POLYMER SCIENCE

Edited by **Faris Yılmaz**

Polymer Science
http://dx.doi.org/10.5772/2749
Edited by Faris Yılmaz

Contributors

Telmo Ojeda, Hale Berber Yamak, Kelly A. Ross, Susan D. Arntfield, Stefan Cenkowski, U. Maver, T. Maver, Z. Peršin, M. Mozetič, A. Vesel, M. Gaberšček, K. Stana-Kleinschek, Kenichi Furukawa, Takahiko Nakaoki, Oleksiy Lyutakov, Jiri Tuma, Jakub Siegel, Ivan Huttel, Václav Švorčík, Petr Slepička, Tomáš Hubáček, Zdeňka Kolská, Simona Trostová, Nikola Slepičková Kasálková, Lucie Bačáková, Zuzana Makajová, Kateřina Kolářová, Alena Řezníčková, Peter James Baker, Keiji Numata

Published by InTech

Janeza Trdine 9, 51000 Rijeka, Croatia

Publishing Process Manager Marina Jozipovic
Technical Editor InTech DTP team
Cover InTech Design team

Additional hard copies can be obtained from orders@intechopen.com

Polymer Science, Edited by Faris Yılmaz
p. cm.
ISBN 978-953-51-0941-9

For EU product safety concerns:
IN TECH d.o.o., Prolaz Marije Krucifikse Kozulić 3, 51000 Rijeka, Croatia,
info@intechopen.com or visit our website at intechopen.com.

Meet the editor

Dr. Faris Yılmaz was born in Palestine on October 07th, 1969. After pre-college education in Jinin-West Bank, he was accepted to Chemical Engineering Department of Middle East Technical University (METU) in Ankara, Turkey. There he obtained his BSc degree in Chemical Engineering, Ms and PhD degree in Polymer Science and Technology. His research has mainly concentrated on the conducting polymers, their solution properties, composites and nanocoposites. He has publications in different polymer journals. He is married and has two children.

Contents

Preface XI

Chapter 1 **Polymers and the Environment 1**
Telmo Ojeda

Chapter 2 **Emulsion Polymerization: Effects of Polymerization Variables on the Properties of Vinyl Acetate Based Emulsion Polymers 35**
Hale Berber Yamak

Chapter 3 **A Polymer Science Approach to Physico-Chemical Characterization and Processing of Pulse Seeds 73**
Kelly A. Ross, Susan D. Arntfield and Stefan Cenkowski

Chapter 4 **Polymer Characterization with the Atomic Force Microscope 113**
U. Maver, T. Maver, Z. Peršin, M. Mozetič, A. Vesel, M. Gaberšček and K. Stana-Kleinschek

Chapter 5 **Absorption Kinetics of Phenol into Different Size Nanopores Present in Syndiotactic Polystyrene and Poly (p-methylstyrene) 133**
Kenichi Furukawa and Takahiko Nakaoki

Chapter 6 **Nonconventional Method of Polymer Patterning 151**
Oleksiy Lyutakov, Jiri Tuma, Jakub Siegel, Ivan Huttel and Václav Švorčík

Chapter 7 **The Properties and Application of Carbon Nanostructures 175**
Petr Slepička, Tomáš Hubáček, Zdeňka Kolská, Simona Trostová, Nikola Slepičková Kasálková, Lucie Bačáková and Václav Švorčík

Chapter 8 **Electrokinetic Potential and Other Surface Properties of Polymer Foils and Their Modifications 203**
Zdeňka Kolská, Zuzana Makajová, Kateřina Kolářová, Nikola Kasálková Slepičková, Simona Trostová, Alena Řezníčková, Jakub Siegel and Václav Švorčík

Chapter 9 **Polymerization of Peptide Polymers for Biomaterial Applications 229**
Peter James Baker and Keiji Numata

Preface

In the last decade, a great number of texts have appeared in the field of polymer science which lead to a most legitimate question: Why another book? We certainly asked ourselves this question when preparing the present text, and we wish to offer certain thoughts in reply.

During the last ten to fifteen years the developments in polymer science has been striking attention and undergone important changes. From rather specialized subject intended for engineers interested in certain definite fields, it has developed into one of the fundamental disciplines common to several branches of engineering and science. To serve this purpose, the subject materials have been prepared to treat a comprehensive aspects of polymer science. Following this trend, a number of rigorous books have discussed different types of polymers with great precision and elegance and at relatively high levels of abstraction, but none is complementary. A position midway between the older, traditional approach in engineering and the recent, somewhat formal expositions seems to be evolving.

In the present text we attempt to do justice to the different topics of polymers and their uses. This text is generally suitable for researchers rather than students. The first chapter of this book discussed sorption mechanism of organic compound in the nanopore of syndiotactic polystyrene crystal. In the second chapter, a discussion was done to illustrate a physico-chemical characterization and processing of pulse seeds. The chemo-enzymatic polymerization for peptide polymers were illustrated in the third chapter. In the fourth chapter, an electrokinetic potential method was used to characterize the surface properties of polymer foils and their modifications. Also, an emulsion polymerizations was discussed in the fifth chapter. Nonconventional methods of polymer surface patterning, polymer characterization using atomic force microscope, biopolymers in the environment, and carbon nanostructure and their properties and applications were discussed in the sixth, seventh, eighth and ninth chapters respectively. Finally, let us point that although many books in the field of polymer science appear, none of them are complementary.

Dr. Faris Yılmaz
Research Laboratory in Floteks Plastics Industry-Bursa,
Turkey

Chapter 1

Polymers and the Environment

Telmo Ojeda

Additional information is available at the end of the chapter

http://dx.doi.org/10.5772/51057

1. Introduction

The traditional polymer materials available today, especially the plastics, are the result of decades of evolution. Their production is extremely efficient in terms of utilization of raw materials and energy, as well as of waste release. The products present a series of excellent properties such as impermeability to water and microorganisms, high mechanical strength, low density (useful for transporting goods), and low cost due to manufacturing scale and process optimization [1]. However, some of their most useful features, the chemical, physical and biological inertness, and durability resulted in their accumulation in the environment if not recycled. Unfortunately, the accumulation of plastics, along with other materials, is becoming a serious problem for all countries in the world. These materials occupy significant volume in landfills and dumps today. Recently, the presence of huge amounts of plastic fragments on the oceans has been observed, considerable part of them coming from the streets, going through the drains with the rain, and then going into the rivers and lakes, and then to the oceans [1]. As a result, there is a very strong and irreversible movement, in all countries of the world, to use materials that do not harm the planet, that is, low environmental impact materials.

In this chapter, alternatives to traditional polymers derived from fossil fuels are commented. Materials derived from plants and microorganisms are presented, as well as biodegradable materials obtained from fossil fuels.

2. Overview of the fate of polymeric wastes

Of course, before we use materials that can accumulate in nature, we must think about reducing their consumption, reusing and recycling (either by reuse of raw materials, or by use of the energy of combustion) [3]. However, certain parts that are formed by small amounts of polymer (ie, a few grams) and may still be contaminated by food are difficult to be collected from nature, cleaned, sorted and recycled, both from the economic and also from the environmental (energy consumption and soil pollution of the process) point of view. This is the case of plastic bags and packaging, especially plastics used in food, in medical and hygiene. In these cases, the use of biodegradable polymer materials may be an excellent solution to the environment [2].

Precisely for this reason, we are now receiving a huge load of information on plastic materials with less impact to the environment. And much of this information is contradictory, not bringing acceptable scientific references on the assertions made. Even the norms for biodegradation tests have been developed under influence of the manufacturers of biodegradable products as a tool to ward off competitors.

Despite the somewhat confusing situation we are currently experiencing, the products on the market are being tested by consumers and the trend is that the most suitable materials in every situation be known over time. Nevertheless something is right: the best product for a given application in a given market may not be the same for another application and/or another market. An important aspect to consider is to know where the polymer material will be disposed, to evaluate the conditions for biodegradation. Some polymers that biodegrade well in industrial composting conditions (high temperatures, high levels of moisture and oxygen) biodegrade much more slowly in the soil at ambient temperatures, PLA (or polylactic acid) being an example. In landfills, all polymers biodegrade very slowly, due to restrictions of oxygen and moisture in the layers below the surface. Biodegradation under anaerobic conditions (deprived of oxygen) produces CH_4 (methane), CO_2 (carbon dioxide), water and biomass (living cells). Methane is a much more potent gas than CO_2 to the greenhouse effect. Biodegradation under aerobic conditions (oxygen abundance), otherwise, does not produces methane, producing mainly CO_2, water and biomass [4].

Biodegradability is a feature that has been highly valued in polymers from the environmental standpoint, but is not the only important one. Sooner or later, all components in a polymer material will be returned to the environment, with the degradation, so it is very important to use pigments, fillers and additives that are not toxic in nature. Furthermore, the environmental impacts should be studied from birth to death of the polymer (or "from cradle to grave"). The use of raw materials from renewable resources (plants) has also been highlighted. However, one point to be considered here is the use of arable land for monocultures in farms that could be producing food and are instead producing raw materials for commodities (like plastics). Likewise, one must consider the possibility of using fertilizers and pesticides in excess, what could impact in eutrophication, acidification, global warming, poisoning of the environment, etc. Another point is to know if the farming practices are conservationists or not, ie, if they seek to preserve the soil or not. For example, the practice of burning crop residues after harvest should be avoided. Another interesting aspect is the one of polymer production. Very complex processes with many steps, which consume much energy and generate much waste tend to be disadvantaged. The need for transportation of raw materials to the factory or of finished products to the consumer market must also be considered. All the characteristics above are usually considered in the life cycle assessment of the product, which, being somewhat complex, can be aided by regulatory standards (e.g. series EN - ISO 14040) [5-7].

During the 1960s percipient environmentalists became aware that the increase in volume of synthetic polymers, particularly in the form of one-trip packaging, presented a potential threat to the environment, what became evident in the appearance of plastics packaging litter in the streets, in the countryside and in the seas [8].

PVC (see Figure 1) is a good example. Although the density of PVC is around 1.4, hollow parts may float in the oceans, which have a density of about 1.03. PVC has a high concentration of chlorine atoms in an organic chemical structure that is new in nature (i.e. a xenobiotic), what renders it very recalcitrant [9]. On the other hand, PVC degrades easily under the action of light or heat, and its decomposition is catalyzed by the HCl released, forming a poly-unsaturated structure which is very degradable. In oceans the HCl might be removed by the moving water and also neutralized by the cations existing in the alkaline medium (pH ~ 8). To increase stability, 1 to 5% of additives based on transition metals, such as salts, derivatives and complexes of Pb, Zn, Cd and Sn are commonly employed [10]. In order to get a more flexible PVC, plasticizers based on phthalate are commonly used, many of them having chronic toxicity to animals, showing body growth problems (ie, teratogenic effects) and reproduction complications in humans. Small fragments of PVC molecules can evaporate. Just as halogenated solvents, these molecules are very inert and can rise to the stratosphere, contributing to the destruction of the ozone layer [9]. In addition to the accumulation in the environment and to the possible toxicity of the additives, it was realized that the incineration of PVC generated many toxic products such as dioxins, due to the high concentration of chlorine atoms present [1].

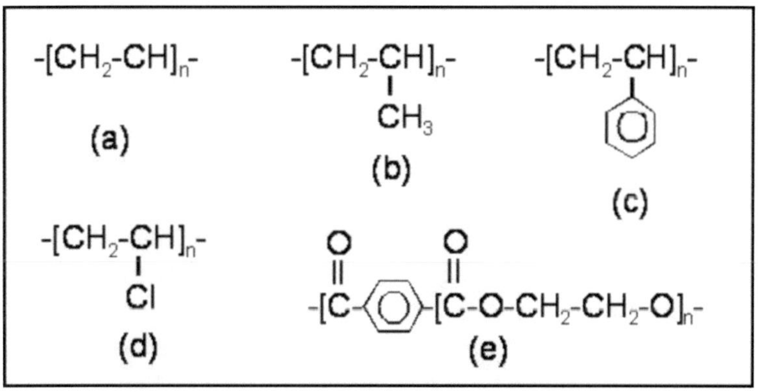

Figure 1. Basic structures of the main thermoplastic polymers in the present: a) polyethylene (PE), b) polypropylene (PP), c) polystyrene (PS), d) polyvinyl chloride (PVC), and e) polyethylene terephthalate (PET).

Polycarbonate (plastic) and **epoxy resins** (coating and adhesive) are normally produced with bisphenol A as one of the monomers. This substance may also be used as an additive for plastics. It is an endocrine disruptor (it can mimic hormones) [1]. Some studies have shown toxicity, carcinogenic effects and possible neurotoxicity at low doses in animals [11-15]. In the case of decomposition of the resin, this toxic monomer might be released into the environment. Polycarbonate can be recycled. Its biodegradation is very slow due to the presence of aromatic rings in the main chain.

Polystyrene (PS, Figure 1) has a density of about 1.05, but hollow parts made with this polymer may fluctuate in the oceans. The presence of aromatic rings at a short distance from the main chain increases its resistance to biodegradation (ie., its recalcitrance). In addition, PS has rigid (although not crystalline) molecules, making difficult the enzymatic action. Although most of the additives used with PS are not toxic, the residual amounts of free styrene, "dimers" and "trimers" (obtained after polymerization) must be kept very low, because they are volatile and can migrate out of the part [16]. This polymer can be recycled.

Polyethylene terephtalate (PET, Figure 1) has a density around 1.4, but again bottles and other parts made with this polymer may float in the ocean until they fracture. PET presents aromatic rings in the main chain, which makes it highly recalcitrant, despite having hydrolysable ester groups. Additionally, catalysts residues employed in their synthesis (either by esterification or transesterification) are present in the polymer. Examples of catalysts are manganese, zinc and cobalt salts (transesterification) and compounds of antimony, germanium, titanium and tin (esterification) at typical concentrations of 50-250 ppm [17]. Phosphorus compounds used to deactivate transesterification catalysts can cause eutrophication of ocean waters. PET has been reused and recycled on a large scale in many countries around the world.

Polyamide 11 (PA 11) is a biopolymer derived from vegetable oil. It is commercialized by Arkema and DSM. PA 11 belongs to the technical polymers family and is not biodegradable. Its properties are similar to those of PA 12, although emissions of greenhouse gases and consumption of nonrenewable resources are reduced during its production. It is used in high-performance applications like automotive fuel lines, pneumatic airbrake tubing, flexible oil and gas pipes, sports shoes and electronic device components [18].

The **nylon 6** and **nylon 66 polyamides** are considered polymers with superior thermal and mechanical properties, and are therefore referred to as engineering polymers. Although they are in the group of the non-biodegradable polymers [19], they may be significantly degraded by white-rot lignin degrading fungi. Its surface erosion suggests that nylons are degraded to soluble monomers [20]. They can be degraded by hydrolysis and also by oxidation. In the latter case, the more reactive hydrogen atoms are those attached to the carbon atom adjacent to the nitrogen atom of the amide group [21].

Polyurethanes (PURs), or carbamates are polyethers or polyesters (with molecular weight of about 200 - 6000 $g \cdot mol^{-1}$) copolymerized by a polyaddition process, with monomers containing isocyanate groups, resulting the characteristic urethane groups in the main chain, which are generally in very low proportions. In most applications, they are thermosetting or thermo-crosslinkable polymers, ie they form a three-dimensional network by chemical reactions under heating, and they do not soften under further heating [1]. The company Cargill produces polyol for polyurethane cushioning, which is soy-based (BiOH polyol), designed especially for flexible foams.

It is believed that most of PUR biodegradation occurs by the action of esterases, however polyester-polyurethane degrading enzymes have been purified and their characteristics have been reported. These enzymes have a hydrophobic binding domain at the surface of the PUR, and a catalytic domain [22]. But there is no evidence that the urethane linkage has been broken.

Polyolefins are polymers produced by the polymerization of alkenes, such as polyethylenes (PEs, Figure 1), polypropylene (PP, Figure 1), polybutene-1 (PB-1) (plastics), polyisobutylene (PIB), ethylene-propylene-rubber (EPR), ethylene-propylene-diene monomer (EPDM) (elastomers), etc. They are a very large class of carbon-chain thermoplastics and elastomers, the most important being polyethylenes and polypropylene. They are extensively used in many different forms and applications. Flexible packaging, included here wrap films, grocery bags and shopping bags made with extruded films and extruded blown films, as well as rigid packaging made by blow moulding and injection moulding represent a considerable amount of the total material consumed [1]. Polyolefins float in the oceans, because they are normally lighter than salty water. They do not normally contain toxic ingredients, although toxic metals may be introduced as pigments. Usual additives are antiacids (e.g., Mg or Ca stearates at ~0.1%) and antioxidants (e.g., hindered phenols and phosphites at ~0.05 – 0.2%). Catalytic residues, such as Ti and Cr compounds, are present at very low levels (ppm). The oxidative degradation of polyolefins in the oceans is favored by the continuous movement of the waves, by the presence of oxygen at the surface, and by the sun exposure. On the other hand, the temperature of plastic materials at sea does not reach that on the ground, due to the effect of heat removal by water. Eventual fouling can limit the exposure of the material to UV radiation. Oxidized residues of polyolefins may sink into the sea due to the change in density that occurs during oxidation. This behavior slows down subsequent degradation/biodegradation, as in deep water there is no UV radiation, the amount of oxygen available is very limited, the temperature is lower (~ 4°C, reaching even 1°C) and there is no agitation by waves. In fact, even the food present in sunk ships degrades and biodegrades very slowly on the sea bottom.

Although the above polymers have a number of environmental impacts from the time of their disposal, their production from oil, natural gas or coal has been optimized through decades of manufacturing. In the case of petroleum, the petrochemical industry uses naphtha, that is a petroleum fraction of approximately 3% of the total. Should naphtha not be used for the petrochemical industry, it would then be burned, what would not improve anything its environmental impact. Moreover, the use of oil to be burned in a combustion engine or in a boiler for heat is becoming an unacceptable luxury to the present day, with the prices of fossil fuels becoming progressively higher. The use of fossil fuels as raw materials, as major carbon sources, appears to be more compatible with the world reality today. The development of renewable forms of energy such as solar thermal and photovoltaic, wind, hydroelectric, wave and tidal, geothermal, biogas and others should allow the replacement of the energy obtained from fossil fuels in a few decades.

Biopolymers: are polymers produced by living organisms. They all have been around for millions of years on our planet, and for this reason microorganisms have had enough time to develop enzymes capable of degrading their structure, so they are biodegradable. In general their end of life environmental impact is low. However, there are at least two cases where this is not true:

1. When the biopolymer is placed in an unsuitable environment for its biodegradation. For example, a landfill does not provide adequate conditions for biodegradation, since oxygen and water are lacking. Under anaerobic conditions (ie in the absence of oxygen), the biopolymer, as well as organic wates in general, will degrade producing biomass, methane (CH_4), carbon dioxide (CO_2) and water, as well as other eventual small molecules (NH_3, N_2, N_2O, H_2S, mercaptans, etc.), depending on its chemical structure. The generated methane is a much more powerful gas than carbon dioxide to global warming, and is not readily reabsorbed by plants, as with CO_2.
2. Once the biopolymer is mixed with other polymers in a recycle stream, it will act as a contaminant, as biopolymers are normally not recyclable, degrading at the recycling processing conditions.

If we look now to the environmental impacts that occur since the extraction/transportation/processing of raw materials to the final production of the biopolymers, we observe that the final balance can be even worse than that of conventional polymers obtained from petroleum. A comparative study of all environmental impacts of a particular product (eg a polymer) can be obtained by a life cycle assessment. The life cycle assessment (LCA) of a product, process or activity is a technique to assess environmental impacts, or the environmental burdens associated with all the stages of its life, from cradle-to-grave, ie: extraction, transportation and processing or raw materials; manufacturing; transportation and distribution; use, reuse and maintenance; recycling; final disposal; material and energy consumption; water consumption and emissions generation; etc. The assumptions and methodologies should be given, and should be clear, consistent and documented, otherwise it may be impossible to compare different LCA studies. Some of the most often evaluated environmental impacts are: toxicity to humans or to other living organsms; fresh water aquatic ecotoxicity; marine aquatic ecotoxicity; terrestrial ecotoxicity; eutrophication, acidification (of rains and soils); global warming potential; ozone depletion; abiotic depletion of mineral resources; depletion of fossil fuels (petroleum, natural gas and coal); visual pollution (litter); photochemical oxidation (smog formation); renewable and non-renewable energy use [5-7]. A difficulty in comparing different types of environmental impacts is the use of a different unit to each type assessed. For example, kg CO_2 equivalent is the required CO_2 mass to produce the same effect (global warming) that the object of study. Then how to compare global warming with, for example, abiotic depletion, which has the unit kg Sb equivalent (resource depletion compared with that of antimony)? The weight of each impact needs to be arbitrated in order to calculate the total impact. An impact that is very significant for some authors may be considered less important for others. Therefore, all assumptions made need to be transparent in the study.

The findings of some LCAs studies of plastic bags are presented below. In a study by Edwards and Fry [5], the authors have concluded that the environmental impacts of all types of carrier bags are dominated by resource use and production stages, whereas transport, secondary packaging and end-of-line management have minimal influence. According to them the key to reducing the impact is to reuse the bags as many times as possible, at least as bin liners. Reuse produces greater benefits than recycle. Recycling or composting generally produce small reductions in global warming potencial and abiotic resource depletion. They found that starch-polyester bags have significant global warming potential and abiotic depletion. The impacts of the oxo-biodegradable high density polyethylene (HDPE) bags are very similar to the conventional HDPE bags, because of the similarity in material content and use. The production of the pro-oxidant additive has minimal impact on most life cycle categories. End-of-life impacts through incineration and landfill are practically identical. The essential difference, although not concluded by the authors, seems to be that oxo-biodegradable HDPE bags do not remain on ground or water as litter, and that they represent a source of carbon, just like humus.

James and Grant [6] have found in their study that polymer based reusable bags have lower environmental impact than all single-use bags evaluated. Degradable bags have similar greenhouse and eutrophication impacts to conventional HDPE bags, because they normally go into landfills. Decisions about degradable polymers should be based on: where and how they will degrade, minimal LCA (not just end-of-life), and commercial benefits.

Tabone et al. [7] assessed plastic bags according to two sets of parameters: green design principles and life cycle assessment – environmental impacts. The first one was related to several general principles, like prevent waste, utilize less material mass, maximize energy efficiency, use non-hazardous inputs, use renewable feedstocks, use local sources, design for recycle, minimize material diversity, degrade after use, maximize cost-efficiency, minimize the potential for accidents, etc. The second one referred to LCA, which was discussed above. They have found that, while biopolymers rank highly in terms of green design, they exhibit relatively large environmental impacs from production. The impacts from biopolymers result from the use of fertilizers, pesticides and arable land required for agriculture production, as well as from the fermentation and other chemical processing steps. Interestingly, HDPE and PLA (polylactic acid) are relatively close in terms of the sum of all environmental impacts. Polyhydroxy alcanoates (PHAs) produced from stover have obtained an excellent environmental position.

As a conclusion of some LCAs, it comes out that there is not a single ideal material or solution adequate for all possible situations on Earth, which always presents the lowest environmental impact. A practical present solution for the plastic bags could be the conventional polyolefin materials formulated with pro-oxidant additives, used as many times as possible. Another interesting solution is the use of agricultural and other organic residues as raw materials for the manufacture of biodegradable polymers. Recycling and composting units should be encouraged in all countries of the world. Renewable energies should substitute the fossil fuels, which should be destinated as a carbon source for the chemical industry.

3. Polymer degradation

There are three main possibilities of degradation of the polymers: enzymatic, hydrolytic and oxidative. The enzymatic degradation, or biodegradation, is the breaking of polymer chains by the action of enzymes, which are natural catalysts of chemical reactions produced by living organisms. For example, cellulose and starch are degraded by specific groups of enzymes known as cellulases and amylases, respectively. Polyesters can be degraded by esterases enzymes [23, 24].

The degradation by hydrolysis consists in breaking certain chemical bonds such as ester, ether and amide, by attack of water molecules. This process can be catalyzed by both acids and bases (saponification). In the case of the ester linkage, a carboxylic acid (or a salt thereof) and an alcohol are produced. The ether linkage is much more resistant to hydrolysis than the ester one, generating two alcohols. Hydrolysis of the amide group results in an amine and a carboxylic acid.

The oxidative degradation consists of several different chemical reactions that take place when free radicals (macrorradicais) are formed in the polymer chains in the presence of oxygen. Free radicals may be formed by the action of ultraviolet radiation from the sun, heat, and mechanical deformation (shear and elongation of the chain during processing, or deformation of the solid material by mechanical action, for example, by the action of water and air) [23, 24].

In most traditional polymer materials, e.g. polyethylenes and polypropylene, the prevailing action of the oxidative degradation is the breakdown of molecular chains into smaller segments containing oxygen incorporated in the form of hydroxyl, ketone, ester, aldehyde, ether, carboxyl , etc. [25, 26]. Unsaturations are also formed in the process. Oxidation of polymer materials is a process that occurs naturally, but may take decades or even centuries to be completed. The presence of certain transition metals (such as V, Mn, Fe, Co, Ni, Cu) accelerates the degradation by a factor of about 10^2 and thus permits the complete degradation within a few years under favorable conditions [27, 28].

3.1. Conventional polymers

The chemical formulas of some polymer materials produced in greater amounts worldwide are shown in Figure 1. The typical average molecular weights (weight average) range from 30,000 to 1,000,000 $gmol^{-1}$ or higher, depending on the polymer applications, therefore "n" can be varied between several tens and hundreds of thousands. It is observed from the figure that all polymers have chemical structures that impart low polarity, that is, low affinity to water. The high molecular weight and the hydrophobicity are two decisive characteristics for the observed recalcitrance (i. e., bio-resistance, persistence in the environment). Additionally, the chlorine atoms and the aromatic rings are structures that further hinder biodegradation.

Polyethylene (Figure 1), according to the process of production, may present short and long branches and at variable levels. Branches generally difficult the ordered packaging of the chains in crystals, thus reducing the degree of crystallinity, what increases the availability of the polymer to the attack by various chemical species, such as free radicals and oxidase enzymes. Moreover, the branching points are constituted by tertiary carbon atoms, which are more easily attacked by free radicals than primary and secondary carbon atoms. On the other hand, although branched molecules are more susceptible to oxidative abiotic degradation, their subsequent biotic degradation by means of β-oxidation of fatty acids (a carboxyl group being assumed at the chain end) may be delayed, because the enzymes involved require straight hydrocarbon chains.

In **polypropylene** (Figure 1) a third of the carbon atoms are tertiary, so the polymer is highly susceptible to degradation by free radicals (even in the absence of oxygen). Its high degradability requires the use of antioxidant additives in high concentrations. However, the presence of a methyl branch per repeat unit impairs the biotic degradation via β-oxidation of fatty acids.

In **polystyrene** (Figure 1), the most vulnerable site to oxidation is the main chain carbon which binds to the phenyl group, that may lose a hydrogen atom, generating a free radical. Furthermore, the phenyl group is UV absorber, and can generate free radicals. The phenyl side groups are distributed unevenly in the chain, preventing PS to crystallize [29], contrary to what occurs with PP, where the distribution of methyl groups is regular (isotactic). Though not crystalline, PS chains are very rigid (the glass transition temperature, or Tg, of PS is well above room temperature, (see Table 2), what hinders the action of degrading enzymes.

The **PVC** resin (Figure 1), as the PS, has the side groups (chlorine atoms) distributed unevenly, being predominantly non-crystalline. The chains are rigid at room temperature, since the Tg is relatively high, but the addition of plasticizers imparts mobility to the chains (lowers Tg). Many of the used phthalates plasticizers present chronic toxicity to animals, showing teratogenic effects, i. e., causing malformations of an embryo or a fetus [9, 10]. PVC degrades easily under the action of light or heat, and its decomposition is self-catalised by the released HCl. To enhance the stability, toxic additives based on transition metals are normally used, as already mentioned [10].

In **PET** (Figure 1), the points that are most susceptible to oxidative attack are the atoms at the alpha position relative to the ester group. Furthermore, hydrolysis of the ester group is also possible. The hydrolysis lowers the pH, what accelerates the degradation. Hydrolysis is also accelerated by temperature, UV radiation and chemicals such as acids, bases and certain transition metals. On the other hand, the ester linkage is highly stabilized by the aromatic rings in the main chain, that also confer rigidity to the chains [17, 30].

4. Biodegradable polymers

Biodegradable polymers are those polymers that, under certain conditions (e. g., in the soil, at room temperature and under aerobic conditions) can be degraded directly by the action of enzymes or after passing through an initial period of hydrolytic or oxidative degradation. The main degrading organisms are fungi, bacteria and archaea, although algae, nematodes, and even insects can also be involved. In aerobic environments, the degradation produces CO_2, H_2O (among other gases) and biomass, i.e., living cells. In anaerobic conditions, CH_4 (methane) is additionaly produced (among other gases). The biodegradable polymer serves as a source of carbon and energy to the microorganisms. But other nutrients are also needed for maintaining microbial activity, such as O, H, N, P, S, Cl, Na, K, Mg, Ca, Mn, Fe, Co, Ni,

	Price	Processing						Equilibrium Moisture	Drying Conditions	Processing Temperature	Pressing (1) Temperature	Shrinkage in the Mold
	US$/kg (2)	Compression	Extrusion	Blown Film	Injection	Blow	Thermoforming	%		------------------------	ºC -----	%
LDPE	1.1	x	x	x	x	x	x	0.02	2h/75ºC	150-250	180	2.0-3.5
HDPE	1.0	x	x		x	x	x	0.01	2h/75ºC	150-250	180	2.0-3.5
PP	1.0	x	x	x	x	x	x	0.01	2h/75ºC	180-250	200	1.5-2.5
PS	1.3	x	x		x		x	0.2	3h/70ºC	180-240	200	0.2-0.6
PVC	0.9-1.0	x	x	x	x	x	x	0.05	3h/65-70ºC	150-205	200	0.1-0.5 (3)
PET	1.5	x	x		x	x	x	0.16	5h/120ºC	250-285	200	0.1-2.0
PBAT	7.0	x	x	x	x	x	x	---	4h/60-70ºC	160	180	---
PBST	7.0	x	x	x	x	x	x	---	4h/60-70ºC	---	180	0.5-1.0
PGA	~ 7.0	x	x		x		x	---	4h/80ºC	230-240	230	---
PLA	2.5	x	x	x	x	x	x	0.4-2.0	4h/80-100ºC	190-210	180-200	0.4
PHB	5.5	x	x		x	x	x	0.2-0.4	4h/80ºC	180-190 (4)	180-190	1.3
PHBV	5.5	x	x		x	x	x	~0.2	4h/80ºC	100-180 (4)	100-180	---
PCL	5-10	x	x		x		x	< 1	8h/45ºC	50-165	80-165	---
PCL/STARCH	5.0	x	x	x	x		x	---	8h/45ºC	120-150 (4)	120-150	---
PVOH	1.0-2.5	x	x		x		x	---	4h/<150ºC	180-230 (4)	180-230	---
Cel. Acetate	4.0	x	x	x	x	x	x	2.2(0.5-8.0)	3h/70ºC	175-225	180-200	0.4-0.6

(1) Used in order to obtain specimens; (2) Rough estimates made by the author, and highly subject to market variations;(3) Plasticized polymers: 1-2%; (4) Polymers subject to strong thermal degradation.
LDPE: low density polyethylene; HDPE: high density polyethylene; PP: polypropylene; PS: polystyrene; PVC: polyvinyl chloride;
PET: polyethylene terephtalate; PBAT: poly(butylene adipate-co-terephtalate); PBST: poly(butylene succinate-co-terephtalate);
PGA: poly(glycolic acid); PLA: polylactic acid; PHB: polyhydroxybutyrate); PHBV: poly(hydroxybutyrate-co-hydroxyvalerate);
PCL: polycaprolactone; PVOH: polyvinyl alcohol; Cel. Acetate: cellulose acetate.

Table 1. Processing characteristics and approximated prices of some biodegradable and conventional polymers.

Polymer	Classification (1)	Crystal-linity %	Tg ºC	Tm ºC	Density 23/4 ºC	Modulus of Elasticity MPa	Break Point Elongation %	Break Point Stress MPa
LDPE	S	50	-30	112	0.910-0.925	50-250	150-600	7-19
HDPE	S	70	-30	135	0.940-0.970	500-1500	10-1000	14-42
PP	S	50	0	163	0.902-0.907	1200-2400	50-300	30-41
PS	S	0	90	-	1.03-1.09	2800-3500	5	40
PVC	S	0	80	-	1.35-1.45	2500	30	55
PET	S	0-50	70	250	1.35	3000	100-300	55
PBAT	SDC	20-35	-30	110-120	1.25-1.27	80	560-710	10-45
PBST	SDC	20-35	45	195	1.25	3000	40-500	40-65
PGA	SDC	50 (46-52)	35	225	1.6	5500	15-35	n. a.
PLA	SDCR	0-42	50-60	175 **(2)**	1.24-1.26	1200-3000	5-10	30-60
PHB	NDCR	60	0-5	170	1.25	2500-3500	2-8	30-40
PHBV	NDCR	0-60	-3 - +5	80-160	1.25	800-2500	15-50	20-30
PCL	SDC	55	-60	57-61	1,145	200-500	700-1000	16-54
PCL/Starch	SDC	n. a.	~30	60-110	1.3	180-200	600-900	25-35
PVOH	SDC	40-50	85	180-230	1.25-1.32	37-45	150-400	44-64
Cel. Acetate	SDC	n. a.	68	170-250	1.22-1.34	2000	35	32

(1) S: synthetic; N: natural; D: degradable; C: compostable; R: renewable resource. (2) Values of 120-180 ºC have been reported; n. a.: information not available. LDPE: low density polyethylene; HDPE: high density polyethylene; PP: polypropylene; PS: polystyrene; PVC: polyvinyl chloride; PET: polyethylene terephthalate; PBAT: poly(butylene adipate-co-terephthalate; PBST: poly(butylene succinate-co-terephthalate); PGA: polyglycolic acid; PLA: polylactic acid; PHB: poly(3-hydroxy-butyrate); PHBV: poly(3-hydroxybutyrate-co-3-hydroxyvalerate); PCL: poly(epsilon-caprolactone); PVOH: polyvinyl alcohol.

Table 2. Physical properties and ecological classification of some biodegradable and conventional polymers.

Cu, Zn and Mo. The presence of water is essential to all known living organisms. Regarding the oxygen requirement, there are the following classes of organisms: obligate aerobes (need O_2), obligate anaerobes (are killed by O_2), microaerophiles (live only at low O_2 concentrations), facultative anaerobes (live with or without O_2) and aerotolerant anaerobes (not affected by O_2 concentration).

There are currently biodegradable polymers with characteristics and properties very similar to those from conventional polymers derived from petroleum (Tables 1 and 2).

Many factors may affect biodegradation. With respect to the material itself, the following factors have influence: chemical structure, molecular weight, degrees of branching and crosslinking (if present), glass transition temperature, crystallinity and solubility, and concentration of additives and pigments. With regard to the environment, the following factors have influence: presence of water, oxygen and other nutrients (in particular, the ratio C : N : P), temperature, pH, osmotic pressure (concentration of ions and solutes in the environment), surface area of the part, and the available microbial population [31].

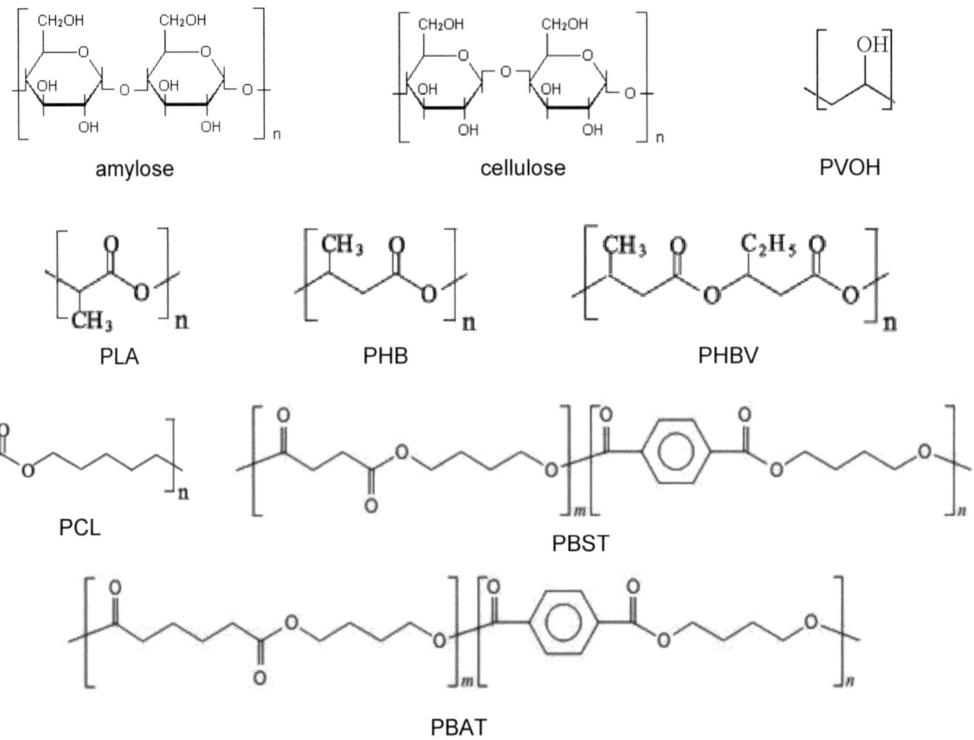

Figure 2. Chemical structures of some biodegradable polymer materials: PVOH: polyvinyl alcohol, PLA: polylactic acid, PHB: poly(hydroxybutyrate), PHBV: poly(hydroxybutyrate-co-hydroxyvalerate), PCL: polycaprolactone, PBST: poly(butylene succinate-co-terephtalate), PBAT: poly(butylene adipate-co-terephtalate).

The vision that a single species will be responsible for the complete degradation of a substrate (polymer) is very common but unrealistic, because xenobiotics are normally degraded by consortia of different species of fungi, bacteria and archaea, among other organisms. Symbiosis, commensalism and co-metabolism are common events. The polymers are degraded by steps by the consortium, each step through one or more enzymes produced by one or more organisms. The final products of degradation are mostly CO_2, H_2O and biomass under aerobic conditions, and additionally, CH_4, under anaerobic conditions.

The first group of polymer materials is comprised of inherently biodegradable materials, i. e. those whose molecules can be biodegraded immediately after coming into contact with microorganisms from soil, composting plants, rivers, etc. Examples of this group are modified starch and cellulose (Figura 2), proteins and their derivatives (i. e., products made of these materials after chemical modifications and/or mixing with other materials). The polymer materials produced by plants or other living organisms are called biopolymers or bioplastics, if they are plastic.

4.1. Group of polysaccharides

4.1.1. Starch

Starch is a polymer of the group of polysaccharides, to which also belong cellulose, hemicellulose and chitin, among others. Its chemical composition depends on the plant, but generally comprises a mixture of 20-30% by weight amylose (Figure 2) and 70-80% by weight of amylopectin. Amylose and amylopectin have molecules formed by glycosidic units bond by α-1,4 ether linkages. Amylose consists of about 200 to 12,000 glycosidic units, forming a straight chain without branching, with secondary (space) helical structure. Amylopectin consists of about 0.6-2.5 millions of glycosidic units, and is strongly branched. The branches are formed by α-1.6 ether linkages, with average length of 20-30 glycosidic units, occurring every 24-30 units of the straight chain [32, 33]. The chain regions next to the branches are included in the amorphous phase of the material. The chains of the crystalline phase normally have the helix conformation. As a consequence of its molecular structure, starch has lower degree of crystallinity than cellulose, what facilitates its biodegradation. Just as cellulose, starch is not a thermoplastic material, due to the intensity of the interaction between the molecules by H bonds. Thus heating and shear forces result in degradation before melting. The mechanical properties are poor, both stiffness and toughness, and related properties.

Starch is usually obtained from corn, potato and cassava, although it may be obtained from other sources as well. The fact that it is obtained from foods, that require fertile soil for cultivation, has been much questioned, considering that more than 10% of the world population is still undernourished today. From the viewpoint of the life-cycle analysis, products from plant or food wastes are highly favored.

In order to prepare thermoplastic starch, the crystal structure of starch has to be destroyed, either by mechanical working, pressure or heat, or by addition of plasticizers, such as water and glycerin. The gelation of starch is the disruption of the semicrystalline structure of its granules during heating in the presence of water over 90% [32]. The gelation process occurs in two steps. The first, at 60-70 °C, is the swelling of the granules, with little leaching, but with loss of chain organization in crystals. The second above 90 °C, causes the complete disappearance of the granular integrity due to events of swelling and dissolution, making the swollen granules vulnerable to shear.

The destructuring of starch is defined as the melting and the disorganization of the molecular structure of the starch granules and subsequent molecular dispersion in water. The thermoplastic starch product has a starch content above 70% (e. g., 85%), being based on gelled and destructured starch, and on the use of plasticizers, allowing the use of conventional equipment for thermoplastic processing. A problem that can also be an advantage is the high hygroscopicity and solubility in water [34].

To improve processability, mechanical properties and moisture resistance while maintaining biodegradability, starch may be mixed with aliphatic or aliphatic-aromatic polyesters, such as polycaprolactone (PCL), polylactic acid (PLA) and poly(butylene adipate-co-terephthalate) (PBAT), suffering complexing with these polymers [33, 35].

The complex formed by amylose with the complexing agent is usually crystalline, characterized by an amylose single helix around the complexing polymer [32]. Amylopectin does not interact with the complexing polymer, remaining in the amorphous phase. Starch can also be blended with polyvinyl alcohol (PVOH), for the production of foamed products, such as trays for food. Starch esters reinforced with natural fibers exhibit properties similar to those of polystyrene (PS).

Among the world's leading suppliers are: Novamont (Mater-Bi products) and its licensees (about 80,000 t per year), Rodenburg (Solanyl products, 40,000 t per year), Corn Products, Japan Corn Starch, Chisson, Biotec, Supol, Starch Technology, VTT Chemical, Groen Granulaat and Plantic. The price of the blends of starch with polyesters is about US$ 5 kg^{-1}, while the price of the modified starch is about US$ 1.0 - 1.5 kg^{-1}.

Starch is biodegraded by amylases in a huge variety of different environments. The biodegradation of starch results from the enzymatic attack of the glycosidic linkages, reducing the chain size and producing mono-, di- and oligosaccharides, easily metabolized by biochemical pathways.

Some manufacturers still mix low levels of starch with polyethylene. The rapid biodegradation of the former increases the available surface of the latter to degradation. Possibly, some of the starch degrading microorganisms may also help with the slow biodegradation of polyethylene [36]. Erlandsson et al [37] have tested a system of starch with LDPE, SBS copolymer, and manganese stearate, after thermo-oxidation at 65 and 95 °C and UV radiation. The starch has stabilized PE regarding thermo-oxidation, but has promoted its photo-oxidation. Among the applications for starch, are films for packaging, shopping bags, garbage bags, mulch films, disposable diapers, foams, foamed trays for food, injection moulded products, blown bottles and flasks, filaments, etc.. The foaming process involves melting (or softening) the polymer and blending it with a foaming agent, typically pentane or carbon dioxide. It is used mainly for polystyrene (PS).

4.1.2. Cellulose

It is the main component of plants, with natural production per year estimated at 7.5 billion tons, and annual human consumption estimated at 200-250 million tonnes. In wood, the cellulose fibrils are joined together by lignin, which is a resin binder.

Cellulose is a polymer made up of about 7,000-15,000 D-glycosidic units (D-glucopyranose residues, Figure 2), joined by β-1.4 ether linkages, that form the cellobiose units (two consecutive glycosidic units), which are repeated along the chain [38]. Each glycosidic unit has three hydroxyl groups, that promote strong interactions by hydrogen bonds. The spatial structure allows molecules crystallize in a horizontal plane, forming fibers. As a consequence, cellulose is sparingly soluble and not processable by thermal and mechanical action, i. e. it is not thermoplastic. Cellulose is a rigid material, whose fibers may be used to reinforce other materials. It presents a small elongation capacity. In order to become thermoplastic, it is necessary that about two of the three hydroxyl groups of the glycosidic units be reacted.

Cellulose is biodegraded by the extracellular cellulase enzyme complex, that is induced in most microorganisms [39]. Only a subgroup of cellulase, known as exogluconase, or β-1,4-gluconase, can attack the terminal glycosidic bond, and is effective in degrading the small crystals (crystallites) in which neither water nor enzymes can penetrate. A manufacturer of cellulose based films is Innovia (NatureFlex products).

Cellulose acetate is a thermoplastic derivative obtained by partial esterification of the hydroxyl groups with acetic acid or anhydride. With an average degree of substitution of up to 2.5 of the 3 available glycosidic hydroxyl groups per unit, the polymer is still biodegradable [40].

The main applications of cellulose are: timber, furniture and fuel; textiles such as cotton; paper, membranes, and explosives. Important cellulose derivatives are cellulose acetate and cellulose acetate butyrate (thermoplastic esters); ethyl, hydroxyethyl and hydroxypropyl cellulose.

4.1.3. Other polysaccharides

Chitin and chitosan: The amount of naturally synthesized chitin is estimated at about 1 billion tons per year, produced by fungi, arthropods, molluscs and some plants. Chitosan is the wholly or partly deacetylated chitin. Because it is biocompatible and biodegradable, it has several biomedical applications: cosmetics, personal care, diet food, treatment for thermal burns, sutures, artificial skin, control of drug delivery, etc.. Its price is between US$ 15 and US$ 50 per kg. Chitosan is biodegraded by enzymatic hydrolysis. The higher the degree of acetylation, the lower the crystallinity and the higher the rate of biodegradation obtained [38, 41].

Pectin: is a polysaccharide obtained from citrus peel and remains of apples or other fruits used to obtain juice. It is a bonding material of plant cells. It is used as thickener, gel-forming agent, emulsion stabilizer, dietary fiber that is not digested by humans, etc..

Xanthan: is a polysaccharide obtained from fermentation of glucose or sucrose by the bacteria Xanthomonas. It is used as food emulsifier additive, rheological modifier (thickener) of oils and cosmetics (with the bentonite), as a stabilizer of aqueous gels, etc.

Pullulan: is a polysaccharide obtained from the fermentation of starch by the yeast *Aurobasidium pullulans*. It is edible and tasteless. It is used as edible films for food packaging with high oxygen permeability, oral care products, adhesives, thickeners, stabilizers, etc..

Alginates: are a family of copolymers of polysaccharides consisting of α-L-guluronic acid blocks (G) and β-D-mannuronic acid (M) [42]. They may bind to certain ions (e. g. Ca^{+2}) through guluronate residues, forming three-dimensional hydrogels. Alginates rich in G blocks bind to a larger number of ions, producing more rigid and resistant gels, whereas alginates rich in M blocks are more flexible and enable higher diffusion rates of solutes through the gel. Alginates may be produced from natural sources (brown algae), by extraction and purification by sterile filtration. But they may also be produced by fermentation of various microorganisms. Chemically modified alginates are also synthesized: by esterification with propylene oxide (for beers and salad dressings), alkylation with alcohols (drug delivery systems), alkylation or allylation of binder groups (in order to obtain photo-crosslinkable gels). There are several methods to fabricate globules, microcapsules, fibers, films and membranes. Alginates are susceptible to degradation: by cleavage of glycosidic linkages through hydrolysis in acidic or basic environments, by free radical oxidation, and by enzymatic degradation.

4.1.4. Hemicellulose

Hemicellulose is a polysaccharide consisting of around 200 monomer units of different sugars, such as xylose (highest contents), mannose, galactose, rhamnose and arabinose, statistically distributed in the chain , which is branched. As a consequence, the material is amorphous, and has low mechanical and hydrolysis resistance. It is easily hydrolyzed by many hemicellulases enzymes from bacteria and fungi. Hemicelluloses are embedded in the cell walls of plants, bond with pectin (another carbohydrate) to cellulose to form a network of cross-linked fibres.

4.2. Lignin

Lignin is a complex and heterogeneous cross-linked polymer, containing aromatic rings, C-C bonds, phenolic hydroxyls, and ether groups, with molar mass higher than 10^4 g mol^{-1}. It is formed in chemical association with cellulose, giving lignocellulose, in the cell walls of plants. Thus lignin is not a polysaccharide, but a complex substance consisting of aromatic structures with alkoxy and hydrocarbon substituents that link the basic aromatic unit into a macromolecular structure through carbon-carbon and carbon-oxygen bonds. It is not heterogeneous both in chemical composition and molar mass.

Lignocellulose is strong and tough, and provides physical, chemical and biological protection to the plant. Lignin is resistant to peroxidation (see oxo-biodegradable polymers), as a result of the presence of many antioxidant-active phenolic groups, which act as protective agents against abiotic peroxidation and biological attack by peroxidase enzymes [43-45]. Lignin is obtained mainly from wood: 25-30% of wood is lignin, filling the spaces among cellulose, hemicellulose and pectin in the cell wall of plants. It is also present in some algae. After the cellulose, it is the second most abundant organic polymer on earth, with about 50 million tons being industrially produced annually. It is the main humus-forming component, which provides nutrients and electric charges to the soils. Humus is slowly biodegradable by oxidase and peroxidase enzymes, produced especially by fungi. Although basidiomycetous white-rot fungi and related litter-decomposing fungi are the most efficient degraders of lignin, mixed cultures of fungi, actinomycetes, and bacteria in soil and compost can also mineralize lignin [46].

The main industrial use of lignin is still the power generation, as biofuel. A biodegradable material based on lignin, obtained as a byproduct from the manufacture of paper, mixed with vegetable fibers, is manufactured by Tecnaro under the trade name of Arboform [47]. This is a hygroscopic thermoplastic material, that can be processed at 140 °C. Its mechanical properties show high rigidity and low deformability.

4.3. Proteins

Proteins are polymers formed by α-amino acids, in which the individual amino acid units, called residues, are linked together by amide (or peptide) bonds. The amide linkages are readily degraded by enzymes, particularly proteases. Soy proteins have been used for edible films and even automotive parts, but proteins have not been consolidated as a thermoplastic of worldwide use [48]. Polyamino acids with free carboxylic groups, such as polyaspartic acid and polyglutamic acid, are excellent candidates for use as water soluble biodegradable polymers [40]. In addition to thermoplastics, other possible applications of proteins are in coatings, adhesives, surfactants and gelatin capsules for pharmaceutical uses.

4.4. Natural rubber

Natural rubber is poly (1,4-cis-isoprene), naturally synthesized by the rubber tree, *Hevea brasiliensis*, present in its milky sap or latex. It is also synthesized industrially by polyaddition of isoprene. The long and flexible molecules (Tg = -73 °C) give elastomeric characteristics to the material, that is usually crosslinked with a curing agent such as elemental sulfur. It partially crystallizes when stretched. Due to the long sequences of double bonds (one per monomeric unit), this rubber has a high reactivity with oxygen, undergoing the peroxidation reactions, yellowing very quickly, and being biodegraded at a relatively high rate [1]. The hydrogen atoms attached to carbon atoms at the alpha position relative to the unsaturations are more reactive, or more labile [21]. Among the main applications are tires and tubes.

4.5. Polyvinyl alcohol (PVOH)

Polyvinyl alcohol is a biodegradable polymer obtained by partial or complete hydrolysis of polyvinyl acetate (PVA, of petrochemical origin) to remove acetate groups (Figure 2). The vinyl alcohol monomer almost exclusively exists as the tautomeric form acetaldehyde, which does not polymerize [49].

PVOH is the only water soluble biodegradable polymer, whose main chain consists only of carbon atoms. Solubility and biodegradability are imparted by the hydroxyl groups, that are capable of establishing hydrogen bonds with water. The partial hydrolysis leaves acetate residues, that allow PVOH solubility in cold water, and decrease the biodegradability. At a hydrolysis level close to 100%, the polymer is soluble only in hot water and is completely biodegradable [49]. Even being an atactic polymer, with non-organized space distribution of the hydroxyl groups in the main chain, PVOH shows crystallinity, because the hydroxyl groups are small enough to accommodate within the crystal, not hindering it [49].

PVOH can not be processed by conventional extrusion, because it decomposes (from about 150 °C up) before reaching its melting temperature of 230 °C, with release of water and formation of terminal unsaturation. Partially hydrolyzed products, with melting temperatures of 180-190 °C, containing internal plasticizer (such as water, glycerol, ethylene glycol and its dimer and trimer, etc.) decompose only slightly during extrusion [50].

Among the leading manufacturers are Hydrolene (Idroplast), Celanese (Celvol), DuPont (Elvanol), Air Products, Kuraray, Hisun, Ranjan and Wacker, as well as some major Chinese producers, working for several years. Annual production exceeds 1 million tons.

Some major applications are: thickener in paint industry; paper coating, hair sprays; shampoos; adhesives; biodegradable products for the feminine hygiene; diapers bottoms; water soluble packaging films (detergents, disinfectants, scouring powder, pesticides, fertilizer, laundry, etc.); fibers for concrete reinforcement; lubricant for eye drops and contact lenses; and material for chemical resistant gloves.

It is believed that the PVOH degrading microorganisms are not spread throughout the environment, and that they are predominantly bacteria and fungi (yeasts and moulds) [50].

Before the start of biodegradation, a period of acclimatization may be required. Acclimatization (natural conditions) and acclimation (laboratory conditions) are the adjustment process of an organism or a colony to an environmental change, normally occurring in short periods of time (days or weeks).

The biodegradation mechanism consists of a random cleavage of 1,3-diketones, that are formed by the enzymatic oxidation of the secondary hydroxyls [51]. The main three types of PVOH-degrading enzymes are polyvinyl alcohol oxidase (secondary alcohol oxidase), polyvinyl alcohol dehydrogenase and β-diketone hydrolase [52].

4.6. Poly(ethylene-co-vinyl alcohol), or EVOH

EVOH is a copolymer of ethylene and vinyl alcohol, obtained from ethylene and vinyl acetate, followed by hydrolysis. It is used as an oxygen barrier film in multilayer films for packaging. Its high cost limits its applications as a biodegradable material.

The second group of biodegradable polymers is formed by hydro-biodegradable materials, i. e. those that need to undergo the chemical process of hydrolysis (breakdown of molecules by reaction with water) before they become biodegradable. Therefore, the decomposition process of the polymer occurs in two stages: first, the molecules break up into small fragments by hydrolysis; and second, these fragments are biodegraded by microorganisms. In both stages, the presence of water is essential, both to chemically fragment the molecules, and to be consumed by microorganisms, that need much water in their cells.

To this group belong the aliphatic (i. e., non-aromatic) and the aliphatic-aromatic polyesters, described below. PCL can also be biodegraded directly by enzymes produced by microorganisms, without the initial stage of hydrolysis [24].

4.7. Group of polyesters

Polyesters are polymers in which the bonds between the monomers occur via ester groups. There are many types of natural esters, and their degrading enzymes - the esterases - are present everywhere, together with the living organisms. The ester bonds are generally easy to hydrolyze [20]. The group of biodegradable polyesters mainly consists of: a) linear aliphatic (i. e., non-aromatic) polyesters, such as polyglycolic acid (PGA) and poly (ε-caprolactone) (PCL); b) aliphatic polyesters with short chain branching, such as polylactic acid (PLA), poly (3-hydroxybutyrate) (PHB) and poly(3-hydroxybutyrate-co-3-hydroxy-valerate) (PHBV), and c) aliphatic-aromatic polyesters such as poly(butylene succinate-co-terephthalate) (PBST) and poly(butylene adipate-co-terephthalate) (PBAT).

4.7.1. Poli(3-hydroxy-butyrate), or PHB, and copolymers

Polyhydroxyalkanoates (PHAs) are polyesters of several hydroxyalkanoates that are synthesized by many microorganisms as a carbon and energy storage material. The hydroxyalkanoates can be synthesized from natural substances such as sucrose (e. g. from sugar cane), carboxylic acids and alcohols. Precisely for this reason, this material is rapidly biodegraded in various environmental conditions by many different organisms. Poly(R-3-hydroxybutyrate) (PHB) is a homopolymer of R-3-hydroxybutyrate, and the best known polymer of the PHA group (Figura 2). The molecular weight generally varies from 50,000 to 2,000,000 $g \cdot mol^{-1}$. The monomers are all optical isomers R, the only ones capable of being hydrolyzed by depolymerases, in the isotactic form [53].

PHAs polymers and copolymers are semicrystalline, with the molecules conformed in helices in the crystalline lamellae, which form spherulites. The native intracellular granules of PHAs with 0.2-0.7 μm in diameter, are amorphous and covered by a protein surface layer of about 2-4 nm, containing phospholipids. PHAs are attacked by intracellular PHAs-depolymerases enzymes, but not extracellular depolymerases [54, 55].

PHB is a biological storage material that is used by archaea, bacteria and fungi as feed source. There are more than 75 bacterial genera capable to synthesize PHAs, that are also produced by archaea, fungi, plants and animals, in the soils and aquatic bodies. In addition to the well known 3-hydroxybutyrate, more than 100 monomeric constituents have already been identified [53, 56].

Bacteria and archae can accumulate PHB up to about 95% by dry weight in their cells in the form of granules. The PHB synthesis can occur, for example, as follows: the bacteria are inoculated in a small batch reactor, along with sucrose, other nutrients and water, pH is adjusted and the temperature is raised. The growing colony is transferred to successively larger reactors. After the initial growth of bacteria, the competition period starts, with bacterial storage of PHB in the cytoplasm. The molecular weight increases continuously up to reactor cooling and addition of solvent, what will cause cell lysis and dissolve PHB, which is then purified and dried. The powder obtained in the extraction process is transformed into pellets in an extruder. At the same time nucleating agents and plasticizers are added to improve processability and mechanical properties.

There are also successful attempts to develop genetically modified plants to produce PHAs [40], but the products obtained are very expensive, not being accepted by the market.

As PHB is a very rigid and brittle polymer, it may be mixed with other polymeric materials that are softer and more strainable, such as PBAT or PCL. Blend-stabilizing copolymers, with an intermediate chemical structure, may be obtained by separated synthesis, by transesterification or by the action of peroxides on the two components. A significant product is the copolymer poly(3-hydroxybutyrate-co-3-hydroxy-valerate), or PHBV (Figure 2), which presents lower crystallinity and rigidity than PHB, increasing the flexibility and the elongation capacity [57].

Some important PHAs manufacturers are Metabolix/ADM (Mirel and Mvera, 50 kt/annum), Biomer, Tianan (PHBV, 1 kT/annum, being expanded) and PHB Industrial (60 t/annum pilot plant).

As for processing, the material depolymerizes at high temperatures, therefore the temperature of 185 °C should not be exceeded. At 195 °C and up decomposition takes place, with emission of inflammable gases and darkening [58].

Some applications are: tubes for seedlings, injection and blown moulded containers, and films [for example, obtained with PHBH, or poly (3-hydroxybutyrate-co-3-hydroxy-hexanoate)].

For medical applications, the price of PHAs is already acceptable, although it is still too high for the commodity market, such as for packaging. Some important aspects to be improved in PHB are: strong degradation during processing[1], excessive brittleness and low elongation capacity[2]. The latter can be solved through the use of copolymers and blends.

The PHAs may undergo simultaneously hydrolytic, oxidative and enzymatic degradation. The PHAs degrading microorganisms are widely distributed in the environment. Just as bacteria and archaea, fungi are also excellent decomposers [59].

---

[1] PHB undergoes β-elimination reactions, which cleave molecules and form chains with terminal unsaturation [60].

[2] This is a consequence of the high crystallinity and the large spherulites formed, since the crystals nucleate slowly but grow fast.

In addition to their high degradative potential, many fungi have remarkable capacity to expand on the substrate surface, surrounding it with their hyphae, which release extracellular enzymes close enough to achieve the substrate [59].

PHAs are biodegradable in windrow composting, soil or marine sediments. The enzymes which are involved in the degradation of PHAs are depolymerases, hydrolases which may be intra- or extracellular and endo- or exoenzymes[3]. They cleave the chain into smaller fragments (hydroxy acids), either monomers or oligomers, used as sources of carbon and energy[4]. The chains must be previously cleaved by extracellular depolymerases, then soluble monomers and oligomers are introduced and metabolized within the cells. Unlike the aquatic environment, the soil environment makes it difficult to transport the enzymes secreted by microorganisms over long distances to the substrate [59]. Biodegradation under aerobic conditions results mostly CO_2, H_2O and biomass, whereas, in anaerobic conditions, it results mainly, in addition to the above components, CH_4 (methane). PHB is abiotically degraded by hydrolysis, with random cleavage of the ester linkages, especially at high temperatures (above 160-170 °C). It is also biotically degraded by many genera of bacteria, archaea and fungi, with biofilm formation. The rate of biodegradation with PHA depolymerase is 10^2-10^3 faster than that of hydrolytic degradation. The degradation of the amorphous regions is faster than that of the crystalline regions. In the PHBV copolymer, the crystallinity maintained constant, the addition of 3-hydroxy-valerate decreases the hydrolysis rate, as well as the enzymatic degradation rate. Longer side chains on the β-carbon further reduce the possibility of enzymatic attack [53].

4.7.2. Poly(E-caprolactone) or poly(6-hydroxy-hexanoate) (PCL)

PCL is a biodegradable polyester obtained from raw materials originating from petroleum, through ring opening polymerization of the lactone with suitable catalysts (Figure 2).

It has good resistance to water and organic solvents. PCL is a polymer stable against abiotic hydrolysis, which occurs slowly with molecular weight decrease. Its melting temperature is low, as its viscosity, facilitating its thermal processing. PCL may present spherulitic structure. It is a soft and flexible polymer, that may be used in blends with other biodegradable polymers, such as starch.

A major global manufacturer is Solvay (Capa, 5,000 t per year). Some applications are foamed food trays, bags, bioabsorbable medical items, replacement of gypsum in the treatment of broken bones, etc.

PCL may be degraded by many microorganisms, including bacteria and fungi, that are spread by soils and water bodies [56]. However, an initial stage of abiotic hydrolysis appears to be necessary [61]. The rates of hydrolysis and biodegradation depend on

---

[3] The enzymes may be classified as intra- or extracellular according their action inside or outside the cell, and also as endo- or exoenzymes, according their action inside or at the end of the substrate molecule.

[4] They are usually induced enzymes whose expression is repressed in the presence of other carbon sources such as glucose and organic acids.

molecular weight and crystallinity [40]. Pronounced biodegradation occurs with molecular weights below about 5,000 g-mol^{-1}. Abiotic and biotic degradation take place preferentially in the amorphous phase. Certain PCL-depolymerases, such as *Pseudomonas* lipase, can hydrolyse both amorphous and crystalline PCL phases. Enzymes from the two major classes of excreted esterases - lipases and cutinases - are able to degrade PCL and its blends [62]. Biodegradation causes surface erosion, without reduction of molecular weight [54].

4.7.3. Polylactic Acid (PLA)

PLA is an aliphatic polyester, derived from renewable resources, e. g. corn starch or sugarcane sucrose. It is a polymer produced from lactic acid (Figure 2), which is obtained from the fermentation of various carbohydrate species: glucose, maltose and dextrose from corn or potato starch; sucrose from beet or sugar cane; and lactose from cheese whey [63].

Lactic acid is chiral and has two isomers: the (S)-lactic acid (or L-(+)-lactic acid) and the (R)-lactic acid (ou D-(-)-lactic acid). The lactic acid monomer may be obtained by fermenting carbohydrate crops such as corn, sugar cane, cassava, wheat and barley, being eventually converted to lactide by means of a combined process of oligomerization and cyclization, with the use of catalysts. The synthesis of PLA (polylactate) was explored by Cargill, Hycail, Neste Oy, Shimadzu and Mitsui. Mitsui used a solvent based process to remove water azeotropically in the condensation polymerization process. Neste has obtained high molecular weight PLLA (i. e., L-PLA, or PLA from L-lactic acid) by joining low molecular weight precursors through urethane links. All the others use the dimer lactide process, with lactide ring opening polymerization. In the process using lactides, the additional step of dimerization of lactic acid increases production costs, but improves the control of molecular weight and end groups of the final polymer [38].

Lactic acid in the L isomeric form may be obtained by fermentation of glucose, while the lactic acid obtained via petrochemistry is in the racemic form (L/D = 1). Through the stereochemical control of lactic acid (ratio of D- and L- optical isomers), one can vary the crystallinity of PLA and also rate of crystallization, transparency, physical properties and even the biodegradation rate. For example, poly(L-lactide), or L-PLA is a semicrystalline polymer with glass transition temperature of 76 °C and melting at 180 °C, while poly(DL-lactide), or DL-PLA, is an amorphous polymer with glass transition at 58 ° C [64]. DL-PLA is used when it is important to have a homogeneous dispersion of the active species in the single-phase matrix, such as in devices for controlled release of drugs (in the same manner that PLAGA copolymers). L-PLA is preferred for applications where mechanical strength and toughness are required, such as in sutures and orthopedic appliances.

The mechanical properties are somewhat higher than those of polyolefins in general. PLA is a hard material, similar in hardness to acrylics (as methyl methacrylate). Because of its hardness, PLA fractures along the edges, resulting in a product that cannot be used. To overcome these limitations, PLA must be compounded with other materials to adjust the hardness [65].

The low glass transition temperature (see Table 2) is the reason for the limited resistance of PLA to heat, making PLA inadequate for hot drink cups, for example. PLA is suitable for frozen food or for packages stored at ambient temperatures.

PLA have been used in films for packaging, thermoformed and injection moulded disposable rigid containers (for example, food containers and trays), blown flasks and bottles, filaments, and biomedical uses (capsules for drug delivery, fibers for tissues and absorbable/degradable surgical sutures, and internal bone fixation implants). It is a polymer with consolidated use in the medical area, due to its biocompatibility and biodegradability in the human body. PLA-based resins may be modified to adapt to many applications, from disposable food-service items to sheet extrusion, and coating for paper [40].

Some of the leading manufacturers of PLA are: NatureWorks (Ingeo, capacity of 140 kt per annum), Teijin (BioFront, 1-5 tons per annum, Hisun Biomaterials (5,000 t per annum), Purac/Sulzer/Synbra, SK Chemicals, Biotech, Futerro, Mitsui Toatsu and Shimazu

The abiotic degradation of PLA takes place in two stages: a) diffusion of water through the amorphous phase, degrading that phase; and b) hydrolysis of crystalline domains, from the surface to the center [61]. The ester linkages are broken randomly. A semicrystalline material such as poly(L-lactate) presents a hydrolysis rate much lower than that from an amorphous material, such as poly (D,L-lactate), with half-lives of, respectively, one or a few years, and a few weeks. The hydrolysis is self-catalyzed by the acidity of the resulting carboxylic groups [66].

PLA can not directly be degraded by microorganisms, but requires first abiotic hydrolytic degradation, so that the microorganisms (mainly bacteria and fungi, which form biofilm) can metabolize the lactic acid and its oligomers dissolved in water. Abiotic hydrolysis takes place at temperatures above the glass transition temperature, i. e., at temperatures above 55 °C. Thus PLA is fully biodegradable in composting conditions of municipal waste plants, although it may need a few months to several years to be degraded under conditions of home composting, soil or oceans [35, 63, 67]. Furthermore, the PLA degrading microorganisms are not widespread in the environment [20, 61, 67].

The polymer passes the tests of compostability, provided that the thickness of the parts do not exceed around 2-3 mm. The extracellular enzymatic degradation consists of two steps: a) the enzyme is adsorbed on the polymer surface, through its binding site; and b) ester bonds are cleaved through the catalytic site of the enzyme [61]. The polymer chain ends are attacked preferentially. The biodegradation rate is a function of the crystallinity and the content of L-monomers [68]. Some enzymes (proteases) that may degrade PLA are proteinase K, pronase and bromelain. Subtilisin, a microbial serine protease, and some mammalian serine proteases, such as α-chymotrypsin, trypsin and elastase, could also degrade PLA [20, 61, 67].

4.7.4. Polyglycolic Acid (PGA)

It is the simplest linear polyester, consisting only of a methylenic group between the ester linkages. It may be synthesized in a way quite similar to that of PLA, by the ring opening polymerization of glycolide, that is the cyclic dimer of glycolic acid. The glass transition temperature of the homopolymer is about 35-40 °C and the melting temperature is about 225-230 °C. Glycolate is copolymerized with lactate in order to obtain a copolymer with appropriate stiffness and elongation capacity (known as PLAGA).

Among the manufacturers are American Cyanamid (Dexon), DuPont (Vicryl) and Kureha (Kuredux). The biodegradation of the PGA is usually faster than that of PLA, although an initial stage of abiotic hydrolysis appears to be necessary, since the polymer has a phase in the crystalline state and another in the amorphous glassy state [61].

PGA and its copolymers with lactic acid have very important medical applications: body absorbable sutures; ligaments reestablishment, through resorbable plates and screws; drugs of controlled release; grafting of arteries; etc. Although the homopolymers PGA and D-PLA are not biodegradable, copolymers of glycolic acid and D-lactic acid, which may still contain L-lactic acid, are usually biodegradable by lipase enzymes [67]. The degradation of PGA seems to follow the same steps of PLA: diffusion of water into the amorphous region, with degradation and erosion; hydrolytic attack of the crystalline region; and biodegradation of monomers and oligomers dissolved in water.

4.7.5. Aliphatic-aromatic polyesters

The aliphatic-aromatic polyesters have petrochemical origin, and are generally produced through traditional polycondensation reactions. The structure of two of them, PBAT and PBST, may be seen in Figure 2. They consist of aliphatic chain segments (residues of 1,4-butanediol, and of adipic or succinic acid), which provide flexibility, toughness, and biodegradability and aromatic segments (residues of terephthalic acid and 1,4-butanediol) , which impart mechanical strength and rigidity. PET, an aromatic polyester, decomposes very slowly in recalcitrant aromatic oligomers [69].

The degradation of the aliphatic-aromatic polymers may be oxidative, hydrolytic and enzymatic. The oxidative degradation occurs in the presence of oxygen gas and heat, ultraviolet radiation and/or mechanical stresses. The hydrolytic degradation occurs in presence of water, and is self-catalyzed by the acidity of the carboxylic acids, being more intense inside the part. The enzymatic hydrolysis uses non-specific enzymes, such as hydrolases and lipases, produced by an enormous variety of organisms, especially the mycelium-forming microorganisms (fungi and actinomycetes) [69]. The amorphous regions are degraded preferentially over the crystalline regions, both chemically and biologically. Interestingly, there is an inverse relationship between the melting temperature of the polyester and its rate of biodegradation, indicating that the crystalline characteristics are a very important factor in its biodegradability [69, 70]. The polyesters that behave like elastomers at the degradation temperature, undergo enzymatic degradation from the moment in which they are placed in the disposal environment, showing surface erosion. The polyesters that behave like glass at the degradation temperature are enzymatically degraded only at the end of the degradation process, from the by-products of the preliminary abiotic degradation. Although the aliphatic-aromatic polyesters present high degradability in industrial composting, their rate of degradation in soil and water bodies is much lower, and their degradability under anaerobic conditions is even lower [69].

Some applications are the same typical for LDPE: transparent blown films, mulch films for agriculture, films for package and bags, and also blown bottles, filaments, injection moulded and thermoformed products [69].

Poly(butylene adipate-co-terephthalate) (PBAT): Some of its main applications are films (mulch, containers, bags), filaments, thermoformed and injection moulded products, and blown bottles. Two products in the market are Ecoflex (BASF, 14,000 t per year) and MaterBi (former EasterBio/Eastman, now Novamont, 15,000 t per year). For some applications, PBAT has a very low stiffness, and may be mixed with PHB or PLA, for example. It may also be mixed with thermoplastic starch [69, 71].

Poly(butylene succinate-co-terephthalate (PBST): Some of its main applications are blown films, filaments, blown and injection moulded containers, thermoformable cups and trays, paper coating, etc. A product in the market is Sorona/Biomax). PBST has good mechanical properties, reasonable processing and biodegrades slowly [69, 72].

4.8. Standards for biodegradation tests - hydrobiodegradable and inherently biodegradable polymers

There is not a standard test of biodegradability of universal validity. For the hydro-biodegradable and inherently biodegradable polymers, it is common the use of patterns for testing compostability. Some standards for compostability are: EN 13432, ASTM D6400, ASTM D5338, ISO 14855 and 17088, and Australian Standard 4736. These are standards for biodegradation in the special conditions found in industrial composting, that require short timescales and rapid CO_2 emissions. There are also standards for biodegradation in soils and aquatic environments but they are less used.

Finally, the third group of biodegradable polymers consists of oxo-biodegradable materials, that is, those that need to undergo the chemical process of oxidation (combination with oxygen, which leads to breakage of the molecules) before they become biodegradable. In general, all the traditional widely used plastic materials are oxo-biodegradable, that is, over time undergo oxidative degradation, what leads to the breakdown of their molecules into smaller fragments, highly oxygenated, capable of being biodegraded. However the time scale to complete degradation and biodegradation is too long: it takes several decades [25, 26, 73]. To accelerate this process, organic salts (such as stearates) of transition metals (such as iron, manganese and cobalt) are added, so reducing the time required for degradation and biodegradation to some years [74]. Such additives are known as pro-oxidants or pro-degrading. Until now, these salts have shown no toxicity to animals, plants or microorganisms, being rather micronutrients to them. To this group belong lignin, lignocellulose, natural rubber and polybutadiene (without the need of pro-oxidant additives) as well as traditional plastics, such as polyethylene, polypropylene, polystyrene and PET, all these formulated with pro-oxidant additives.

4.9. Oxo-biodegradable polymers

Oxo-biodegradable polymers are the polymer materials that present in their formulation pro-oxidant and antioxidant additives, so as to provide a planned period of useful life, after which the materials begin to degrade oxidatively, the residues being inherently biodegradable. It is also possible that the polymer contains pro-oxidant chemical structures, such as double bonds and atoms susceptible to attack by free radicals. The oxidation process is called peroxidation, and occurs through a free radical mechanism [1]. The first step is the formation of a free radical in the polymer (i. e., a macrorradical), through the homolytic cleavage of a CH or a CC bond, that could take place because of the heat, the UV radiation or the mechanical stress (e. g., shear or elongation during the processing of the material, or the wind or ocean waves action). Then the polymer radical formed may capture an oxygen molecule, generating a peroxide radical, which after capturing a hydrogen atom bound to the polymer will form a hydroperoxide bond. The hydroperoxide decomposes over time, generating an alkoxy and a hydroxyl radical. The decomposition can be accelerated by about 10^2 times with the use of catalysts based on organic salts of Fe, Mn, Co, etc [27, 28]. These salts also help to carry oxygen to the polymer molecules. The hydroxyl radical may capture hydrogen atoms, generating new macrorradicais. The alkoxy radical can recombine, generating a ketone group, or breaking the molecule, generating a new radical. The ketone group is susceptible to degradation by UV, which can cause rupture of the chain, by the mechanisms of Norish I and II. The free radical reactions involving organic polymer and oxygen generate many different molecules, which may contain the groups hydroxyl, carbonyl, ether, ester, carboxyl, etc., and also insaturations [25, 26]. Therefore, the final product from the polymer abiotic degradation generally consists of small and strongly oxygenated molecules, capable of crossing the cell wall (if present) and membrane, and that are metabolyzed in the cytoplasm of microorganisms with the help of the available enzymes, which depend on the chemical structure of the oligomers and the genetic potential of the organism [4].

The antioxidant additives present in the formulation of a polymer resin may have the function of protecting it against degradation by deactivating the free radicals formed (primary antioxidants) or by decomposing hydroperoxides formed (secondary antioxidants). The former are useful during the service life of the material at ambient temperatures, whereas the latter are most useful when processing the material at elevated temperatures [75].

Molecular weight reduction is generally a consequence of oxidative degradation (being e. g. the case of PE, PP and PS), what causes the collapse of the mechanical properties, and consequent disintegration (fragmentation) of the part [27, 28, 76]. Oxidative degradation also causes the incorporation of oxygen atoms in the chains and the rise of double bonds. The residue from abiotic degradation of a plastic material is no longer plastic, but a complex mixture of unsaturated and oxygenated oligomers, showing some hydrophilicity and being biodegradable by a large number of genera of naturally occurring microorganisms [25, 26, 28].

Characteristics	Days of exposure						
	0	54	80	136	190	242	299
Average molar mass[a]							
Mn, g mol^{-1}	10500	10700	8630	3450	3310	3300	2620
Mw, g mol^{-1}	183000	112000	80500	13000	15000	9210	7850
Polydispersity (Mw/Mn)	17.43	10.47	9.33	3.77	4.53	2.79	3.00
Carbonyl index[b]	0.000	–	0.187	0.427	0.580	0.675	–
Degree of crystallinity (%)[c]	58	59	–	63	66	70	–
Mechanical properties[d]							
Stress at fracture (MPa)	52 ± 4	29 ± 6	16 ± 2	0	0	0	0
Strain at fracture (%)	400±40	230±40	60 ± 9	0	0	0	0

[a] From SEC. Mn, Mw = number and weight average molar masses, respectively.
[b] From FTIR spectroscopy.
[c] From DSC, second heating runs.
[d] From tension tests. Values expressed with their standard errors.

Table 3. Characteristics of oxo-biodegradable polyethylene films subjected to weathering for different periods of time [28].

In Table 3, it is possible to observe the changes in molecular weight and carbonyl concentration, as well as the consequent changes in mechanical properties, that occur with the outdoor weathering of films of a HDPE/LDPE blend for several months. The increase in crystallinity can be explained by the higher freedom of motion of smaller polymer chains, that could be rearranged in more crystalline structures [28].

The rates of biodegradation of the residues from oxo-biodegradable materials are usually lower than those of most hydro-biodegradable materials. It generally takes a few years to quantitative biodegradation of the oxo-bio materials, depending on resin type, environmental conditions and formulation of additives used. Figure 3 shows the mineralization curves for an oxo-biodegradable HDPE/LLDPE blend biodegraded in composting conditions at two different temperatures.

After a certain level of oxidative degradation, biofilms can be observed on the oxidized polymer residues [25-27]. These biofilms mainly consist of fungi and bacteria, although archaea, algae and protozoa may also be present. The oxo-bio materials may be recycled with conventional polymer materials, provided that the resins still contain antioxidant additives in concentration sufficient to prevent oxidative degradation during processing and service life. Some people consider that the residues of oxo-biodegradable polymers contain toxic metals, but so far there is no evidence of toxicity of them to plants or animals. Instead, the cations of Fe, Co and Mn are micronutrients, acting as cofactors of enzymes. At very high concentrations, these cations may damage the plants, even because they increase the osmotic pressure of the environment and may dehydrate the root cells.

Conventional materials, such as polyolefins and polystyrene, can be converted to oxo-biodegradable materials by adding 1-5% (typically) of additives, what tend to increase the total cost of the resin in 5-15%. Some of the leading manufacturers of oxo-bio additives are Symphony Environmental (d2w), EPI (TDPA), Wells (reverse), Willow Ridge (PDQ) and others.

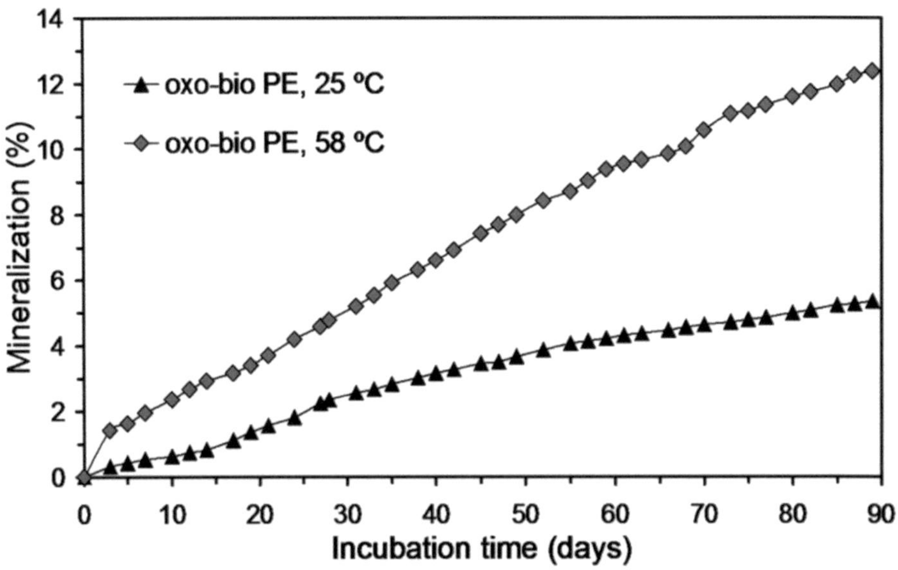

Figure 3. Biodegradation of polyethylene films as a function of incubation time at 25 and 58 °C in compost/perlite, at 50% relative humidity [28].

4.10. Standards for biodegradation tests - oxo-biodegradable polymers

Standards for slowly biodegradable products are a challenge, because it becomes very difficult to simulate in a laboratory the biodegradation that takes place in real environments. In laboratory microcosms, the environment is isolated, without the possibility of free mass exchange with the surroundings. So there may be accumulation of metabolites produced by microorganisms, to the point where they become toxic and disrupt microbial growth. Thus, it is possible that only the beginning of biodegradation can be observed, but this does not mean that biodegradation would be interrupted in real conditions.

It is not possible to provide a specific timescale in a general standard for oxo-biodegradable polymers, (as distinct from a standard for industrial composting) because the conditions found in industrial composting are specific and the conditions found in the open environment are variable. Also, the time taken for oxo-biodegradable plastic to commence and complete the processes of degradation and biodegradation can be varied.

Some standards for the analysis of oxo-biodegradable products are: ASTM 6954-04 (USA), BS8472 (UK), SPCR 141 (Sweden) and UAE standard 5009:2009 (United Arab Emirates). They usually require three test levels: 1) abiotic degradability (decrease of molecular weight and mechanical properties, low gel formation); 2) biodegradability (biofilm formation, release of CO_2, tested with the residue obtained at level 1); and 3) ecotoxicity (to animals, plants and microorganisms) of the residues from levels 1 and 2. The ecotoxicity tests may follow the OECD standards, for example, to algae (OECD 201), microcrustaceans (OECD 202), fish (OECD 203), earthworms (OECD 207) and plants (OECD 208). A highly sensitive method for measuring low rates of biodegradation was developed by Chiellini et al. [77], and used with modifications by Fontanella et al. [27] and Ojeda et al. [28].

4.11. Blends

In order to maintain a good compromise among biodegradability, chemical and physical properties, and costs, mixtures or blends of polymers of any of the three groups mentioned above (i.e., inherently, hydro- and oxo-biodegradable polymers) have appeared on the market. Some biodegradable polymers are very rigid and brittle (e. g., PHB and PLA), while others are very soft and flexible (e. g., PCL and PBAT). The mixture (blend) of two or more different polymers may lead to a blend of interesting intermediate mechanical properties. The processability of the final material may also be improved, included here the resistance to chemical decomposition during processing, among other features. Examples of commercial blends are: modified starch + PBAT (e. g., Novamont - MaterBi; Corn Products/Basf - Ecobras) for films, thermoformed and injection moulded parts; PBAT + PLA (e. g., BASF - Ecovio) for films, thermoformed and extruded parts; PLA + starch (e. g., Cereplast - Cereplast Compostables) for bags and packaging, injection moulded and extruded articles for food, pens, etc.

5. Non-biodegradable polymers derived from renewable resources

Some polymer materials are produced wholly or partly from renewable (the term *renewable* here is very limited, as was previously explained) raw materials. Some examples are listed below.

Braskem - *green polyethylene*: is the traditional polyethylene, but derived from ethylene produced with ethanol from sugar cane. DuPont - *PTT* - poly(trimethylene terephthalate): one of its monomers, 1,3-propanediol, is obtained from corn or sugar beets. Coca-Cola Co: *PlantBottle*: PET bottle made from ethylene glycol obtained from alcohol derived from sugar cane and molasses. In addition, the PP cap is slightly smaller. The proportion of raw materials obtained from fossil fuels (oil, coal and gas) and obtained from plants can be found through analysis of the proportion of carbon-12 to carbon-14 present in the polymer, since fossil fuels contain virtually no carbon-14 whose half-life is about 5730 years (see ASTM D6866-11).

6. Conclusion - Future perspectives of environmental friendly polymer materials

The environmental impacts produced by conventional polymers on the planet are now clearly observed. As a consequence, these materials, especially plastic bags, have suffered many attacks in several countries, and alternative solutions to their use have been encouraged. However, to date, no definitive and universal solution for the replacement of conventional polymer materials has emerged. And it is very likely that the path to be taken is exactly this: to promote the diversity of materials available, according to the local diversities of the planet. Depending on environmental conditions, available raw materials, local cultures, industrial parks, etc., different polymeric or not polymeric materials may be elected as the most appropriate for a given population at a given time in history. The effects of globalization and the current facility of transportation from a continent to another can not be neglected.

The continuous worldwide implementation of renewable forms of energy might permit the use of petroleum as a petrochemical feedstock for many more years. However, mechanical, chemical and energy recovery would need to be improved greatly, and the products difficult to recycle should be mixed with pro-oxidant additives. Another interesting solution is the production of biodegradable polymer materials from agricultural and other organic wastes, such as PHAs produced from stover. Anyway, composting units should be encouraged in all countries around the world.

Author details

Telmo Ojeda
Environmental Sciences – Federal Institute for Education, Science and Technology (IFRS) - Porto Alegre, Brazil

7. References

[1] Scott G (1999) Polymer and the environment. Cambridge: RSC Paperbacks. 132p.
[2] Stevens ES (2003) What makes green plastics green? BioCycle 24: 24-27
[3] Al-Salem SM, Lettieri P, Baeyens J (2009) Recycling and recovery routes of plastic solid waste (PSW): a review. Waste management 29: 2625-2643.
[4] Atlas RM, Bartha R (1998) Microbial ecology: fundamentals and applications, 4th. Menlo Park: Benjamin/Cummings. 694p.
[5] Edwards C, Fry JM (2011) Life Cycle Assessment of Supermarket Carrier Bags. Environmental Agency, Bristol, UK. Available: http://www.biodeg.org/files/uploaded/Carrier_Bags_Report_EA.pdf. Accessed 2012 May 25.
[6] James K, Grant T. LCA of degradable plastic bags. Centre for design at RMIT university. Available: www.europeanplasticfilms.eu/docs/Jamesandgrant1.pdf. Acessed 2012 May 25.

[7] Tabone MD et al. (2010) Sustainability metrics: life cycle assessment and green design in polymers. Environ Sci Technol. 44:8264-8269.
[8] Scott G. (2002) Why biodegradable polymers? In: Scott G, editor. Degradable polymers. Dordrecht/Boston/London: Kluwer Academic Publishers. pp. 1-15.
[9] Thornton J (2002) Environmental impacts of polyvinyl chloride building materials. Healthy building network. Available: http://www.healthybuilding.net/pvc/Thornton_Enviro_Impacts_of_PVC.pdf. Acessed 2012 May 25.
[10] Titow WV (1984) *PVC technology*, 4th. ed., Elsevier. pp. 263-286.
[11] Okada H et al. (2008) Direct evidence revealing structural elements essential for the high binding ability of bisphenol A to human estrogen-related receptor-gamma. Environ. Health Perspect. 116(1):32-38.
[12] Vom Saal S, Myers JP (2008) Bisphenol A and risk of metabolic disorders. JAMA. 300(11):1353-1355.
[13] Vogel S (2009) The politics of plastics: the making and unmaking of bisphenol A' safety. American journal of publich health 99(S3):559-566.
[14] Ogiue-Ikeda M et al. (2008) Rapid modulation of synaptic plasticity by estrogens as well as endocrine disrupter in hippocampal neurons. Brain research reviews 57(2):363-375.
[15] Soto AM, Sonnenschein C (2010) Environmental causes of cancer: endocrine disruptors as carcinogens. *Nature Reviews Endocrinology* 6 (7): 363–370.
[16] Yanagiba Y et al. (2008) Styrene trimer may increase thryroid hormone levels via down-regulation of the aryl hydrocarabon receptor (AhR) target gene UDP-glucuronosyltransferase. Environmental health perspectives 116(6):740-745.
[17] Schumann H-D, Thiele UK (1996) Polyester producing plants - principles and technology. Landsberg/Lech: Moderne Industrie. 72 p.
[18] Mark JE (1999) Polymer data handbook, Oxford: Oxford University Press. 1012 p.
[19] Alexander M (1973) Nonbiodegradable and other recalcitrant molecules. Biotechnology and bioengineering. 15: 611-647.
[20] Shimao M (2001) Biodegradation of plastics. Current opinion in biotechnology. 12: 242-247.
[21] Gugumus F (1992) Stabilization of plastics against thermal oxidation. In: Gächter H, Müller F, editors. Plastics additives handbook, 3rd ed. Munich: Hanser Publishers.
[22] Nakajima-Kambe T et al. (1999) Microbial degradation of polyurethane, polyester polyurethanes and polyether polyurethanes. Applied microbiology and biotechnology. 51(2): 134-140.
[23] Scott G, Wiles DM (2001) Programmed-life plastics from polyolefins: a new look at sustainability. Biomacromolecules. 2(3): 615-622.
[24] Scott G (1997) Abiotic control of polymer biodegradation. Trip. 5(11): 361-368.
[25] Khabbaz F, Albertsson A-C, Karlsson S (1999) Chemical and morphological changes of environmentally degradable polyethylene films exposed to thermo-oxidation. Polymer degradation and stability. 63:127-138.
[26] Albertsson A-C et al. (1995) Degradation product pattern and morphology changes as means to differentiate abiotically and biotically aged degradable polyethylene. Polymer. 36(16): 3075-3083.

[27] Fontanella S et al. (2010) Comparison of the biodegradability of various polyethylene films containing pro-oxidant additives. Polymer degradation and stability. 95: 1011-1021.

[28] Ojeda TFM et al. (2009) Abiotic and biotic degradation of oxo-biodegradable polyethylenes. Polymer degradation and stability. 94: 965-970.

[29] Azapagic A, Emsley A, Hamerton I. (2003) Polymers: the environment and sustainable development. Wiley. pp. 22-23.

[30] Valle MLM et al. (2004) Degradação de poliolefinas utilizando catalisadores zeolíticos. Polímeros: ciência e tecnologia. 14: 17-21.

[31] Jacques RJS et al. (2005) Anthracene biodegradation by *Pseudomonas* sp. isolated from a petrochemical sludge landfarming site. International biodeterioration & biodegradation 56: 143-150.

[32] Bastioli C (2004) Starch-polymer composites. In: Scott G, editor. Degradable polymers. Dordrecht/Boston/London: Kluwer academic publishers. pp. 133-161.

[33] Lu DR, Xiao CM, Xu SJ (2009) Starch-based completely biodegradable polymer materials. Express polymer letter. 3(6): 366-375.

[34] Bastioli C (2005) Starch-based technology. In: Bastioli C, editor. Handbook of biodegradable polymers. Shawbury, Shrewsbury, Shropshire: Rapra Technology. pp.257-286.

[35] Australian government - department of the environment, water, heritage and the arts. Degradable plastics. Available: http://www.environment.gov.au/settlements/ publications/waste/degradables/biodegradable exec-summary.htm). Accessed: 2012 May 25.

[36] Nakamura EM et al. (2005) Study and development of LDPE/starch partially biodegradable compounds. Journal of materials processing technology. (162-163): 236-241.

[37] Erlandsson B, Karlsson S, Albertsson A.-C (1997) The mode of action of corn starch and pro-oxidant system in LDPE: influence of thermo-oxidation and UV-irradiation on the molecular weight changes. Polymer degradation and stability. 55: 237-245.

[38] Chiellini E, Chiellini F, Cinelli P (2002) Polymers from renewable resources. In: Scott, G, editor. Degradable polymers. Dordrecht/Boston/London: Kluwer. pp. 165-178.

[39] Lynd RL et al. Microbial cellulose utilization: fundamentals and biotechnology. Microbiology and molecular biology reviews. 66(3): 506-577.

[40] Gross RA, Kalra B (2002) Biodegradable polymers for the environment. Science. 297: 803-807.

[41] Piskin E (2002) Biodegradable polymers in medicine. In: Scott G, editor. Degradable polymers, Dordrecht/Boston/London: Kluwer Academic. pp. 347-349.

[42] Chandra R, Rustgi R (1998) Biodegradable polymers. Prog. Polym. Sci. 23: 1273-1335.

[43] Martone PT et al. (2009) Discovery of lignin in seaweed reveals convergent evolution of cell-wall architecture. Current biology. 19(2): 169–75.

[44] Sjöström E (1993) Wood chemistry: fundamentals and applications. Academic Press. (ISBN 0-12-647480-X)

[45] Boerjan W, Ralph J, Baucher M (2003) Lignin bios. Ann. rev. plant biol. 54(1):519–549.

[46] Tuomela et al. (2000) Biodegradation of lignin in a compost environment: a review. Bioresource technology. 72(2): 169-183.
[47] Tecnaro - Arboform. Available: http://www.tecnaro.de/english/arboform.htm. Accessed: 2012 May 25.
[48] Slater S et al. (2004) Evaluating the environmental impact of biopolymers. In: Steinbüchel A, editor. Biopolymers, v. 10. Wiley Interscience. pp. 473-480.
[49] Vinyl alcohol polymers. Encyclopedia of polymer science and technology, John Wiley & Sons, v.8, p.399-436.
[50] Rudnik E (2008) Compostable polymer materials. Elsevier. 211p.
[51] Swift G (2002) Environmentally biodegradable water-soluble polymers. In: Scott, G. Degradable polymers – principles and applications, Kluwer Academic Publishers. pp. 379-412.
[52] Yubin LV et al. (2011) Research development of biodegradable modification of poly(vinyl alcohol) film. Advanced materials research. 380: 234-237.
[53] Reddy CSK, Ghai R, Rashami, Kalia VC (2003) Polyhydroxyalkanoates: an overview. Bioresource technology. 87: 137-146.
[54] Jendrossek D (2001) Microbial degradation of polyesters, Advances in biochemical engineering/biotechnology. 71: 293-325.
[55] Braunegg G (2002) Sustainable poly(hydroxyalkanoate) (PHA) production. In: Scott G, editor. Degradable polymers – principles and applications. Dordrecht/Boston/London: Kluwer Academic Publishers. pp. 235-293.
[56] Suyama T et al. (1998) Phylogenetic affiliation of soil bacteria that degrade aliphatic polyesters available commercially as biodegradable plastics. Applied and environmental microbiology. 64(12): 5008-5011.
[57] Sudesh K, Doi Y (2005) Polyhydroxyalkanoates. In: Bastioli C, editor. Handbook of biodegradable polymers. Shawbury, Shrewsbury, Shropshire: Rapra Technology. pp. 219-256.
[58] Biomer biopolyesters. Available: http://www.biomer.de. Accessed: 2012 May 25.
[59] Sang BI, Hori K, Unno H (2002) Fungal contribution to in situ biodegradation of poly(3-hydroxybutyrate-co-3-hydroxyvalerate) film in soil. Appl. microbiol. biotechnol. 58: 241-247.
[60] Chodak I (2002) Polyhydroxyalkanoates: properties and modification for high volume applications. In: Scott G, editor. Degradable polymers – principles and applications. Dordrecht/Boston/London: Kluwer Academic Publishers. pp. 295-319.
[61] Li S, Vert M (2002) Biodegradation of aliphatic polyesters. In: Scott, G. Degradable polymers – principles and applications. Dordrecht/Boston/London: Kluwer Academic Publishers. pp. 71-131.
[62] Tucker N, Johnson M, editors (2004) Low environmental impact polymers, UK: Rapra. 360p.
[63] Garlotta DA (2001) Literature review of poly(lactic acid). Journal of polymers and the environment. 9(2): 2001.
[64] Reeve MS et al. (1994) Polylactide stereochemistry: effect on enzymatic degradability. Macromolecules. 27: 825-831.

[65] Briassoulis D (2004) An overview on the mechanical behaviour of biodegradable agricultural films. Journal of polymers and the environment. 12(2): 65-81.

[66] Hakkarainen M (2002) Aliphatic polyesters: abiotic and biotic degradation and degradation products. Advances in polymer science. 157: 113-138.

[67] Tokiwa Y et al. (2009) Biodegradability of plastics. International journal of molecular sciences. 10: 3722-3742.

[68] Auras R, Lim L-T (2010) *Poly(lactic acid): synthesis, structures, properties, processing and applications*. Wiley, 2010.

[69] Müller R-J (2005) Aliphatic-aromatic polyester. In: Bastioli C, editor. Handbook of biodegradable polymers. UK: Rapra Technology. pp. 303-337.

[70] Marten E, Müller R-J, Deckwer W-D (2005) Studies on the enzymatic hydrolysis of polyesters. II. Aliphatic-aromatic copolyesters. Polymer degradation and stability. 88(3): 371-381.

[71] Witt U et al. (2001) Biodegradation of aliphatic-aromatic copolyesters: evaluation of the final biodegradability and ecotoxicological impact of degradation intermediates. Chemosphere 44: 289-299.

[72] Ki HC, Park OO (2001) Synthesis, characterization and biodegradability of the biodegradabale aliphatic-aromatic random copolyesters. Polymer 42: 1849-1861.

[73] Ojeda T et al. (2011) Degradability of linear polyolefins under natural weathering. Polymer degradation and stabilility. 96: 703-707.

[74] Scott G (2002) Degradation and stabilization of carbon-chain polymers. In: Scott G, editor. Degradable polymers – principles and applications. Dordrecht/Boston/London: Kluwer Academic Publishers. pp. 27-50.

[75] Schwarzenbach K et al. (2009) Antioxidants. In: Zweifel H, Maier RD, Schiller M, editors. Plastics Additives Handbook. Cincinnati: Hanser. pp.1-137.

[76] Ojeda T et al. (2009) Abiotic and biotic degradation of oxo-biodegradable foamed polystyrene. Polymer degradation and stability. 94: 2128-2133.

[77] Chiellini E, Corti A, Swift G (2003) Biodegradation of thermally-oxidized, fragmented low-density polyethylenes. Polymer degradation and stability 81: 341-351.

Chapter 2

Emulsion Polymerization: Effects of Polymerization Variables on the Properties of Vinyl Acetate Based Emulsion Polymers

Hale Berber Yamak

Additional information is available at the end of the chapter

http://dx.doi.org/10.5772/51498

1. Introduction

Emulsion polymerization is scientifically, technologically and commercially important reaction. It was developed during the World War II because of the need to replace the latex of natural rubber. The synthetic rubbers were produced through radical copolymerization of styrene and butadiene [1-5]. Today, emulsion polymerization is the large part of a massive global industry. It produces high molecular weight colloidal polymers and no or negligible volatile organic compounds. The reaction medium is usually water and this facilitates agitation and mass transfer, and provides an inherently safe process. Moreover the process is environmentally friendly. Other domains justifying also a big production are that of the versatility of the reaction and the ability to control the properties of the emulsion polymers produced. Because of these unique properties, the industry including waterborne polymers produced by emulsion polymerization continues to expand incrementally. The wide range of the products which include synthetic rubbers, toughened plastics, paints, adhesives, paper coatings, floor polishes, sealants, cement and concrete additives, and nonwoven tissues can be produced by emulsion polymerization. More sophisticated applications are also found in cosmetics, biomaterials and high-tech products.

In the emulsion polymerization products, poly(vinyl acetate) emulsion homopolymer and vinyl acetate based emulsion copolymers have a great importance in industrial aspect as well as scientific aspect. They account for 28% of the total waterborne synthetic latexes. Poly(vinyl acetate) emulsion homopolymer was the first synthetic polymer latex to be made on a commercial scale [6-9]. Its production is growing steadily in both actual quantities and different applications. The largest volume applications are in the area of coating and adhesive. It offers good durability, availability at low cost, compatibility with other materials, excellent adhesive characteristic, and ability to form continuous film upon drying of the emulsions. In addition, vinyl acetate can mostly be copolymerized with ethylene, acrylic esters, versatic ester, or vinyl chloride. So it is possible to overcome some poor properties of the vinyl acetate homopolymer such as weak resistance against alkaline and water, being hydrolysis, and impractical values of glass transition temperature and minimum film forming temperature for many applications by these copolymerizations.

Otherwise, the reaction variables play a determinative role on the emulsion copolymerization reactions and the properties of the resulting copolymers due to the significant differences between the properties of vinyl acetate and other comonomers. The emulsion polymerization of vinyl acetate possesses the rather typical properties in comparison the emulsion polymerizations of the comonomers. Vinyl acetate has high water solubility, a high monomer-polymer swelling ratio, and a high chain transfer constant. Thus, the type of emulsion polymerization process (batch, semi-continuous or continuous) is a very important factor affecting the polymerization mechanism and the final properties of the copolymers. There are also many different variables such as agitation speed, initiator type and concentration, emulsifier type and concentration, feeding policy, feeding rate and temperature. Eventually, the production of vinyl acetate based copolymer latexes in a wide range of molecular, particle-morphological, colloidal, physical and film properties can be possible for use in wide variety of applications by change in molecular structure of the comonomer, copolymer composition and the emulsion polymerization variables.

2. Emulsion polymerization

Emulsion polymerization is a complex process in which the radical addition polymerization proceeds in a heterogeneous system. This process involves emulsification of the relatively hydrophobic monomer in water by an oil-in-water emulsifier, followed by the initiation reaction with either a water-soluble or an oil-soluble free radical initiator. At the end of the polymerization, a milky fluid called "latex", "synthetic latex" or "polymer dispersion" is obtained. Latex is defined as "colloidal dispersion of polymer particles in an aqueous medium". The polymer may be organic or inorganic. In general, latexes contain 40-60 % polymer solids and comprise a large population of polymer particles dispersed in the continuous aqueous phase (about 10^{15} particles per mL of latex). The particles are within the size range 10 nm to 1000 nm in a diameter and are generally spherical. A typical of particle is composed of 1-10000 macromolecules, and each macromolecule contains about 100–10^6 monomer units [10-16].

The earliest literature references to produce synthetic latex (the term first referred to the white, sticky sap of the rubber tree) are patents originated from Farbenfabriken Bayer in the years 1909 to 1912 [1-3]. These studies involved polymerization of dien monomers in the form of aqueous emulsions which are stabilized by gelatin, egg white (protein), starch, flour, and blood serum as protective colloids to produce something resembling natural rubber latex. Initiation of polymerization depended on aerial oxygen. But these attempts and other similar studies that followed them were substantially different from what is known today as "emulsion polymerization". In 1929, Dinsmore, who was working for The Goodyear Tire &

Rubber Company, was the first to be granted a patent to produce a synthetic rubber in the presence of soap as emulsifying agent [4]. This was followed by addition of a free radical initiator (water- or monomer-soluble peroxides), which led to polymerization of the emulsified monomer [5]. The reasons for regarding it as an "emulsion polymerization" were the addition of soap, presumably added as an emulsifying agent in the first instance, as well as of a protein-aceous protective colloid, and the assuming that polymerization took place in the emulsified monomer droplets. Later, the practice of emulsion polymerization grew rapidly and industrial-scale production started in the mid-1930. The major developments in emulsion polymerization took place around the Second World War as a result of the intensive collaborative efforts between academia, industry and government laboratories. During and after World War II, the production of many types of latex both in homopolymers and copolymers of different composition was achieved by using different monomers such as butadiene, styrene, acrylic esters, acrylonitrile and vinyl acetate. A wide variety of initiating systems were used. Conversions of the polymerization reactions were increased. Later, the works of that period were published in reports and books [7-9,17-18]. Otherwise very few papers on the subject were published in the scientific journals during the period 1910-1945, in comparison with the patents [19-23]. From 1945 to the present, numerous books including the literature of emulsion polymerization, its mechanism, kinetics and formulation, and many other topics related to the emulsion polymerization have been published [10-18, 24-31]. In addition, many conferences on the emulsion polymerization have been organized by different institutes since 1966, and the proceedings/books of them have been published. There are also an excessive number of papers on the emulsion polymerization and related subjects in literature. The number of publications on this subject continues to increase steadily.

Looking at the historical development of the emulsion polymerization, it is seen that the trigger factor in this development was the necessity for synthetic rubber in the wartime. The production of styrene/butadiene rubber (SBR) satisfied this requirement. Today, millions of tons of synthetic latexes are produced by the emulsion polymerization process for use in wide variety of applications. In the synthetic latexes, the most important groups are styrene/butadiene copolymers, vinyl acetate homopolymers and copolymers, and polyacrylates. Other synthetic latexes contain copolymers of ethylene, styrene, vinyl esters, vinyl chloride, vinylidene chloride, acrylonitrile, cloroprene and polyurethane.

Styrene/butadiene latexes account for 37% of the total waterborne synthetic latexes. They are widely used for tires and molded foam. They mostly consist of 70-75% butadiene by weight and 30-25% styrene by weight for use as general-purpose rubbers. Their carboxylated forms contain acrylic, methacrylic, maleic, fumaric or itaconic acid whose carboxylic groups provide stabilization of the polymer particles and a good interaction with fillers and pigments. They are used in carpet-backing and paper-coating applications. Styrene/butadiene ratios are commonly 50/50 and 60/40 (by weight). In these compositions, these copolymers are still rubbery at normal ambient temperatures, and the latex particles readily integrate to form coherent films as the latex dries. Styrene/butadiene copolymers become non-rubbery by increasing the styrene content in the copolymer composition, e.g., 85/15 and 90/10 by weight, and are used as organic stiffening and reinforcing fillers. When styrene is replaced by acrylonitrile, elastic and solvent resistant emulsion copolymers are obtained, which are used for dipping goods. Additionally, polychloroprene rubbers obtained by emulsion polymerization offer great resistance to chemicals and atmospheric ozone.

Acrylic latexes include pure acrylics and styrene acrylics, which are about 30% of produced waterborne synthetic latexes. Acrylic monomers comprise the monomeric alkyl esters of acrylic acid and methacrylic acid, and also their derivatives. The most used acrylic monomers in the emulsion polymerizations are methyl-, ethyl-, butyl- and 2-ethylhexyl-acrylate, methyl methacrylate, and acrylic- and methacrylic-acid. Homopolymer latexes of these monomers are used as exterior or interior coatings, binder for leather, textiles and paper, and as adhesives, laminates, elastomers, plasticizer and floor polishes. These latexes are stable, have good pigment binding and durability. The copolymerizations of these esters with styrene in an enormous range of accessible copolymer compositions offer almost unlimited opportunities to choose for the glass transition temperature, the minimum film forming temperature, the hydrophilic/hydrophobic properties and morphology design.

Vinyl chloride/vinylidene chloride monomers can be polymerized by emulsion polymerization. Poly(vinyl chloride) (E-PVC) product is mostly applied as the dried form. It is spray-dried and milled to form fine powders ("crumbs") which is mixed with plasticizer to form a plastisol, i.e., dispersion of poly(vinyl chloride) particles in liquid organic media. The plastisol is poured into molds to make rubber dolls, shower curtains, embossed wall coverings and many of other common objects. In packing materials, especially for food packaging, the films of poly(vinylidene chloride) latexes are used, which are highly impermeable for both, oxygen and water vapor.

In both of the large volume and the small volume applications, this variety of emulsion polymers and the widespread use of them are caused by emulsion polymerization which offers many kinetic and technological advantages over other polymerization methods. The dispersion medium is water that provides inexpensive, nonflammable, nontoxic and relatively odorless systems. This polymerization has relative simplicity of the technological process. It is possible to produce high molecular weight polymer at a high reaction rates, and the viscosity of latex is independent of the molecular weight. Thus the producing of high solids content emulsions with low viscosity can be achieved in contrast to solutions of polymers. This method offers better temperature control during polymerization due to more rapid heat transfer in the low viscosity emulsion. There are possibilities of feeding the ingredients at any stage of reaction and the achievement of many copolymerizations that consist of different monomers in wide variety physical properties. The control of undesirable side reactions such as chain transfers, and the range and distribution of particle size can also be obtained. In addition, the dry form of emulsion polymers can be used in many applications as well as the use of latex itself (in wet form). For the formation of dry emulsion polymer, the polymer is isolated by coagulating the latex, filtering off the aqueous medium, and washing the derived crumb. The dried crumbs of polymers may be used as molding resins, or in some cases. Nevertheless, there are some disadvantages of the emulsion polymerization. The presence of emulsifiers and other ingredients in the system constitutes unavoidable contamination to the polymer. The separation of the polymer from dispersion medium requires additional operations.

Moreover, many applications of emulsion polymers such as paints, floor polishes, inks, varnishes, carpet backing, paper coatings, and adhesives, lead to the isolation of the polymer by the removal of water. By this way, latex is transformed into a polymer film. The film formation process of latex occurs in three major steps: first, the polymer particles become into close contact with each other by evaporation of water. Second, as more water evaporates, the particles undergo deformation to form a void-free solid structure which is still mechanically weak. Last, fusion occurs between adjacent particles to generate mechanically strong film. In many applications, the key stage is the transition between wet, dispersed polymer and dry film. The application temperature should be above the minimum film forming temperature (MFFT) of the latex which commonly corresponds to the glass transition temperature (T_g) of the latex polymer in the presence of water [32].

2.1. Main ingredients of emulsion polymerization

A typical emulsion polymerization formulation comprises four basic ingredients: 1) monomer, 2) dispersion medium, 3) emulsifier, 4) initiator. Further auxiliaries, such as chain transfer agents, buffers, acids, bases, anti-aging agents, biocids, etc., can be used. In emulsion polymerization process, a monomer or a mixture of monomers is emulsified in the presence of an aqueous solution of an emulsifier in a suitable container. The monomer is thus present almost entirely as emulsion droplets dispersed in water. The initiator causes the monomer molecules to polymerize within a certain temperature range. When the polymerization is complete, a stable colloidal dispersion of polymer particles in an aqueous medium (the latex) will remain.

2.1.1. Monomer

Emulsion polymerization requires free-radical polymerizable monomers which form the structure of the polymer. The major monomers used in emulsion polymerization include butadiene, styrene, acrylonitrile, acrylate ester and methacrylate ester monomers, vinyl acetate, acrylic acid and methacrylic acid, and vinyl chloride. All these monomers have a different structure and, chemical and physical properties which can be considerable influence on the course of emulsion polymerization. The first classification of emulsion polymerization process is done with respect to the nature of monomers studied up to that time. This classification is based on data for the different solubilities of monomers in water and for the different initial rates of polymerization caused by the monomer solubilities in water. According to this classification, monomers are divided into three groups. The first group includes monomers which have good solubility in water such as acrylonitrile (solubility in water 8%). The second group includes monomers having 1-3 % solubility in water (methyl methacrylate and other acrylates). The third group includes monomers practically insoluble in water (butadiene, isoprene, styrene, vinyl chloride, etc.) [12].

2.1.2. Dispersion medium

In emulsion polymerizations, the dispersion medium, for monomer droplets and polymer particles, is generally water as well as liquids other than water. Water is cheap, inert and environmentally friendly. It provides an excellent heat transfer and low viscosity. It also acts as the medium of transfer of monomer from droplets to particles, the locus of initiator decomposition and oligomer formation, the medium of dynamic exchange of emulsifier between the phases, and the solvent for emulsifier, initiator, and other ingredients.

2.1.3. Emulsifier

These materials perform many important functions in emulsion polymerizations [11,13,30], such as *(i)* reducing the interfacial tension between the monomer phase and the water phase so that, with agitation, the monomer is dispersed (or emulsified) in the water phase. *(ii)* Micelle generating substances. If these substances are used above the critical micelle concentration (CMC), they will form micelles which are ordered clusters of emulsifier molecules, with the oil-soluble part of the molecule oriented toward the center of the cluster and the water-soluble part of the molecule toward the water. *(iii)* Stabilizing the monomer droplets in an emulsion form. *(iv)* Serving to solubilize the monomer within emulsifier micelles. *(v)* Stabilizing the growing latex particles. (vi) Also, stabilizing the particles of the final latex. (vii) Acting to solubilize the polymer. (viii) Serving as the site for the nucleation of particles. (ix) Acting as chain transfer agents or retarders.

Emulsifiers (also referred to as surfactant, soap, dispersing agent, and detergents) are surface-active agents. These materials consist of a long-chain hydrophobic (oil-soluble) group (dodecyl, hexadecyl or alkyl-benzene) and a hydrophilic (water-soluble) head group. They are usually classified according to the nature of this head group. This group may be anionic, cationic, zwitterionic or non-ionic [30]. Anionic emulsifiers having negatively charged hydrophilic head group are the sodium, potassium and ammonium salts of higher fatty acids, and sulfonated derivatives of of aliphatic, arylaliphatic, or naphtenic compounds. Sodium lauryl (dodecyl) sulfate, [$C_{12}H_{25}OSO_3^-Na^+$], sodium dodecyl benzene sulfonate, [$C_{12}H_{25}C_6H_4SO_3^-Na^+$] and sodium dioctyl sulfosuccinate, [$(C_{18}H_7COOCH_2)_2SO_3^-Na^+$] are commonly used in emulsion polymerizations as anionic emulsifiers. Quaternary salts such as acetyl dimethyl benzyl ammonium chloride and hexadecyl trimethyl ammonium bromide may be given examples for cationic emulsifiers. Zwitterionic (amphoteric) emulsifiers can show cationic or anionic properties depending on pH of the medium. They are mainly alkylamino or alkylimino propionic acids. Non-ionic emulsifiers carry no charge unlike ionic emulsifiers. The most used type of these emulsifiers is that with a head group of ethylene oxide (EO) units. Polyoxyethylenated alkylphenols, polyoxyethylenated straight-chain alcohols and polyoxyethylenated polyoxypropylene glycols (i.e., block copolymers formed from ethylene oxide and propylene oxide) are the most commonly three classes of non-ionic emulsifiers used for emulsion polymerization formulations. Polyoxyethylenated alkylphenol type of emulsifiers includes two main members: nonylphenol polyoxyethylene glycol, [$C_9H_{17}C_6H_4O\text{-}(CH_2CH_2\text{-}O)_nH$],and octylphenol polyoxyethylene glycol, [$C_8H_{15}C_6H_4O\text{-}(CH_2CH_2\text{-}O)_nH$]. The number of

EO units, (n), may be diversified from a few toabout 100 (typically from 1 to 70 EO units), which characterize the distribution of polyEO chain lengths for each specific emulsifier. A typical example for polyoxyethylenated polyoxypropylene glycols is polyethylene oxide-polypropylene oxide-polyethylene oxide triblock copolymer, [H-$(OCH_2CH_2)_a$-$(OCH_3CH$-$CH_2)_b$-$(OCH_2CH_2)_a$ -OH], in which the polyEO portion constitutes between 10 and 80% of the copolymer.

In general, the anionic emulsifiers are extensively preferred in many emulsion polymerization systems. They serve as strong particle generators and stabilize the latex particles via electrostatic repulsion mechanism. But latexes stabilized with this type of emulsifiers are often unstable upon addition of electrolytes and in freeze-thaw cycles. Furthermore, these emulsifiers have limited stabilizing effectiveness at high solids (e.g., > 40%) and present high water sensitivity. To overcome these problems, non-ionic emulsifiers can be used to nucleate and stabilize the particles in the course of emulsion polymerization. In this case, it is the steric stabilization mechanism that protects the interactive particles from coagulation. In addition, the use of non-ionic types improves the stability of latex product against electrolytes, freeze-thaw cycles, water and high shear rates. As a result of them, in many emulsion polymerization recipes (particularly in industry), mixtures of anionic and non-ionic emulsifiers have been widely used together in a synergistic manner to control the particle size and to impart enhanced colloidal stability [33-35]. The cationic and zwitterionic emulsifiers are used infrequently in emulsion polymerization applications.

Besides all these types of emulsifiers, polymeric and reactive emulsifiers can be used in emulsion polymerizations. Polymeric emulsifiers are often non-ionic water-soluble polymers such as poly(vinyl alcohol), hydroxyethyl cellulose and poly(vinyl pyrrolidone), and called sometimes as a *"protective colloid"*. They are used to increase the particle stability in latexes against coagulation. Reactive emulsifiers (*"surfmers"*), which have polymerizable reactive group, can copolymerize with the main monomer and be covalently anchored onto the surface of latex particles. When these compounds used in emulsion polymerizations, the emulsifier migration is reduced. Furthermore, surfmers improve the water resistance and surface adhesion as well as resistance against electrolytes and freeze-thaw cycles in comparison to conventional emulsifiers. Surfmers can be anionic with sulfate or sulfonate head groups (sodium dodecyl allyl sulfosuccinate), cationic (alkyl maleate trimethylamino ethyl bromide), or non-ionic (functionalized poly(ethylene oxide)-poly(butylenes oxide)copolymer). The reactive groups can be in different types, for example, allylics, acrylamides, (meth)acrylates, styrenics, or maleates [36-37].

2.1.4. Initiator

Emulsion polymerization occurs almost entirely following the radical mechanism. The function of the initiator is to generate free radicals, which in turn lead to the propagation of the polymer molecules. The free radicals can be commonly produced by two main ways: *(i)* thermal decomposition, or *(ii)* redox reactions. In addition, the free-radical initiators can be either water or oil-soluble.

The most commonly used water-soluble initiators are persulfates (peroxodisulfates). For example, potassium-, sodium-, and ammonium-persulfate. Persulfate ion decomposes thermally in the aqueous phase to give two sulfate radical anions which can initiate the polymerization. Hydrogen peroxide and other peroxides are thermal decomposition type initiators and they are soluble in both the aqueous and monomer-swolen polymer phases. Besides of these, oil-soluble compounds such as benzoyl peroxide and azobisisobutyronitrile (AIBN) can be employed as thermal initiators in emulsion polymerizations. The other initiation system consists of redox initiators (such as persulfate-bisulfite system) which produce free radicals through an oxidation-reduction reaction at relatively low temperatures.

The main types of free radicals which are produced by thermally or redox system are:

a. Persulfates

$$S_2O_8^{-2} \rightarrow SO_4\bullet^{-1} + SO_4\bullet^{-1} \tag{1}$$

b. Hydrogen peroxide

$$HO-OH \rightarrow HO\bullet + HO\bullet \tag{2}$$

c. Organic peroxides

$$RO-OR' \rightarrow RO\bullet + R'O\bullet \tag{3}$$

d. Azo compound

$$RN=NR' \rightarrow R\bullet + R'\bullet + N_2 \tag{4}$$

e. Persulfate-Bisulfite

$$S_2O_8^{-2} + HSO_3^{-1} \rightarrow SO_4\bullet^{-1} + SO_3\bullet^{-1} + HSO_4^{-1} \tag{5}$$

There is also surface active initiators which are called as *"inisurfs"*, for example; bis[2-(4'-sulfophenyl)alkyl]-2,2'-azodiisobutyrate ammonium salts and 2,2'-azobis(N-2'-methylpropanoyl-2-amino-alkyl-1-sulfonate)s. The initiators of this type carry stabilizing groups in their structures, and emulsion polymerization can be successfully carried out in the presence of them, without additional stabilizers up to more than 50% in solid content [37]. Moreover, the free radicals needed to initiate the emulsion polymerization can be produced by ultrasonically, or radiation-induced. ^{60}Co γ radiation is the most widely used as radiation-induced initiation system in the emulsion polymerizations.

2.1.5. Other ingredients

The formulations of emulsion polymerization may include a wide variety of ingredients: *chain transfer agents:* are added to a latex formulation to help regulate the molar mass and molar mass distribution of the latex polymer. The mercaptans are the most common type of chain transfer agents. The surface active transfer agents, *"transurfs"*, are also used in emulsion polymerizations. *Buffers:* are often added to a latex formulation to regulate the pH of the polymerization system. Generally, for this purpose, sodium bicarbonate has been chosen. In addition, coalescing aids, plasticizers, thickening agents, antimicrobial agents, antioxidants, UV-absorbers, pigments, fillers, and other additives can take place in a recipe of emulsion polymerization.

2.2. Kinetic and mechanism of emulsion polymerization

Emulsion polymerization is a type of free-radical addition polymerization. Such reactions are comprised of three principal steps, namely initiation, propagation and termination. In the first stage an initiator is used to produce free-radicals which react with monomer containing unsaturated carbon-carbon bonds (its general structure; $CH_2{=}CR^1R^2$, where R^1 and R^2 are two substituent groups) to initiate the polymerization. When the radical reacts with a monomer molecule a larger free-radical (active center) is formed which, in turn, reacts with another monomer molecule, thus propagating the polymer chain. Growing polymer chains are finally terminated (free electrons coupled) with another free radical, or with chain transfer agents, inhibitors, etc.

The three stages of the free-radical polymerization are shown in the following steps:

Initiation: The reaction of initiation can be described as a two-stage process. In the first stage the initiator is decomposed to free-radicals, in the second stage the primary radicals react with the monomer, converting it to a growing radical.

The first stage where free-radicals can be generated by two principal processes: (1) homolytic scission (i.e. homolysis) of a single bond which can be achieved by the action of heat or radiation, and (2) chemical reaction involving electron transfer mechanism (redox reactions).

The most common method used in emulsion polymerizations is thermal initiation in which the initiator (*I*) dissociates homolytically to generate a pair of free-radicals ($R\bullet$) as shown below:

$$I \xrightarrow{k_d} 2R\bullet \tag{6}$$

where k_d is the rate constant for the initiator dissociation. The rate of this dissociation, R_d, is given by,

$$R_d = 2fk_d[I] \tag{7}$$

where $[I]$ is the concentration of the initiator and f is the initiator efficiency. The initiator efficiency is the fraction of primary free radicals ($R\bullet$) which are successful in initiating polymerization, and is in the range 0.3-0.8 due to wastage reactions. The factor of 2 enters because two primary free radicals are formed from each molecule of initiator.

In the second stage, the free radicals generated from the initiator system attack the first monomer (M) molecule to initiate chain growth:

$$R\bullet + M \xrightarrow{k_i} RM\bullet \tag{8}$$

where k_i is the rate constant for the initiation. The rate of initiation, R_i, is equal to the rate of dissociation of an initiator. Because the primary radical adds to monomer is much faster than the first stage, and so the dissociation of the initiator is the rate-determining step in the initiation sequence. According to this, R_i is given by

$$R_i = 2fk_d[I] \tag{9}$$

Propagation: The propagation step is only one which produces polymer. This involves essentially the addition of a large number of monomer molecules (n) to the active centers ($RM\bullet$) for the growth of polymer chain as shown below.

$$RM\bullet + nM \xrightarrow{k_p} P_{n+1}\bullet \tag{10}$$

where k_p is the rate constant for propagation.

The rate of polymerization, R_p, is known as the rate of monomer consumption. Monomer is consumed by the propagation reactions as well as by the initiation reaction. The corresponding rate of polymerization is then:

$$R_p = -\frac{d[M]}{dt} = k_i[R\bullet][M] + k_p[M\bullet][M] \tag{11}$$

where $[R\bullet]$ is the primary free-radicals concentration, [M] is the monomer concentration and $[M\bullet]$ is the total concentration of every size of chain radicals. The amount of monomer consumed in the initiation step can be neglected due to the number of monomer molecules reacting in the initiation step is far less than the number in the propagation step for a process producing high polymer, and a very close approximation of the polymerization rate can be given simply by the rate of propagation. Then, the polymerization rate can be writen:

$$R_p = k_p[M\bullet][M] \tag{12}$$

Termination: In last step of the polymerization, the growing polymer chain is terminated. There are two main mechanisms, recombination and disproportionation, for termination reactions. In these mechanisms, the growing polymer chain react with another growing chain or another free radical of some kind.

Recombination;

$$P_n\bullet + P_m\bullet \xrightarrow{k_{tc}} P_{n+m} \tag{13}$$

in which two growing chains constitute the coupling with each other resulting in a single polymer molecule.

Disproportionation;

$$P_n \bullet + P_m \bullet \xrightarrow{k_{td}} P_n + P_m \qquad (14)$$

in which one growing chain abstracts a hydrogen atom from another, leaving it with an unsaturated end-group. This mechanism occurs more rarely than recombination. It results in the formation of two polymer molecules, one saturated and one unsaturated. In the above equations, k_{tc} and k_{td} are the rate constants for termination by recombination and disproportionation, respectively. The overall rate constant for termination reaction is given as $k_t=k_{tc}+k_{td}$.

In addition to these main termination reactions, there are some other reactions which can terminate the growing chain radical. These reactions can be occurred by removal of an atom from some substances present in the reaction mixture to give a new radical which may or may not start another chain (chain transfer reactions), or by addition to some substance (such as retarder or inhibitor) into the reaction mixture to give a new radical having little or no ability to continue the propagation of the chain [10].

In the chain transfer reactions, some substances such as polymer, monomer, solvent, additives, impurities, or initiator can act as a chain transfer agent. An example of these reactions is given:

$$P_n \bullet + T - A \longrightarrow P_n T + A \bullet \qquad (15)$$

where *T-A* is a chain transfer agent. The chain radical abstracts $T \bullet$ (often a hydrogen or halogen atom) from *T-A* molecule to yield a terminate polymer molecule and a new free radical, $A \bullet$ which can initiate a new chain. The main effect of chain transfer is to reduce the molecular weight of the polymer. If the new radical $A \bullet$ is as reactive as the primary radicals, $R \bullet$, there will be no effect on the rate of polymerization.

In polymerization kinetic, *steady state conditions* must obtain, i.e. where the rate of generation of free radicals (initiation) is equal to the rate at which they disappear (termination). This implies a constant overall concentration of propagating free radicals, $[M \bullet]$. The equation for the steady state conditions is:

$$R_i = R_t = -\frac{d[R\bullet]}{dt} = 2k_t[M\bullet]^2 \qquad (16)$$

In practice, most free-radical polymerizations operate under steady state conditions after an induction period wich may be at most a few seconds. When Equation 2.16 is rearranged,

$$[M\bullet] = \left(\frac{R_i}{2k_t}\right)^{1/2} \qquad (17)$$

and a general expression for the rate of polymerization can be obtained by combining Equation 2.12 and 2.17,

$$R_p = k_p[M]\left(\frac{R_i}{2k_t}\right)^{1/2} \tag{18}$$

This equation show that the polymerization rate depends on the square root of the initiation rate. If we make an arrangement on this equation by using Equation 2.9, we can say that the polymerization rate depends on the square root of the initiator concentration:

$$R_p = k_p[M]\left(\frac{fk_d[I]}{k_t}\right)^{1/2} \tag{19}$$

In the emulsion polymerizations, the free-radical mechanism is very closely connected with the heterogeneous nature of the emulsion polymerization in which the micellar phase, the aqueous phase, the monomer droplet phase and the particle phase exist. Therefore, a number of mechanisms have been proposed for latex particle formation. The most important qualitative mechanism of emulsion polymerization was proposed by Harkins in 1945 and 1946 [38-39]. The following main premises of this mechanism were taken by Smith and Ewart. They managed to obtain first quantitative theory, which consist of equations for determining the rate of polymerization and the number of latex particles, for emulsion polymerization [40]. These theories have been applied only to the emulsion polymerization of "hydrophobic monomers" (such as sytrene) in the presence of water-soluble initiators and nonspecific, micelle-forming emulsifiers at concentrations significantly exceeding the critical micelle concentration. A schematic representation of the Harkins theory is illustrated in Figure 1 [11,41]. Such a system contains the emulsified monomer droplets (ca. 1-10 μm in diameter) dispersed in the continuous aqueous phase with the aid of an emulsifier at the very beginning of polymerization. Monomer-swollen micelles (5-10 nm in diameter) also exist in this system provided that the concentration of emulsifier in the aqueous phase is above (CMC). Only a small fraction (approximately 1%) of the total monomer is actually solubilized by the micelles and only an insignificant amount of monomer (with hydrophobic monomers such as styrene; about 0.04%) is dissolved by water. Most of the monomer molecules are in the monomer droplets which act as a reservoir of monomers during polymerization. Thus, the system prior to initiation contains mainly three parts: the water phase, large monomer droplets dispersed throughout the water phase, and the emulsifier micelles containing solubilized monomer. In addition, a very small amount of molecularly dissolved emulsifier and monomer-free micelles may exist (Figure 1.a).

After the emulsion of the monomer phase in the water phase and the presence of the emulsifier micelles established, the polymerization is initiated by the addition of initiator. According to the theories proposed by Harkins and Smith and Ewart, conventional emulsion polymerization mechanism occurs into three intervals including the initial (particle formation or nucleation) stage, the particle growth stage and the completion stage.

The initial stage (Interval I): This stage is also called as *"particle formation"* or *"nucleation"*. With the addition of initiator to the reaction mixture, the free-radicals which initiate the polymerization are generated in the aqueous phase and diffuse into monomer-swollen micelles. These micelles are the principal locus for the initiation of polymer particle nuclei. In other words, they act as a meeting place for the hydrophobic monomer and the water-soluble initiator. Since they exhibit an extremely large oil-water interfacial area for diffusing of free-radicals and have high monomer concentration. On the other hand, a small amount of particle initiation can occur within the continuous aqueous phase. Monomer molecules dissolved in this phase are first polymerized by waterborne free-radicals. This would result in the increased hdrophobicity of oligomeric radicals. When a critical chain length is achieved, these oligomeric radicals become so hydrophobic that they show a strong tendency to enter the monomer-swollen micelles and then continue to propagate by reacting with those monomer molecules. But this nucleation becomes less significant as the amount of micellar emulsifier in the system increases. The amount of polymerization occurring in the monomer droplets is regarded as being a very minor proportion of the whole because of their small surface area for diffusing of the free-radicals. As a result, monomer-swollen micelles are favored as the sites of the nucleation of polymer particles. Therefore, this nucleation mechanism, proposed by Harkins and Smith and Ewart and modified by Gardon, is called as **"micellar" or "heterogeneous" nucleation** [42].

After nucleation, monomer-swollen micelles are transformed into polymer particles swollen with monomer. The monomers are then acquired to the particles continuously from monomer droplets by diffusion through the aqueous phase, because monomer has been consumed with in the reaction loci by the polymerization process. Thus, the micelles grow from tiny groups of emulsifier and monomer molecules to larger groups of polymer molecules held in emulsion by the action of the emulsifier molecules located on the exterior surfaces of the particles. This action of the emulsifiers for providing colloidal stability of the growing particle nuclei occurs by emulsifiers supplied from the aqueous phase; this in turn tends to leads to dissociation of micelles containing monomer in which polymerization has not yet started. With the continued adsorption of micellar emulsifiers on to growing particles, the micelles start to disappear (Figure 1.b). The particle nucleation stage (Interval I) ends with this disappearance of the micelles at relatively early in the reaction (e.g. between 10% and 20% conversion). During Interval I, the rate of polymerization increases with the increasing time of reaction and only one out of every 100-1000 micelles becomes a polymer particle. The number of particles nucleated per unit volume of water (N_p) is proportional to the emulsifier concentration and initiator concentration to the 0.6 and 0.4 powers, respectively according to the Smith-Ewart theory. After the particle nucleation process is completed, this number remains relatively constant toward the end of polymerization.

The particle growth stage (Interval II): After the particle nucleation process is completed, polymerization proceeds homogeneously in the polymer particles as the monomer concentration in the particles is maintained at a constant concentration by diffusion of monomer from the monomer droplets. The rate of polymerization in this stage is constant. In addition, during this stage, the number of monomer-swollen polymer particles and the monomer/polymer ratio remain constant. The monomer droplets decrease in size as the size of the polymeric particles increase. When monomer droplets completely disappear in the polymerization system (at 50-80% conversion), the particle growth stage (Interval II) ends (Figure 1.c). In this situation, the polymer particles contain all the unreacted monomer and essentially all of the emulsifier molecules are also attached to the surface of polymer particles.

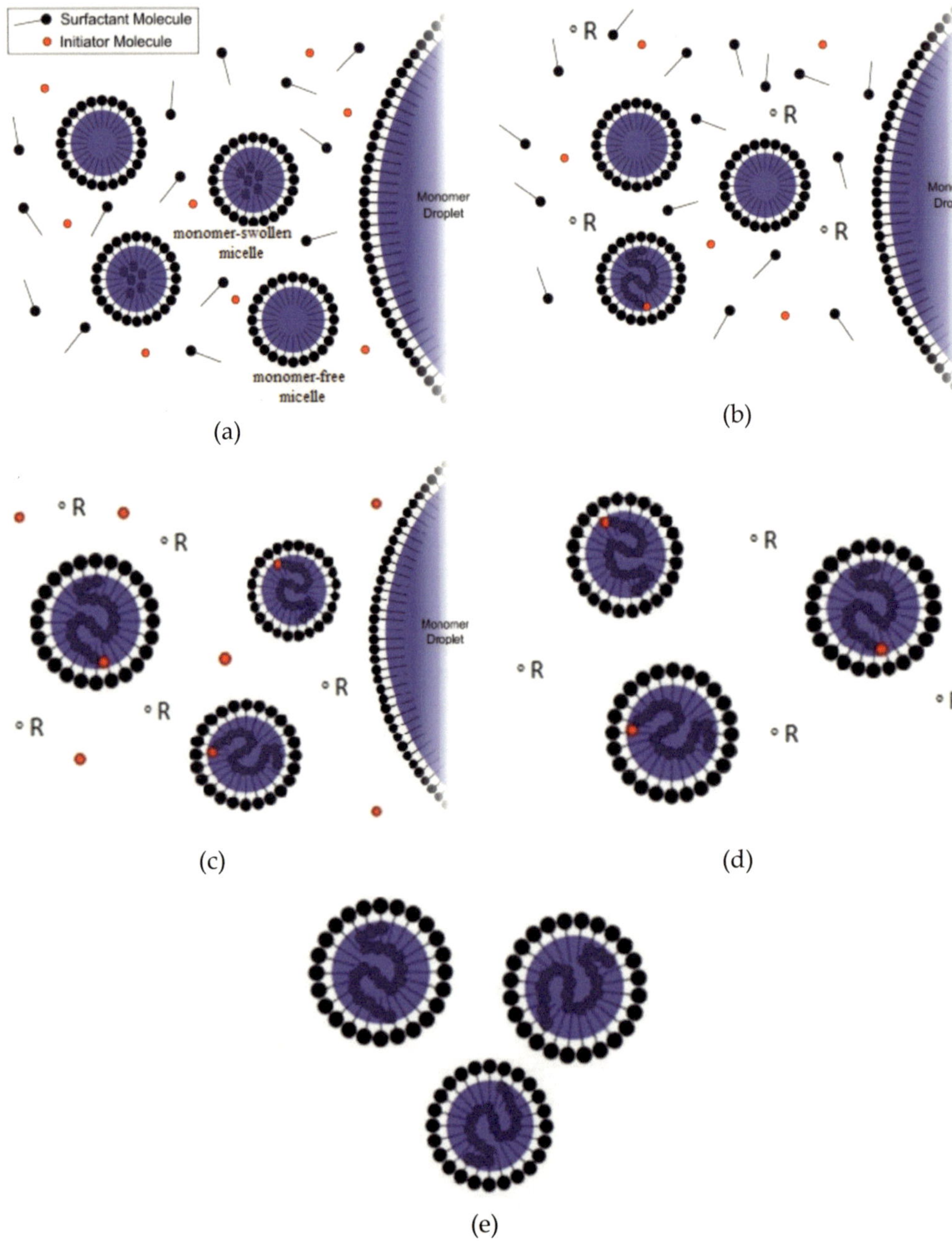

Figure 1. Schematic representation for the mechanism of emulsion polymerization [11,41]

The completion stage (Interval III): This is the final stage of the reaction. In this stage, polymerization continues within the monomer-swollen polymer particles which were formed during Interval I, and persisted and grew during Interval II (Figure 1.d). In the ideal case, the number of reaction loci during this stage is essentially fixed at the number which had become formed at the end of Interval I. Whereas, the concentration of monomer in the reaction loci and the polymerization rate continues to decrease toward the end of polymerization. Finally, the polymerization is complete and the conversion of essentially 100% is usually achieved. The system now comprises a dispersion of small polymer particles stabilized with the molecules of the original emulsifiers (Figure 1.e).

To calculate the rate of emulsion polymerization (R_p) of relatively water-insoluble monomers such as styrene and butadiene, Smith Ewart case 2 kinetics has been widely used [40]:

$$R_p = k_p \left[M\right]_p (nN_p / N_A) \quad (20)$$

where k_p is the rate constant of propagation, $[M_p]$ the concentration of monomer in the particles, n the average number of free radicals per particle, and N_A the Avogadro number.

Other than micellar nucleation, many mechanisms have been proposed to explain the particle nucleation stage. The best-known alternative theory for particle nucleation is that of **"homogeneous nucleation"** which includes the formation of particle nuclei in the continuous aqueous phase. This theory is proposed by Priest, Roe and Fitch and Tsai, and extended by Hansen and Ugelstad (HUFT) describes the emulsion polymerization of water-solubble monomers such as vinyl acetate and acrylonitrile, their water solubility though low (< 3%) is much in excess of the amount of monomer which may be solubilized by the emulsifier [43-48]. It is also the only mechanism which can apply to monomers of low water-solubility, such as styrene, in emulsifier-free reaction system, and also in reaction system which contain a micellizing emulsifier but at such a concentration that is below the CMC. When the monomers are somewhat soluble in the continuous phase, emulsifier micelles have little influence on particle formation. Emulsifier may be required, however, to ensure colloidal stability of the product as it is formed and subsequently "on the shell".

Initially, free-radicals are generated in the aqueous phase by the thermal decomposition of initiator and they can grow by polymerization with those monomer molecules dissolved in the aqueous phase. When a growing oligomer reaches to critical chain lengths, it becomes water-insoluble. The water-insoluble radical then collapses upon itself and becomes effectively a separate phase. The primary particles thus formed provide a potential reaction locus which has within it an active growing free-radical if monomer molecules are available. Replenishment of monomer to the particles takes place by diffusing from the monomer droplets through the continuous phase. Indeed, the collapsed growing oligomer can be regarded as being similar to a emulsifier micelle. The oligomer is to some extent surface-active, because it incorporates both a hydrophilic moiety derived from the water-soluble initiator and hydrophobic units derived from the monomer molecules. The particles also may grow by limited coagulation of the relatively unstable primary particles [48-49]. The coagulation of these particles can occur with already-existing particles or other primary particles in the earliest (the first few tens of seconds) stage of the reaction. The particles formed by coagulation are now a potential reaction locus where the polymerization continues in the presence of further free-radical as well as monomer molecules. Colloidal stability of the collapsed oligomers and the growing particles is regulated by the amount of emulsifier adsorbed on their surfaces as well as by any other stabilizing group, ionic or non-ionic, introduced by functional comonomers, initiator fragments or by chain transfer to molecules. The emulsifiers come from those dissolved in the aqueous phase and those adsorbed on the monomer droplet surfaces. If emulsifier micelles are present, then these are regarded as reservoirs of emulsifier molecules, dissociating as required to provide stabilizer.

The basic principle of this nucleation theory is that the formation of primary particles takes place up to the point where the rate of formation of the radicals in the aqueous phase is equal to the rate of their disappearing via capture of radicals by swollen micelles-if present-initially and by particles already formed (Figure 2, [14]), and via coagulation.

According to this principle, the qualitative kinetic equation developed by Fitch and Tsai is:

$$\left(\frac{dN}{dt}\right) = bR_{iw} - R_c - R_f \tag{21}$$

where t is the reaction time, N the number of particles, b a parameter that takes into account the aggregation of oligomeric radicals, R_{iw} the effective rate of initiation or free-radical generation in the aqueous phase, R_c the rate of capture of free-radicals, and R_f the rate of coagulation of the particles.

Feeney, Napper and Gilbert proposed a different particle nucleation theory, which resembles homogeneous nucleation [50]. They showed that particle size distributions during particle formation (Interval I) were positively skewed, confirming the role of coagulation. They called this phenomenon as "**coagulative nucleation**" mechanism. They also have provided evidence that there can be coagulation even when the initial emulsifier concentration is above the CMC. It is possible that so many particles called “precursor” could be formed probably by homogeneous nucleation. Their size could be too small, and hence their surface area so much during growth, that they ‘run out of’ sufficient stabilizer providing their colloidal stability. This would result in coagulation in which the volume growth of the precursor particles by coagulation is much faster than that by propogation reactions. When these particles are sufficiently large to absorb appreciable amounts of monomer, “mature” primary latex particles occur. Monomer within the mature primary particles polymerizes more quickly than it those within the precursor particles. Because, the higher surface area of the precursor particles reduce the equilibrium swelling of the polymer by monomer, and hence the monomer concentration decreases. The smaller particles also can cause to exit of the free-radicals from the particles more easily. Consequently, the polymerization rate of the precursor particles decreases with increasing time due to these two effects. In the other hand, the positively skewed implies that the rate of particle nucleation increases with increasing time, and Feeney, Napper and Gilbert have explained that this skewed is a result of the coagulative nucleation.

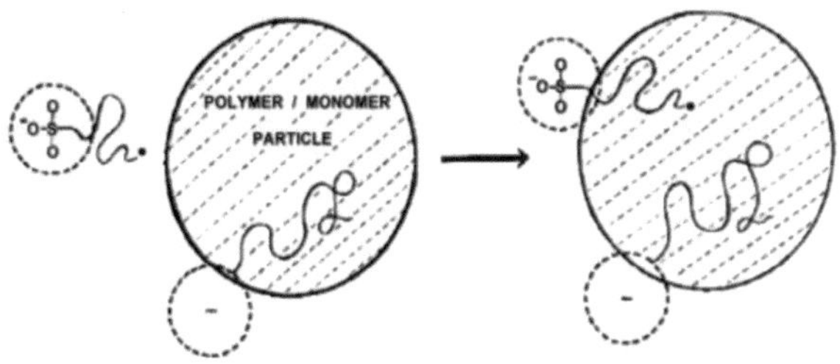

Figure 2. Capture of radicals by a particle in homogeneous nucleation [14]

2.3. Other types of emulsion polymerization

Oil-in-water emulsion polymerization systems are typically classified as possessing the characteristics of one of three types of emulsions: *macro-emulsions*, *mini-emulsions* or *micro-emulsions*. These emulsions are the initial systems for emulsion polymerization. There are quite differences between these systems in some aspects such as the size of the droplets (i.e. the discontinuous or dispersed phase), the interfacial area of the droplets, the particle nucleation mechanism and the stability of the emulsion.

Macro-emulsion polymerizations: The conventional emulsion polymerization is referred as macro-emulsion polymerization. In this chapter, all explanations included so far relate to macro-emulsion polymerization. It is initially composed of a monomer emulsion of relatively large (1-100 μm) monomer droplets and significant free or micellar emulsifier. This emulsion is thermodynamically unstable, but kinetically stable. Phase separation is rapid unless the system is well agitated. In macro-emulsion polymerizations, the nucleation takes place outside the monomer droplets which generally do not contribute to the particle nucleation due to their very small droplet surface area. The monomer droplets act as only monomer reservoirs which supply the monomers to the polymerization loci through the aqueous phase.

Micro-emulsion polymerization: In micro-emulsion polymerization, the initial system is micro-emulsion which consist of monomer droplets (varying from 10 to 100 nm) dispersed in water with the aid of a classical emulsifier (e.g. sodium dodecyl sulfate, SLS) and a "co-surfactant" such a low molar mass alcohol (pentanol or hexanol). Micro-emulsions are thermodynamically stable and optically one-phase solution. There is an excessive amount of emulsifier in these emulsions. Therefore, they are concentrated systems of micelles and the micelles exist throughout the reaction. One of the most interesting aspects of these micelles is their ability to accommodate monomer molecules. Furthermore, their high total surface area relative to nucleated particles implies the monomer-swollen micelles preferentially capture primary radicals generated in the continuous aqueous phase. Then the probability of capturing radicals by the micelles remains essentially one during the entire process and new particles thus are formed continuously. A growth in micellar size is always observed during the reaction due to the internal dynamics of micro-emulsions. This takes the form of either coagulation active and inactive micelles or the diffusion of monomer from the unreacted micelles to the nucleated particles. Reaction rates can be very fast. Finally, the small latex particles less than 50 nm are obtained from micro-emulsion polymerization. They on the average contain only one polymer molecule with average molecular weight exceeding one million [14-15,51].

Mini-emulsion polymerization: Mini-emulsions are produced by dispersing monomer in water by means of vigorous mechanical agitation or homogenization using a mixed emulsifier system including a classical emulsifier and a water-insoluble "co-surfactant" such as a long chain fatty alcohol or alkane (e.g. cetyl alcohol or hexadecane). The stability of mini-emulsions can continue for as little as days and as long as months. The polymerization of these emulsions can begin with submicrometre monomer droplets (from 50 to 500 nm). The long chain co-surfactants penetrate less the oil-water interface than the small monomer molecules, thus the emulsifier molecules move closer together, decreasing the interfacial area. Hence, the monomer droplets formed smaller size can become predominant particle nucleation loci provided that the total monomer droplet surface area becomes large enough to compete effectively with the continuous aqueous phase, in which particle nuclei are generated, for capturing waterborne free-radicals (monomer droplet nucleation). For propagation, the monomer is provided from within the polymerizing particles. Not all monomer droplets become polymer particles. Monomer droplets can be disappeared by another mechanism different from nucleation such as monomer diffusion to growing particles or by collision with polymer particles. No free, monomer-swollen micelles are present in this system since excess emulsifier has been adsorbed onto the large droplet-water interfacial area. The final polymer particles have almost the same size as the initial monomer droplets. Their size is often larger and their particle size distribution is broader than those obtained by conventional means of homogeneous or micellar nucleation [15,52-54].

In contrast to these *oil-in-water emulsions*, it is possible that the emulsion polymerization can also be carried out with inverse emulsions. *Inverse (water-in-oil) emulsion polymerization* in which an aqueous solution of a water miscible hydrophilic monomer such as acrylamide, acrylic acid, or methacrylic acid is dispersed in a continuous hydrophobic oil phase with the aid of a water-in-oil emulsifier such as sorbitan mono-oleate or -stearate. The emulsifier is ordinarily above the CMC. Polymerization can be initiated with either oil-soluble or water-soluble initiators. If an oil-soluble initiator is used, the system is an almost exact 'mirror-image' of a conventional emulsion polymerization system. The final latex is a colloidal dispersion of submicroscopic, water-swollen particles in oil. This type of emulsion polymerization enables the preparation of high molecular weights water-soluble polymers at rapid reaction rates. It is also possible that the water-swollen polymer particles produced by this emulsion polymerization transfer to aqueous phase rapidly by inversion of the latex.

As well as *"inverse emulsion polymerization"*, there is also *"inverse micro-emulsion polymerization"* [15,51,55].

2.4. Types of emulsion polymerization process

Three types of process are commonly used in carrying out the emulsion polymerization: *batch, semi-continuous,* and *continuous*. This classification is made according to the way in which the Interval II and Interval III reactions are effected [29].

Batch process: The batch-type process is the simplest method for effecting emulsion polymerization. All ingredients are placed in a reactor at the beginning of the reaction. The system is agitated, and heated to reaction temperature. Polymerization begins as soon as the initiator is added. Then, the reaction system is kept there by heating or cooling, as needed, and by agitating until the samples removed indicate the desired conversion of monomer to polymer. The only significant changes which can be made in such cases are to the reaction temperature, reactor design and the type and speed of agitation. Emulsion polymerization behavior in batch process can be divided into three intervals, as clearly explained in section 2.2: Interval I, free-radicals generated in the aqueous phase enter micelles and form new polymer particles. In Interval II, particles grow by continuously feeding of monomer from monomer droplets. In Interval III, polymerization continues in the monomer-swollen particles. During Interval III of a batch process, monomer concentrations in the polymer particles and in the aqueous phase decrease with time.

It is commonly used in the laboratory to study reaction mechanism and kinetics, but most commercial latexes are not manufactured by this process because of their undesirable properties. This process has the important disadvantages that limited control is exerted over either monomer/polymer ratio in the reaction loci, or over heat transfer in the reaction, or over copolymer composition. Nevertheless, batch process has an important role, particularly in more fundamental studies.

Semi-continuous (or semi-batch) process: This process is very versatile and is widely used, both industrially and in academic laboratories. Semi-continuous emulsion polymerization offers great degree of operational flexibility than batch or continuous processes. It allows one great control over the course of the polymerization, the rate of heat generation, and the properties and the morphology of the polymer latex particles. It also makes it possible to achieve relatively high polymer quality such as homogeneous chemical composition and particle size distribution.

The semi-continuous emulsion polymerization processes is characterized by continued addition of reaction ingredients such as monomer, emulsifier, initiator, or water to the reaction system throughout the polymerization. In this emulsion polymerization process, two major types of feeds are used for the introduction of ingredients to the reactor; neat monomer feed (M) or monomer emulsion feed (ME). In M feed method, the feed contains only monomer and all the other ingredients are initially in the reactor. Otherwise, the major components of the ME feed are a monomer, a part from the emulsifier, and water. But it contains other ingredients. The main difference between the two types is the emulsifier concentration throughout the polymerization. The initiator can initially be charged or/and continuously fed during polymerization. By the continuous addition of the emulsifier, monomer, and initiator to the reaction mixture, the mechanisms (especially particle formation) and kinetics of the semi-continuous emulsion polymerization process become more complicated in comparison with the batch counterpart. Both types of feed can be added to the reactor at any desired time or feeding rate during the feed addition stage. Generally, two types of feed addition strategy have been applied: monomer-flooded and monomer-starved. In the monomer-flooded type, the rate of monomer addition is higher than the maximum polymerization rate attainable by the system. The monomer accumulates in the system as monomer droplets and the polymer particles are saturated with the monomer, as in Interval II of a batch process. In the monomer-starved type, the rate of addition is lower than the rate of polymerization and the polymerization reaction occurs in Interval III where particles are not swollen to their maximum size, leading to the production of small sized, high solid content latexes [56-66].

A semi-continuous process generally contains three successive operations: seeding batch stage (or preliminary batch), feed addition semi-continuous stage and finishing batch stage. Generally, a semi-continuous process starts with a seeding batch stage which is the most critical stage of a semi-continuous process. It controls the particle formation in the whole course of the reaction. Therefore, the distributions of ingredients between initial reactor charge and feed have a great effect on the particle formation. The time interval between initiation and the start of feeding, which is called the "pre-period" or "seeding time", will determine the duration of seeding batch. Further growth of polymer particles is achieved by absorption of the monomer from the feed. After the pre-period time, particle nucleation may be complete but, in some conditions, nucleation can continue during the feeding period. If no pre-period is allowed, the seeding proceeds during the feed addition stage. The feeding stage is generally known as the "growth stage" and is the only stage within the three stages defined above which is carried out semi-continuous wise. The finishing batch is carried out to reduce the amount of unreacted monomer remaining in the reaction mixture to a minimum [61,63-64].

Consequently, semi-continuous emulsion polymerization is an important process, which overcomes the disadvantages of the batch and continuous processes, for the manufacture of a variety of latex products because of its operational flexibility and, providing control on the polymerization and the properties of polymer and latex.

Continuous process: In this process, the reaction system is continuously fed to, and removed from, a suitable reactor at rates such that the total volume of system undergoing reaction at any instant is constant. There are two basic type reactors for effecting continuous emulsion polymerization: the continuous stirred-tank reactor (CSTR) and tubular reactor. In tubular system, the composition is constant with time at any given position in the reactor. They are generally unsatisfactory for industrial production because of the very long tube lengths required, the poor degree of mixing and the difficulties of cleaning. They also require plug-flow in which the reacting mixture passes through the reactor without any forward or backward mixing. For the production of large amounts of the same product, the use of a CSTR may be preferable. In CSTR system, the reaction system is fed continuously to tank reactor, and from this reactor it is removed continuously at the same rate, the mixture in the tank being agitated to such an extent that it remains effectively homogeneous at all stage of the process. These reactors are characterized by isothermal, spatially uniform operation. There may be only one CSTR, or multiple CSTRs in a chain, the product from one being the feed for the next in the chain. The greater the numbers of CSTRs in a chain, the closer the properties approach those from a batch reaction. In such a system, residence time, which is defined as a finite time for remaining of the reaction system in the reactor, is important matter. Because, the performance of continuous flow reactor for emulsion polymerization in terms of conversion, particle concentration and particle size distribution strongly depends on the residence time distribution. Generally, conversion and particle number are lower and the particle size distribution is broader as compared to batch process. In particular, the particle nucleation stage is sensitive to residence time distribution [14,29,67]. Continuous emulsion polymerization process is useful for the production of commercial latexes. This process enables economical production of large volume, formation a highly uniform and well regulated product, and fewer problems with wall polymer buildup and coagulation. But, it allows less operational flexibility and less control on the product characteristics such as specific particle size distribution or particle morphologies [15].

Besides of this classification including the above three processes, another classification is also made according to the way in which the reaction loci are formed. By this way, two types of reaction system are defined: *"ab initio emulsion polymerization"* and *"seeded emulsion polymerization"* [29].

ab initio emulsion polymerization: In this type reactions, the particle nucleation (the generation of reaction loci) proceed in a significant period (during the early stage) of the reaction by mechanisms described as in section 2.2. No reaction loci are initially present.

Seeded (multi-stage) emulsion polymerization: In this reaction type, reaction loci are present in the initial system. The loci are formed by a separate reaction called as first stage of the emulsion polymerization. Thus, the particle nucleation stage of the reaction is eliminated by use of a pre-formed latex. Then, the pre-formed latex is introduced to the reaction mixture at the beginning. There must be enough seed particles to avoid new particles being subsequently nucleated. If the particle number is sufficiently high (typically, $\geq 10^{16}$ particles per liter is satisfactory), then the seed particles efficiently capture free-radical species from the aqueous phase and all primary particles that form by homogeneous nucleation (second stage of the emulsion polymerization). It is also possible that coagulation of tiny primary particles onto seed particles (heterocoagulation strongly affected by the surface charge on the seed particles) in the absence of emulsifier. When emulsifier is in the formulation, the amount of emulsifier must be just enough to maintain the colloidal stability of the seed latex particles as they grow through polymerization and/or coagulation with primary particles, but not so high as to generate new particles. In the second stage of the emulsion polymerization, monomer can be added to reaction system including seed particles by three different methods: dynamic swelling, batch and semi-continuous. By dynamic swelling method, the second stage is carried out after the seed particles are swollen with the second monomer. In batch method where seed latex has a second monomer added in the second stage of the emulsion polymerization. For semi-continuous method, the second monomer is continuously added in monomer-starved conditions to the reaction system including the seed latex during the second stage. In addition, one or more types of monomer can be used in both stages of the seeded emulsion polymerization. For example, in ABS (acrylonitrile/butadiene/styrene) polymer, polybutadiene (PB) seed latex is polymerized with a mixture of styrene and acrylonitrile monomers added in the second stage of the emulsion polymerization. As a result, it is possible to produce polymer particles with a wide variety of morphologies such as 'dumb-bell' shaped, 'ice-cream cone' shaped, 'raspberry' shaped, or core-shell due to the multi-stage emulsion polymerization process enables changing the experimental conditions in a wide range. Furthermore, this process provides many advantages in the production of latex particles being uniform and having excellent monodispersity in size. It also has high production yields. Such features make these latexes more practical in final uses for a wide variety of application areas [14,29.68-69].

3. Emulsion polymerization of vinyl acetate

The production of poly(vinyl acetate), PVAc latexes using poly(vinyl alcohol) as an emulsifier began in Germany during the mid-1930 [6]. Vinyl acetate, VAc emulsion polymerization recipes in use in Germany before and during World War II were originally disclosed in the BIOS (British Intelligence Objectives Subcommision) and the US FIAT (Field Intelligence Agency Technical) reports [7-8]. On the basis of these report information, manufacture of PVAc latex paints was begun in England in 1948. Afterwards, the production of PVAc latexes has continued to the present day, growing steadily over the years. During this period, too many studies have been done to explain the mechanism and kinetics of emulsion polymerization of VAc, to develop its polymerization conditions, and to improve the properties of PVAc latexes and their films. These subjects are still very important in both industrially and scientifically. Furthermore, VAc copolymers have been developed by using many kinds of comonomers such as ethylene, acrylates, maleates, vinyl chloride and versatic acid esters.

The industrial importance of VAc latexes necessitates the understanding of the mechanism and kinetics of VAc emulsion homopolymerization and copolymerizations with other comonomers by scientific communities as well as industrial communities. VAc is polymerized exclusively via free-radical polymerization. The emulsion homopolymerization mechanism of VAc is different from that of other vinyl monomers like styrene and butadiene. Although Smith-Ewart theory and related theories successfully explain the emulsion polymerization of nearly water-insoluble monomers, they are not fit the emulsion polymerization of VAc [4,70-71]. The most characteristic feature of VAc is its polarity and relatively high solubility in water in comparison with other hydrophobic monomers such as styrene. For example, the water-solubility value of VAc is 2.5% by weight at 20 °C, while this value for styrene is $3.6x10^{-2}$ % [72]. Therefore, the initiation occurs in the aqueous phase and is explained by the *"homogeneous nucleation"* (as described in section 2.2). Another characteristic feature of VAc that affects the polymerization mechanism is its low reactivity and the correspondingly high reactivity of PVAc radicals in comparison with other vinyl monomers. This establishes the participation of the majority of components in the system in chain transfer during the polymerization of VAc. This has a substantial influence on the rate of polymerization, the molecular structure of the polymers, their branching, the tendency toward graft copolymerization, and also the tendency toward formation of significant amounts of gel fraction. In particular, the presence of a poly(vinyl alcohol), PVOH protective colloid in VAc emulsion polymerizations greatly affects branching and reaction mechanism. This agent is used as the principal surface-active substances in the system. More conventional emulsifiers may also be added to the system, but their role seems to be secondary that of this protective colloid. The chain transfer of PVAc to polymer, monomer and solvent (methanol) during emulsion polymerization in the absence of the PVOH is also a very important subject. As a result of these significant branching reactions of PVAc, there are considerable research results in literature on this subject [27,30,73-78].

Indeed, VAc has different features in comparison with other monomers in other respects than its water-solubility and the strong tendency of its radicals to the chain transfers. It has a high equilibrium monomer-polymer swelling ratio (values as great as 7:1). The heat of polymerization of VAc monomer is 21±0.5 kcal/mole [10]. Removing of the polymerization heat is of great importance in providing the control of the polymerization reaction. If not, a runaway exothermic reaction causes to uncontrollable boiling and consequent foaming or destabilization.

All these specific features of VAc monomer require a controlled emulsion polymerization process to be useful this monomer in particular industrial practice. The most commonly used process in the industrial practice of VAc emulsion polymerization is the delayed addition method in semi-continuous process. In this method, 5-15% of the VAc monomer is added into the reactor with some of the ingredients such as water, emulsifier and initiator at start. After the particle-generating period is completed, the remaining ingredients are incrementally during the polymerization. In some cases there is also gradual addition of remaining neat monomer and initiator when all other water phase ingredients are in the initial reactor charge. Additionally, many variations can be made in this process (see also section 2.4, semi-continuous process). The delayed addition process avoids the coalescence of the monomer-swollen latex particles, thus attaining greater colloidal stability. Emulsion prepared in this manner also show smaller particle size than emulsions prepared by the batch process. In the other hand, the semi-continuous process produces a broader molar mass distribution than the batch process and a high molar mass fraction which is attributed to chain transfer reactions due to monomer-starved conditions [15].

In the emulsion polymerization of VAc, the most commonly used initiators are the water-soluble, thermally decomposed, free-radical producing persulfates (peroxodisulfates) such as potassium-, sodium-, and ammonium-persulfate. Non-ionic and anionic emulsifiers (generally, mixtures of them) are the most widely preferred types because of improved compatibility with negatively charged PVAc particles as a result of persulfate initiator fragments. The use of protective colloids such as PVOH and hydroxyethyl cellulose (HEC) is common in the industrial emulsion polymerization of VAc. These water-soluble polymers can be used in the presence of emulsifier or in the emulsifier-free polymerization for the production of PVAc latexes and stable latexes can be made with only these protective colloids. The grafting reactions also take place between these protective colloids and VAc radicals, which influence the final product properties such as molar mass, viscosity and particle morphology. The VAc emulsion system is sensitive to changes of pH and the optimum pH range (pH=4.5—5.5) can be generally obtained by using sodium bicarbonate buffer system.

As a result of the emulsion polymerization of VAc monomer, an amorphous, non-crystalline and thermoplastic polymer is obtained. Poly(vinyl acetate) is soluble in alcohols, ketones, aromatic hydrocarbons and ethers. It can be also solubilized in emulsifier solutions. It absorbs 3 to 6 wt% water between 20 to 70 °C within 24 hrs. Moreover, in the presence of water, it can hydrolyze to form vinyl alcohol units and acetic acid, or the acetate of the basic cation. The minimum film forming temperature (MFFT), the glass transition temperature (T_g) and the mechanical properties of the PVAc depend on the molecular weight. For medium molecular weight, its T_g is approximately 30 °C and its MFFT is approximately 20 °C. Its Young modulus is 600 N/mm^2 at 25 °C with 50% relative humidity. The tensile strength and the elongation at break of PVAc are between 29.4—49.0 N/mm^2 and 10—20% at 20 °C, respectively [30]. These properties depend on not only the molecular weight of the polymer, but also the emulsion polymerization type, the ingredients of the polymerization and the particle morphology of the polymer. The thermal degradation of PVAc occurs at 150-220 °C. PVAc is brittle on cooling below room temperature, i.e., at 10-15 °C [30].

Nevertheless, some of these properties of the PVAc make the VAc homopolymer latexes and its films unsuitable for many applications. For example, its films have low resistance against water, alkaline, and hydrolysis. They show poor mechanical properties. T_g and MFFT values of this homopolymer also constitute the major problems in its applications because the MFFT is too high for this polymer to perform as an effective binder for pigments and fillers at normal ambient temperatures. While the T_g is too low for the polymer to be useful as a rigid plastic at normal ambient temperatures, the same value is too high for application as a binder and an adhesive at the same temperatures. Due to these properties of the VAc homopolymer, VAc is copolymerized with other monomers such as ethylene, n-ethyl acrylate, n-butyl acrylate, 2-ethylhexyl acrylate, methyl methacrylate, vinyl chloride, vinyl propionate, vinyl versatate, maleates (dibutyl malate), fumarates, and acrylonitrile to improve the poor properties of this polymer. Terpolymers of VAc are also widely produced. The selection of comonomers to be used to produce VAc copolymer latexes for any given application depends principally upon the functional suitability of the comonomer and its coast. As well as the VAc homopolymer latex, The VAc based copolymer/terolymer latexes are widely used for exterior and interior architectural coatings, adhesives, textile and paper industry, and numerous other applications.

4. Effects of polymerization variables on the properties of vinyl acetate based emulsion polymers

VAc monomer is a polar and reactive, and has the characteristic features. The significant differences between VAc and the comonomers lead to difficulties during copolymerization when the VAc is copolymerized with less polar and more reactive monomers. The emulsion copolymerization of VAc with the comonomers requires the special treatments. Other properties such as the copolymer structure, the particle morphology, and the colloidal, physical and film properties of the copolymer latex are also affected from this situation. In order to prevent potential problems and produce copolymer emulsion polymers in homogeneous copolymer structure and desired properties, the various processes and different ingredient systems are applied for VAc copolymerization systems. The major copolymerization systems of VAc and, the effects of copolymerization variables on the course of polymerization and the properties of emulsion copolymers are given in detail below.

One of the most important industrial latexes is the vinyl acetate/n-butyl acrylate (VAc/BuA) copolymer latex. This copolymer latex is utilized extensively in interior and exterior paints and is superior to either homopolymer alone. The main features of these monomers are: reactivity ratios (r_{VAc}=0.05 and r_{BuA}=5.5), water solubilities (25 g/L for VAc and 1-1.5 g/L for BuA, at 20 °C), and glass transition temperatures (T_g (PVAc) =32 °C and T_g (PBuA) = –54 °C). The presence of BuA in VAc emulsion copolymerizations results in internally plasticized copolymers with MFFTs in the ambient temperature range. BuA also imparts softness and tackiness, and thus the durability of VAc copolymer increases. Although polyacrylates and polymethacrylates are less susceptible to hydrolysis than the PVAc, VAc/BuA emulsion copolymers show sensibility to hydrolysis. In industry, 15-25% by weight of BuA is preferred.

The emulsion polymerization process types, especially batch and semi-continuous processes, are very effective on the VAc/BuA copolymerization and the obtained copolymers. In batch and semi-continuous emulsion processes of VAc/BuA comonomer mixtures having various compositions, homogeneous copolymer compositions are obtained only by semi-continuous process. In contrast, batch process produces the VAc/BuA latexes in heterogeneous copolymer composition. Due to the significantly differences in water solubility and reactivity ratio of this monomer couple, the semi-continuous process provides better control over compositional heterogeneity than batch process for this monomer couple [27,79-82]. The copolymer latexes produced by batch process have a larger average particle size (PS) and narrower particle size distribution (PSD) than the latexes produced by semi-cotinuous process. By increasing of BuA content in comonomer mixture the average particle size of semi-continuous latexes decreases, whereas it is independent of comonomer composition for batch-latexes [27,80,82-86]. In addition, the particle structures of batch-latexes have a larger BuA-rich core surrounded by a PVAc-rich shell, indicating heterogeneous structure, than the semi-continues latex particles [80]. Also, the ratio of each monomer on particle surface depending on the process type and copolymer composition effects hydrolysis and electrolyte resistance of the latexes [27,82]. The broader molecular weight distribution (MWD) with bimodal character is observed for semi-continuous copolymers. The MWD of batch copolymers are broad, but they are narrower and in uni-modal character in comparison to those of semi-continuous copolymers. This results for both processes indicates the excessive branching reactions [27,82-85,87]. When the film properties of VAc/BuA copolymer latexes of various copolymer composition are examined, it is seen that the semi-continuous process provide better properties to copolymer films such as filming ability and film homogeneity (examined according to the T_g and surface analyzes) than the batch process [27,88]. The homogeneity influences hardness and adhesive properties of latexes as well as copolymer composition, and hardness decreases with increasing homogeneity [79]. The thermodynamic work of adhesion decreases with increasing BuA content for semi-continuous copolymer films [86]. In contrast, tensile properties of the batch copolymer films show a higher ultimate tensile strength, higher Young's modulus, and lower percent elongation to break compared to semi-continuous latex films [27,88]. In addition, by the increase of BuA content in the copolymer, the film abilities increase and T_g values decrease for copolymers produced by both process [27,85-86,88].

The semi-continuous copolymerization of VAc/BuA monomer system can be carried out by applying different addition policy of copolymerization ingredients. In the copolymerization of VAc/BuA system having a certain monomer ratio, the monomer addition type directly affects the course of copolymerization. For seeded-semi-continuous copolymerizations in which a certain part of monomer present in the initial reactor charge as a remaining part in the feeding, the polymerization rate approaches a constant value depending on the monomer addition rate when the monomer-starved feed addition strategy is applied (for a neat monomer feed method). It increases with increasing monomer addition rate, and approaches the rate observed for batch polymerization when the feed rates exceed the maximum values for monomer –starved conditions. In the other hand, the polymerization rate is at maximum and independent upon the monomer feed rate when monomer-flooded condition is applied [87,89]. Furthermore, for the semi-continuous process of VAc/BuA, micro-phase separation in particle structure is revealed, and the feeding rate of comonomers influences the change in this particle morphology of copolymers. The faster feeding rate causes to greater separation while a multi-core-shell structure is formed with the slowest feeding [90]. The initiator addition policy also affects the polymerization rate, the instantaneous monomer conversion, the average PS and PSD, and the polymerization stability of the VAc/BuA semi-continuous emulsion systems. Increasing the amount of ammonium persulfate initiator in the initial reactor charge leads to increase the rate of polymerization and the conversion. But it is observed that this amount is increased above a certain value, destabilization occurs and the average PS increase [91-92].

The stabilizing system significantly affects many colloidal and film properties of the VAc/BuA emulsion copolymers. Berber and co-workers studied the semi-continuous emulsion copolymerization of this comonomer system in the presence of conventional non-ionic emulsifiers and different protective colloids which were water-soluble polymer

(PVOH) and polymerizable oligomeric N-methylol acrylamide (o-NMA). The VAc/BuA copolymer latexes obtained by o-NMA at different copolymer compositions were found to have lower viscosity, finer particle size, better latex stability and lower MWD and their films showed the higher T_g, better film forming behavior and better water resistance, compared to those synthesized by PVOH. Figure 3 shows the effect of the BuA content and the stabilizing system on the T_g of the VAc/BuA latex films [85]. They also compared the effects of two different initiator systems which consist of ammonium persulfate and potassium persulafte, on the properties of VAc/BuA copolymers synthesized in presence of o-NMA. It was seen that the particle size of copolymers varied by the increasing BuA content in the copolymer regardless of the type of initiator. Otherwise, for the all copolymer compositions, molecular weights (MWs) of the copolymers synthesized in the presence of potassium persulfate were found to be higher and the latex viscosities of them were lower, than those of VAc/BuA copolymers synthesized in the presence of ammonium persulfate [83].

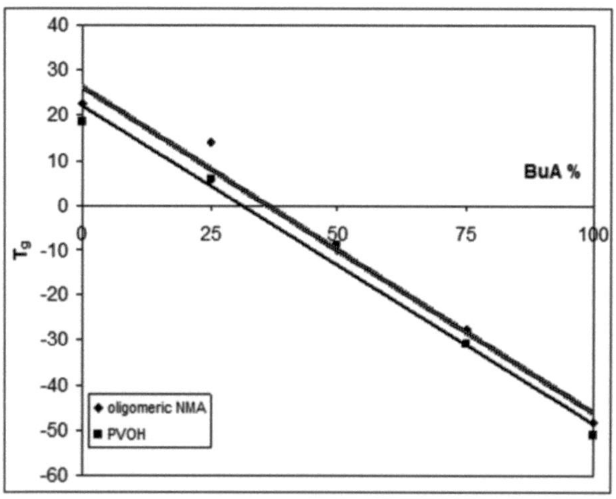

Figure 3. Changes in T_g with copolymer composition for VAc/BuA copolymer latexes obtained by o-NMA and PVOH [85]

The seeded emulsion polymerization process can be applied for VAc/BuA comonomer system. Seeded reactions allow for the production of mono-dispersee latexes. Key factors in producing narrow PSD latexes are concentration of seed latex and ionic strength. The increasing of seed latex concentration (55% by weight) causes to the largest mono-dispersee particles [93].

The emulsion copolymerization of VAc/BuA monomer system can be performed by other emulsion polymerization methods, mini-emulsion and micro-emulsion polymerizations. When the mini-emulsion copolymerization carries out with the 50/50 molar ratio of VAc/BuA in the presence of sodium hexadecyl sulfate emulsifier, hexadecane co-surfactant and ammonium persulfate initiator, the results of this polymerization conducted in a batch process are different from that of conventional batch polymerization of this monomer couple [94]. For the mini-emulsion polymerization the polymerization rate is slowed done, the final particle size is larger and the coagulation during the polymerization occurs, compared to the conventional batch process. There is also less mixing between the BuA-rich core and VAc-rich shell. The presence of co-surfactant makes significant effects to both of course of the copolymerization and the resulting latex and films [94].

The emulsion copolymerizations of vinyl acetate with maleic acid diesters such as dibutyl maleate (DBM) and dioctyl maleate (DOM) are also possible. These diesters are easily copolymerized but can not homopolymerize. The main features of these monomer couples are: reactivity ratios (r_{VAc}=0.171 and r_{DBM}=0.046; r_{VAc}=0.195 and r_{DOM}=0.945) and water solubilities (25 g/L for VAc, 0.35 g/L for DBM and almost zero for DOM, at 20 °C) [95-96]. Because of the significantly different monomer properties between VAc and these maleic acid diesters, copolymer latexes having a wide range of properties can be obtained by varying the comonomer composition, thus influencing the glass transition temperature and minimum film forming temperature of the latexes. The use of these comonomers provides internal plastification to VAc polymers and they reduce the Tg value of them. The hardness, flexibility, and alkali, water and ultra-violet light resistance of their films can be improved. Furthermore, these diesters gain more stability to the copolymer latex by being strongly associated to the latex particles. It is then expected that the final properties of the copolymers will be enhanced and will not be altered during time due for instance to the migration of the plasticizer. The copolymers of VAc/Maleic acid diester are used for paints and adhesives.

Although these copolymers are industrially very useful, there are no more scientific researches about them. The earliest studies in the literature about the VAc/Maleic acid diester copolymers belong to Donescu and co-workers [57,97]. They investigated the semi-continuous emulsion polymerization of VAc and DBM (VAc/DBM: 60/40 by volume). They used PVOH, 20 moles ethoxylated cetyl alcohol and potassium persulfate as protective colloid, emulsifier and initiator respectively. The seeded-semi-continuous process was carried out by applying monomer emulsion feed method. The effects of emulsifier distribution between initial reactor charge and continuously introduced monomer, and agitation speed on the monomer conversion and copolymer structure were also investigated. They observed decrease in the polymerization rate with an increase of agitation speed. On the other hand, the transfer reactions were higher for the smaller speed. When the influence of the emulsifier distribution on conversion was examined, it was seen that the conversion increased and then decreased with the increasing initial reactor charge concentration of emulsifier. This behavior has been explained that the chain transfer effect of emulsifier is more pronounced over a certain emulsifier concentration. The emulsifier and DBM act as chain transfer agent in this type emulsion copolymerization of VAc/DBM [57]. In addition, Donescu and Fusulan copolymerized VAc with DBM using a semi-continuous process for different copolymer compositions. They also noticed a decrease in particle size with increasing DBM monomer. The retarding effect of DBM on the VAc polymerization was also determined [98]. In another study, Donescu and co-workers copolymerized VAc with another type of maleic acid diesters, 2-ethylhexyl maleate which is synonym for DOM, by semi-continuous emulsion polymerization [97]. They used a different stabilizing system including hydroxylethyl cellulose (HEC) as protective colloid and sodium sulfosuccinate of 6 moles ethoxylated nonyl phenol as emulsifier compared to the VAc/DBM system. They also used n-butanol as chain transfer agent. The comonomer addition was applied in flooded conditions for a certain comonomer ratio, VAc/2-ethylhexyl maleate: 80/20 by volume). They investigated the effect of HEC on the particle number, conversion, and latex surface tension during the copolymerization. The oscillations occurred in these investigation parameters due to the flocculation and nucleation of new particles. The authors suggested that HEC splits its chain in the presence of the initiator, thus continuously stabilizing the new particles formed. In theVAc/2-ethylhexyl maleate copolymerization system, it was seen that the copolymerization rate and conversion decreased with the increasing agitation rate. The higher the agitation rate also affected the average PS of copolymers, and the smaller the PS found up to 600 rpm agitation rate. In addition, the presence of n-butanol in that reaction system decreased the average particle diameter and cross-linked copolymer content.

Wu and Schork also studied the copolymerization of VAc/DOM for different compositions of monomer mixture. They compared the conventional (macro-) and the mini-emulsion polymerization by applying both of batch and semi-continuous process [95]. In their study, a certain difference in PSs and particle numbers between macro-emulsion and mini-emulsion feed semi-continuous operation is seen. The PS is smaller and the particle number is higher. The size decreases with a decrease in feed rate for both macro- and mini-emulsion polymerization. The PSs in semi-continuous operations are smaller than in batch mini-emulsion polymerizations when the feed rate of monomer emulsion is sufficiently low. For batch copolymerizations of VAc and DOM, the rate of macro-emulsion polymerization is faster than that of mini-emulsion polymerization while the particle number increases with increasing conversion throughout the reaction for both macro- and mini-emulsion processes. Due to the extremely water-insoluble DOM monomer, a significant deviation in the composition of the VAc/DOM copolymer is observed. The most homogeneous compositions are obtained by mini-emulsion copolymerizations and the coagulation resistance of this emulsion polymerization system is better than that of macro-emulsion process, as well. For both of macro- and mini-emulsion polymerizations, the increasing concentration of DOM in the monomer mixture results in a reduction of the total polymerization rate and reduction of the copolymer MW.

Berber investigated the effects of some polymerization variables such as process type, initiator addition type, emulsifier type and its distribution between the initial reactor charge and feed, and feeding time on the monomer conversion, the latex stability and the average particle size for semi-continuous emulsion copolymerizations of VAc with DOM (VAc/DOM: 60/40, by weight) [99]. The emulsion copolymerizations were carried out in the presence of potassium persulfate as initiator and o-NMA as polymerizable co-stabilizer. A mixture of monomers, such as including VAc and DOM, should be added to reaction medium by the monomer emulsion feed method because of the large water-solubility differences between the two monomers. Thus, the selected emulsifier system should constitute "pre-emulsion" for monomer emulsion feed of the monomer couple. For the mixture of VAc and DOM monomers, anionic and non-ionic emulsifiers such as sodium dodecylbenzene sulfonate (Maranil A 25), dioctyl sodium sulfosuccinate (Disponil SUS 87), sodium lauryl ether sulphate (SLES), 30 and 10 moles ethoxylated nonyl phenoles (NP 30 and NP 10), and their combinations were used. The non-ionic emulsifiers system consisting of equal amounts of NP 30 and NP 10 which can form stable pre-emulsion and can produce stable copolymer latex was selected as suitable emulsifier system for VAc/DOM emulsion copolymerization (Table 1). The change of emulsifier concentration in the initial reactor charge (from 0% to 33%, by weight) affected the stability of polymerization and the average PS of copolymer latexes. The very high or the very low concentrations of emulsifier in the initial charge caused to polymerization instability and increase in average PS (Table 2).

Emulsifiers	Stable Pre-emulsion	Stable Latex
Maranil A 25	-	-
Disponil SUS 87	-	-
SLES	-	-
NP 30	-	-
NP 10	-	-
% 70 SLES + % 30 NP 10	✓	-
% 30 NP 30 + % 70 NP 10	✓	-
% 50 NP 30 + % 50 NP 10	✓	✓

Table 1. The effects of emulsifier type on the stability of pre-emulsion and resulting latex for VAc/DOM system

Emulsifier concentration in initial reactor charge (%, w/w)	Appearance of latex	Particle size (nm)
33	Creaming and incrustation on the latex surface	284
25	Partially incrustation on the latex surface	221
0	Coagulation and incrustation on the latex surface	482

Table 2. The effects of emulsifier concentration in initial reactor charge on the stability and the particle size of latex for VAc/DOM system

In addition, the presence of seeding stage in the semi-continuous process (by 10 wt % monomer), the increasing of the feeding time and better control of the initiator feed increased the monomer conversion, the average particle size, and the stabilities of the polymerization and latex of VAc/DOM monomer system. The effect of these reaction variables on the conversion and the particle size of the VAC/DOM latex are given in Table 3 [99].

Monomer content in initial reactor charge (%, w/w)	Seeding stage	Feeding stage	Completion stage	Initiator feed	Conversion (%)	Particle size (nm)
-	-	70 °C, 2.5 h.	70 °C, 30 min.	as 5 portion during feeding time	82	145
-	-	70 °C, 3 h.	70 °C, 1 h.	as drop wise during feeding time	86	160
10	30 min. at 65 °C + 30 min. at 70 °C	70 °C, 5 h.	80 °C, 1h.	half of the total amount at beginning of the reaction + the remaining amount as drop wise during feeding time	95	221

Table 3. The effect of some polymerization variables on conversion and particle size of VAc/DOM latex

Vinyl versatate monomer (vinyl ester of versatic acid) is a commercial product of Shell Chemicals and called as VeoVA. It is a branched vinyl neodecanoate, and this branched vinyl ester (VeoVA 10) is copolymerized with VAc. The reactivity ratios of VAc and vinyl neodecanoate are respectively 0.99 and 0.92. The water solubilities are 25 g/L for VAc and 0.1 g/L for vinyl neodecanoate at 20 °C. The glass transition temperatures are 32 °C for PVAc and –3 °C for polyVeoVa. VeoVA can act as internal plasticizer in VAc emulsion copolymers. The most distinguishing feature of these branched vinyl esters is their resistance to hydrolysis, both as monomers and in polymers. Polymerization of branched vinyl esters with VAc results in polymers whose hydrolytic stability improves with increasing concentrations of branched vinyl ester. Higher molar mass vinyl branched esters offer excellent hydrophobic properties. These esters are strongly resistant to saponification, water absorption and UV degredation. The properties of VAc polymer such as water and alkali resistance and exterior durability improve with these esters. These all advantages make the VAc/branched vinyl ester copolymers more suitable for applications. They are used in architectural paints, interior and exterior applications, and general markets, Typically, about 15-30% VeoVa-10 is used to optimize the cost-performance properties of VAc polymers. Interior latexes tend to contain 15-20% monomer. Exterior latexes usually contain 20-30% of the branched monomer [15,30]. The high solids emulsion copolymerization of VAc and VeoVA 10 can be carried out in a continuous industrial reactors, continuous loop reactor and continuous stirred tank reactor (CSTR). When the heat generation rate is high a thermal runaway occurred in the CSTR reactor, whereas the loop reactor is easily controlled [100].

For the emulsion copolymerizations of VAc, methyl methacrylate (MMA) comonomer is added to increase the T_g of the VAc polymer. The homopolymer of the MMA has a T_g of approximately 106 °C. The other features of these monomers are: reactivity ratios (r_{VAc}=0.03 and r_{MMA}=22.21) and water solubilities (25 g/L for VAc and 16 g/L for MMA, at 20 °C) [15]. This comonomer also imparts hardness and no tackiness. VAc/MMA emulsion copolymers have unlimited uses. VAc can also be copolymerized with other acrylates such as methylacrylate (MA), 2-ethylhexyl acrylate (2-EHA) by emulsion polymerization. The main features of these monomers are: reactivity ratios (r_{VAc}=0.04 and r_{MA}=7.5 ; r_{VAc}=0.029 and r_{EHA}=6.7), water solubilities (25 g/L for VAc, 0.1 g/L for 2-EHA and 52 g/L for MA, at 20 °C) and glass transition temperatures (T_g (PVAc) =32 °C and T_g (PMA) =22 °C) [30]. 2-EHA is a softer monomer and adds durability to the copolymer. The tack of the final copolymer films increases when this comonomer is used in high concentrations. The terpolymerizations of these comonomers are also possible with VAc and other acrylates [15,30].

Moreover, some water-soluble functional monomers can be used in emulsion copolymerizations or terpolymerizations. Acrylic acid, methacrylic acid, fumaric acid, crotonic acid, maleic acid, itoconic acid, N-methylol acrylamide and some polymerizable monomer containing amines, amides and acetoacetates are used alone and in combination with each other to improve the stability and adhesion properties of VAc emulsion polymers.

The presence of acid comonomers decreases the rate of reaction and icreases PS. Also, it leads to the splitting reactions of the potassium persulfate initiator [101]. Acid comonomer groups in the copolymer structuture act as effective thickeners and impart good blending properties with inorganic additives. Carboxylic groups located on the particle surface impart high pH and high stability The minor amount of acid comonomers is used to modify VAc/acrylates copolymer emulsions [102]. Acrylic acid (AA) and methacrylic acid (MAA)-modified VAc/BuA or VAc/2-EHA copolymer latexes are used in the pressure-sensitive adhesive market. The feeding of the acid comonomers in the monomer feed produces a greater viscosity response upon pH adjustment compared to a water solution feed of them [70]. Minor amounts of N-methylol acrylamide are sometimes used impart cross-linking and heat-set properties to the VAc homopolymer and VAc/acrylate copolymers, especially in textile printing binder applications and adhesive applications for wood.

5. Conclusion

This chapter basically focused on the scientific and industrial importance of the emulsion polymerization and vinyl acetate based emulsion polymers from past to present. Firstly, the basic issues of conventional emulsion polymerization were given. Its ingredients, kinetics, and mechanisms were explained in detail. Other emulsion polymerization methods including micro-, mini- and inverse-emulsion polymerization were mentioned, followed by the description of main emulsion polymerization processes comprising batch, semi-continuous, continuous and seeded, and their application types. Secondly, the emulsion polymerization of vinyl acetate was given. The characteristic feature of vinyl acetate monomer, its emulsion polymerization conditions, and the main properties of its homopolymer latex were summarized. Finally, the emulsion copolymerizations of vinyl acetate with other monomers having specific features and industrially importance were discussed briefly. The effects of the copolymerization variables and the components on the course of emulsion polymerization and the properties of the resulting copolymers were explained, which was supported by the literature results.

Author details

Hale Berber Yamak
Department of Chemistry, Yildiz Technical University, Istanbul, Turkey

6. References

[1] Hoffman F and Delbrück K (1909) German Patent. 250,690 (to Farbenfabriken Bayer A.G).
[2] Hoffman F and Delbrück K (1912) German Patent. 254,672 (to Farbenfabriken Bayer A.G).
[3] Hoffman F and Delbrück K (1912) German Patent. 255,129 (to Farbenfabriken Bayer A.G).
[4] Dinsmore R.P (1929) U.S. Patent. 1,732,975 (to Goodyear Tire & Rubber Co.).
[5] Luther M and Hück C (1932) U.S. Patent. 1,864,078 (to I.G Farbenindustrie A.G).
[6] Starck W and Freudenberg H (1940) U.S. Patent. 2,227,163 (to I.G Farbenindustrie A.G).
[7] B.I.O.S (British Intelligence Objectives Subcommision) (1947) Manufacture of Vinyl Acetate Polymers and Derivatives. Final Report 744. London: H.M.S.O.
[8] US FIAT (Field Intelligence Agency Technical) (1947) Polymerization of Vinyl Acetate. Final Report 1102. London: H.M.S.O.
[9] B.I.O.S (British Intelligence Objectives Subcommision) (1954) The German Plastic Industry During the Period 1939-1945. Surveys Report 34. London: H.M.S.O.
[10] Bovey F.A, Kolthoff I.M, Medalia, A.I, Meehan, E.J (1965) Emulsion Polymerization. High Polymer Series Vol. IX. New York: Interscience Publishers Inc.
[11] Blackley D.C (1975) Emulsion Polymerization, Theory and Practice. London: Applied Science Publishers.
[12] Eliseeva V.I, Ivanchev S.S, Kuchanov S.I, Lebedev A.V (1981). Emulsion Polymerization and Its Applications in Industry. New York: Plenum Publishing Corporation.
[13] Piirma I (1982) Emulsion Polymerization. New York:Academic Press.
[14] Fitch M.R (1997) Polymer Colloids: A Comprehensive Introduction. USA: Academic Press.
[15] Lovell P.A, El-Aasser M.S (1997) Emulsion Polymerization and Emulsion Polymers. England: Wiley.

[16] Urban D, Takamura K (2002) Polymer Dispersions and Their Industrial Applications. Germany: Wiley-VCH Verlag GmbH & Co.
[17] Whitby G.S, Davis C.C, Dunbrook R.F (1954) Synthetic Rubber. New York: Wiley.
[18] Hoelscher F (1969) Dispersionen Synthetischer Hochpolymeren, Part I. New York: Springer-Verlag.
[19] Ostromislensky I.I (1916) Condensation of Alcohols and Aldehydes in Presence of Dehydrating Agents. Mechanism of the Process J. Russ. Phys. Chem. Soc. 48:1071.
[20] Whitby G.S, Katz M (1933) Synthetic Rubber. Ind .Eng. Chem. 41: 1204-1211, 1338-1348.
[21] Berezan K, Dobromyslova A, Dogadkin B (1938) Synthetic Rubber From Alcohol. Bull. Acad. Sci. U.S.S.R. 7:409.
[22] Fikentscher H (1938) Emulsionspolymerisation und Technische. Angew. Chem. 51:433.
[23] Hohenstein W.P and Mark H (1946) polymerization of Olefins and Diolefins in Suspension and Emulsion. Part II. J. Polym. Sci. 1: 549-580.
[24] Talalay A, Magat B (1945) Synthetic Rubber from Alcohol: A Survey Based on the Russian Literaure. New York: Interscience.
[25] Williams H.L (1956) Polymerization in Emulsion In: Schildknecht C.H editor. Polymer Processes. New York: Interscience. pp. 111-171.
[26] Warson H (1972) Applications of Synthetic Resin Emulsions. London: Benn.
[27] El-Aasser M.S, Vanderhoff, J.W (1981) Emulsion Polymerization of Vinyl Acetate. London: Applied Science Publishers.
[28] Gilbert R.G (1995) Emulsion Polymerization-A mechanistic Approach. New York: Academic Press.
[29] Blackley D.C (1997) Polymer Latices Science and Technology. Volume 2. London: Chapman & Hall.
[30] Erbil H.Y (2000) Vinyl Acetate Emulsion Polymerization and Copolymerization with Acrylic Monomers. Florida: CRC Pres.
[31] Herk A (2005) Chemistry and Technology of Emulsion Polymerization. Oxford: Blackwell Publishing.
[32] Winnik M.A (1997) The Formation and Properties of Latex Films. In: Lovell P.A, El-Aasser M.S editors. Emulsion Polymerization and Emulsion Polymers. England: Wiley. pp. 468-470.
[33] Chern C.S (2006) Emulsion Polymerization Mechanisms and Kinetics. Prog. Polym. Sci. 31:443-486.
[34] Lijen C, Shi Y. L, Chorng S.C, Shuo C.W (1997) Critical Micelle Concentration of Mixed Surfactant SDS/NP$(EO)_{40}$ and Its Role in Emulsion Polymerization. Colloids and Surfaces A. 122:161-168.
[35] Chern C.S, Lin S.Y, Chang S.C, Lin J.Y, Lin Y.F (1998). Effect of Initiator on Styrene Emulsion Polymerisation Stabilised by Mixed SDS/NP-40 Surfactants. Polymer. 39:2281-2289.
[36] Javier I, Amalvy M.J, Unzué H.A.S, José M.A (2002) Reactive Surfactants in Heterophase Polymerization: Colloidal Properties, Film-Water Absorption and Surfactant Exudation. J. Polym. Sci. 40:2994–3000.

[37] Guyot A, Tauer K, Asua J.M, Van S, Gauthier C, Hellgren A.C, Sherrington D.C, Goni A.M, Sjoberg M, Sindt O, Vidal F, Unzue M, Schoonbroad H, Shipper E, Desmazes P.L (1999) Reactive Surfactants in Heterophase Polymerization. Acta Polym. 50:57-66.
[38] Harkins W.D (1945) A General Theory of The Reaction Loci in Emulsion Polymerization. J. Chem. Phys. 13:381.
[39] Harkins W.D (1946) A General Theory of Reaction Loci in Emulsion Polymerization. J. Chem. Phys.14:47-48.
[40] Smith W and Ewart V (1948) Kinetics of Emulsion Polymerization. J. Chem. Phys. 16: 592-607.
[41] Available: en.wikipedia.org/wiki/Emulsion_polymerization.
[42] Gardon J. L (1968) Emulsion Polymerization I. Recalculation and Extension of Smith-Ewart Theory. J. Polym. Sci. A. 6:623-641.
[43] Priest W.J (1952) Particle Growth in the Aqueous Polymerization of Vinyl Acetate. J. Phys. Chem. 56:1077-1083.
[44] Roe C.P (1968) Surface Chemistry Aspects of Emulsion Polymerization. Ind. Eng. Chem. 60:20-33.
[45] Fitch R.M, Tsai C.H (1970) Polymer Colloids: Particle Formation in Nonmicellar Systems. Polym Lett. 8:703-710.
[46] Fitch R.M, Tsai C.H (1971) Particle Formation in Polymer Colloids. III. Prediction of the Number of Particles by Homogeneous Nucleation Theory. In: Fitch R.M editor. Polymer Colloids. Newyork: Plenum Press. pp. 73-102.
[47] Hansen F.K, Ugelstad J (1978) Particle Nucleation in Emulsion Polymerization. I. A Theory for Homogeneous Nucleation. J. Polym. Sci. A. 16:1953-1979.
[48] Alexander A.E (1962) Some Aspects of Emulsion Polymerization. J. Oil Colour Chem. Assoc. 45:12-15.
[49] Fitch R.M and Watson R.C (1979) Coagulation Kinetics in Polymer Colloids Determined by Light Scattering. J. Colloid İnterface Sci. 68:14-20.
[50] Feeney P.J, Geissler E, Gilbert R.G, Napper D.H (1988) SANS Study of Particle Nucleation in Emulsion Polymerization. J. Colloid İnterface Sci. 121:508-513.
[51] Hunkeler D, Candau F, Pichot C, Hemielec A.E, Xie T.Y, Barton J, Vaskova V, Guillot J, Dimoine M.V, Reichert K.H (1994) Heterophase Polymerizations: A Physical and Kinetic Comparison and Categorizarion. Adv. Polym. Sci. 112:115-133.
[52] Ugelstad J, El-Aasser M.S, Vanderhoff J.W (1973) J. Polym. Sci. Polym. Lett. 11: 503.
[53] Lopez de Arbina L and Asua J.M (1992) High-solids-Content Batch Miniemulsion Polymerization. Polym. 33: 4832-4837.
[54] Asua J.M (2002) Miniemulsion Polymerization. Prog. Polym. Sci. 27: 1283-1346.
[55] Vanderhoff J.W (1993) Recent Advances in the Preparation of Latexes. Chem. Eng. Sci. 48:203-217.
[56] Donescu D, Gosa K, Ciupitoiu A (1985) Semicontinuous Emulsion Polymerization of Vinyl Acetate. Part I. Homopolymerization with Poly(vinyl alcohol) and Nonionic Coemulsifier. J Macromol. Sci. Part A. 22:931-941.

[57] Donescu D, Gosa K, Languri J and Ciupiţoiu A (1985) Semicontinuous Emulsion Polymerization of Vinyl Acetate. Part II. Copolymerization with Dibutyl Maleate J Macromol. Sci. Part A. 22:941.

[58] nuparek J, Bradna P, Mrkvıckova L, Lednicky F, Quadrat O (1995) Effect of Initial Polymerization Conditions on the Structure of Ethyl Acrylate-Methacrylic Acid Copolymer Latex Particles. Collect. Czech. Chem. Commun. 60:1756-1764.

[59] Sajjadi S and Brooks B.W (1999) Semibatch Emulsion Polymerization of Butyl Acrylate. I. Effect of Monomer Distribution. J.Appl. Polym. Sci. 74:3094-3095.

[60] Sajjadi S and Brooks B.W (2000) Semibatch Emulsion Polymerization of Butyl Acrylate. II. Effects of Emulsifier Distribution. J.Appl. Polym. Sci. 79:582-584.

[61] Sajjadi S, Brooks B.W (2000) Semibatch Emulsion Polymerization Reactors: Polybutyl Acrylate Case Study. Chem. Eng. Sci. 55:4757-4781.

[62] Sajjadi S, Brooks B.W (2000) Unseeded Semibatch Emulsion Polymerization of Butyl Acrylate: Bimodal Particle Size Distribution. J. Polym. Sci.Part A. 38:528-545.

[63] Sajjadi S (2000) Particle Formation and Coagulation in the Seeded Semibatch Emulsion Polymerization of Butyl Acrylate. J. Polym. Sci. Part A. 38: 3612-3630.

[64] Sajjadi S (2001) Particle Formation Under Monomer-Starved Conditions in the Semibatch Emulsion Polymerization of Styrene. I. Experimental. J. Polym. Sci. Part A. 39:3940-3952.

[65] Al-Bagoury M, Yaacoub E.J (2003) Semicontinuous Emulsion Copolymerization of 3-O-methacryloyl-1,2:5,6-di-O-isopropylidene-α-D-glucofuranose (3-MDG) and Butyl Acrylate (BA). Monomer Feed Addition. J.Appl. Polym.Sci. 90: 091-2102.

[66] Al-Bagoury M, Yaacoub E.J (2004) Semicontinuous Emulsion Copolymerization of 3-O-methacryloyl-1,2:5,6-di-O-isopropylidene-α-D-glucofuranose (3-MDG) and Butyl Acrylate (BA) by Pre-emulsion Addition Technique. Europ. Polym. J. 40:2617-2627.

[67] Mayer M.J.J, Meuldijk J, Thoenes D (1994) Emulsion Polymerization in Various Reactor Types: Recipes with High Monomer Content. Chem. Eng. Sci. 49:4971-4980.

[68] Zhao K, Sun P, Liu D, Dai G (2004) The Formation Mechanism of Poly(vinyl acetate)/Poly(butyl acrylate) Core/Shell Latex in Two-satge Seeded Semi-continuous Starved Emulsion Polymerization. Europ. Polym. J. 40:89-96.

[69] im J.W, Suh K.D (2008) Monodisperse Polymer Particles Synthesized by Seeded Polymerization Techniques. J. Ind. Eng. Chem. 14:1-9.

[70] O'Donnell J.T, Mersobian R.B, Woodward A.E (1958) Vinyl Acetate Emulsion Polymerization. J. Polym. Sci. 28:171-177.

[71] French D.M (1958) Mechanism of Vinyl Acetate Emulsion Polymerization. J. Polym. Sci. 32: 395-411.

[72] Lindemann M.K (1967) Vinyl Polymrization. New York: Dekker. Vol 1. Pt 1. 260.

[73] Okaya T, Tanaka T, Yuki K (1993) Study on Physical Properties of Poly (vinyl acetate) Emulsion Films Obtained in Batchwise and in Semicontinuous Systems. J. Appl. Polym. Sci. 50:745.

[74] Gavat I, Dmonie V, Donescu D, Hagiopol C, Munteanu M, Gosa K and Deleanu T.H (1978) Grafting process in vinyl acetate polymerization in the presence of nonionic emulsifiers. J. Polym. Sci. Polym Symp. 64:125.

[75] Gilmore C.M, Poehlin G.W, Schork F.J (1993) Modeling poly(vinyl alcohol)-stabilized vinyl acetate emulsion polymerization. I. Theory. J. Appl. Polym. Sci.48: 1449,1461.
[76] Gonzalez G.S.M, Dimonie V.L, Sudol, E.D, Yue H.J, Klein A, El-Aasser M.S (1996) Characterization of Poly(vinyl alcohol) During the Emulsion Polymerization of Vinyl Acetate Using Poly(vinyl alcohol) as Emulsifier. J. Polym. Sci. A: Polym. Chem. 34:849-862.
[77] Suzuki A, Yano M, Saiga T, Kikuchi K, Okaya T (2003) Study on the Initial Stage of Emulsion Polymerization of Vinyl Acetate Using Poly(vinyl alcohol) as a Protective Colloid. Colloid Polym. Sci. 281:337-342.
[78] Carra S, Sliepcevich A, Canevarolo A and Carra S (2005) Grafting and Adsorption of Poly(vinyl) Alcohol in Vinyl Acetate Emulsion Polymerization. Polymer. 46:1379-1384.
[79] Chujo K, Harada Y, Tokuhara S, Tanaka, K (1969) The Effects of Various Monomer Addition Methods on the Emulsion Copolymerization of Vinyl Acetate and Butyl Acrylate. J. Polym. Sci. Part C. 27: 321-332.
[80] Mısra S.C, Pichot C, El-Aasser M.S, VanderHoff, J.W (1979) Effect of Emulsion Polymerization Process on The Morphology of Vinyl Acetate-Butyl Acrylate Copolymer Lateks Films. J. Polym. Sci. Polym. Lett. Edd.17:567-572.
[81] Pichot C, Lauro M, Pham Q (1981) Microstructure of Vinyl Acetate–Butyl Acrylate Copolymers Studied by 13C-NMR Spectroscopy: Influence of Emulsion Polymerization Process J.Polym. Sci. Polym. Chem. Ed. 19:2619.
[82] El-Aasser M.S, Makgawınata T, VanderHoff, J.W (1983) Batch and Semicontinuous Emulsion Copolymerization of Vinyl Acetate-Butyl Acrylate. 1. Bulk, Surface and Colloidal Properties of Copolymer Latexes. J. Polym. Sci. 21:2363-2382.
[83] Sarac A, Berber H, Yıldırım H (2006) Semi-continuous Emulsion Copolymerization of Vinyl Acetate and Butyl Acrylate Using a New Protective Colloid. Part 2. Effects of monomer ratio and initiator. Polym. Adv. Technol. 17: 860–864.
[84] Berber H, Sarac A, Yıldırım H (2011) Synthesis and Characterization of Water-based Poly(vinyl acetate-co-butyl acrylate) Latexes Containing Oligomeric Protective Colloid. Polym. Bull. 66:881–892.
[85] Berber H, Sarac A, Yıldırım H (2011) A Comparative Study on Water-based Coatings Prepared in the Presence of Oligomeric and Conventional Protective Colloids. Prog. Org. Coat. 71:225–233.
[86] Erbil H.Y(1996) Surface Energetics of Films of Poly(vinyl acetate-butyl acrylate) Emulsion Copolymers. Polymer. 24:5483-5491.
[87] VanderHoff J.W (1985) Mechanism of Emulsion Polymerization. J.Polym. Sci. PolyM. Symp. 72:161-198.
[88] El-Aasser M.S, Makgawınata T, VanderHoff J.W (1983) Batch and Semicontinuous Emulsion Copolymerization of Vinyl Acetate-Butyl Acrylate. II. Morphological and Mechanical Properties of Copolymer Latex Films. J. Polym. Sci. 21:2383-2395.
[89] Dimitratos J, El-Aasser M.S, Georgakis C, Klein A (1990) Pseudosteady States in Semicontinuous Emulsion Copolymerization. J. Appl. Polym. Sci. 40:1005-1021.

[90] Sun P.Q, Liu D.Z, Zhao K, Chen G.T (1998) Development of Particle Morphology Simulating of Emulsion Copolymerization of Vinyl Acetate and Butyl Acrylate. Acta Polym. Sinica. 5:542-548.

[91] Lazaridis N, Alexopoulos A.H, Kiparissides C (2001) Semi-batch Emulsion Copolymerization of Vinyl Acetate and Butyl Acrylate Using Oligomeric Nonionic Surfactants. Macromol.Chem.Phys. 202:2614-2622.

[92] Vandezande G.A, Rudin A (1992) In: Daniels E.D, Sudol E.D, El-Aasser M.S, editors. Polymer Latexes. Vol. 492. Washington: ACS Symp. Series. pp. 114.

[93] Vandezande G.A, Rudin A (1992) In: Daniels E.D, Sudol E.D, El-Aasser M.S, editors. Polymer Latexes. Vol.492. Washington: ACS Symp. Series. pp. 134.

[94] Delgado J, El-Aasser M.S, VanderHoff J.W (1986) Miniemulsion Copolymerization of Vinyl Acetate and Butyl Acrylate. I. Differences Between the Miniemulsion Copolymerization and The Emulsion Copolymerization Processes J. Polym. Sci. Polym.Chem. Ed. 24: 861-874.

[95] Wu, X.Q., Schork, F.J., (2000). "Batch and Semibatch Mini/Macroemulsion Copolymerization of Vinyl Acetate and Comonomers", Ind Eng Res. 39: 2855-2865.

[96] Brandrup J, Immergut E.H (1989). Polymer Handbook. New York: Wiley.

[97] Donescu D, Goşa K, Languri J (1990) Semicontinuous Emulsion Polymerization of Vinyl Acetate. VIII. Copolymerization with Di-2-ethylhexyl Maleate. Acta Polym. 41: 210-214.

[98] Donescu D, Fusulan L (1994) Semicontinuous Emulsion Polymerization of Vinyl Acetate X. Kinetics of Homopolymerizations, Copolymerizations, and Initiator Decomposition in the Presence of Sulfosuccinate-type Surfactants. 15:543-560.

[99] Berber H (2012) The Synthesis, Modification and Characterization of Nano-Sized Emulsion Polymers Using Functional Vinyl Monomers. Phd Thesis. İstanbul: Yildiz Technical University.

[100] Abad C, De La Cal J.C, Asua J.M (1995) Core-shell Structured Latex Particles. III. Structure–properties Relationship in Toughening of Polycarbonate with Poly(*n*-butyl acrylate)/Poly(benzyl methacrylate–styrene) Structured Latex Particles. J. Appl. Polym. Sci. 56:419-455.

[101] Donescu D, Fusulan L, Gosa K, Ciupitoiu A (1994) Semicontinuous Emulsion Polymerization of Vinyl Acetate. 12. Comopolymerization with Acid Monomers. Revue Roamanie de Chimie. 39:843-849.

[102] Garjria C, Vijayendran B.R (1983) Acid Distribution in Carboxylated Vinyl-Acrylic Latexes J. Appl. Polym. Sci. 28:1667-1676.

Chapter 3

A Polymer Science Approach to Physico-Chemical Characterization and Processing of Pulse Seeds

Kelly A. Ross, Susan D. Arntfield and Stefan Cenkowski

Additional information is available at the end of the chapter

http://dx.doi.org/10.5772/46145

1. Introduction

A polymer science approach in the physico-chemical characterization of food systems has been highlighted in the literature as concepts of polymer science have been applied to understanding the effect of the glass transition on various food properties. The importance of the glass transition with respect to the processing and the stability of foods has long been recognized in food processing unit operations such as agglomeration, freezing, dehydration, flaking, sheeting, baking, and extrusion (Abbas et al., 2010; Campanella et al., 2002; Rhaman, 2006; Roos, 2010). However, the concept of the glass transition with respect to the processing of pulse seeds has been largely ignored. Successful value-added utilization of pulses, such as beans, for human consumption involves whole seed processing. Adequate hydration represents a significant requirement for processing of pulse products because only those pulse varieties for which water can consistently be absorbed to an acceptable level will be used in the processing of whole seed products and pulse based products (An et al., 2010). There is agreement in the literature that the seed coat is the structure that is primarily responsible for controlling water uptake, thus the seed coat is the principle barrier to water uptake (Arechavaleta-Medina & Synder, 1981; DeSouza & Marcos-Filho, 2001; Ma et al., 2004; Meyer et al., 2007; Ross et al., 2008; Zeng et al., 2005). The permeability of the seed coats of pulses have been studied extensively as poor hydration behaviour is commonly observed in pulse seeds. Although much research has been devoted to understanding the cause for differences in water uptake behaviour of pulse seeds, the approach has either been from a food science perspective in terms of characterizing the physical and chemical differences between impermeable and permeable seeds or from a botanical perspective concerned with defining the differences between the morphology and anatomy of permeable and impermeable seeds (Ross et al., 2008). A polymer science approach to understanding water uptake is necessary. An innovative explanation of water uptake behaviour in pulse seeds can be achieved by merging concepts from food science and polymer science. Key chemical differences in pulse seed coats have been studied with respect to their influence on water uptake behavior and the glass transition (Ross et al., 2010a). The effect of the glass transition temperature of seed coats on water uptake behaviour was reported by Ross et al. (2008). The mechanism of water uptake for pulse seeds possessing seed coats with a glass transition above ambient conditions was explained based on an analogy between a temperature driven glass transition (T_g) and a solvent driven glass transition (a_g) presented in polymer science literature (Ross et al., 2010b). Alternatively, a hypothesis for the mechanism of water uptake in seeds possessing seed coats with a glass transition near ambient conditions implicating the time required to reach saturated surface concentration upon exposure to solvent was based on work reported in the field of polymer science (Ross et al., 2010b).

"Milling", by definition, is a process by which materials are reduced from a larger size to a smaller size (Wood & Malcolmson, 2011). In the case of pulse processing there are several operations that can be characterized in this way, including: 1) dehulling which is defined as loosening and removal of the seed coat to produce polished seed (i.e. footballs); 2) splitting which is defined as loosening and cleavage of the two cotyledons to produce split seeds (i.e. splits); and 3) flour milling or grinding which is defined as reducing whole seed or cotyledons to flour (Wood & Malcolmson, 2011). Variation in the ease of "milling" between the different pulse species primarily explains the wide variation in methodologies and pre-treatments that have been developed to optimize yields (Wood & Malcolmson, 2011). Dehulling is an important aspect of pulse processing as the dehulling efficiency (yield of dehulled seeds) is an important quality characteristic for pulse breeders, processors, and exporters as it ultimately dictates whether a dehulling operation is economically feasible (Wang et al., 2005; Ross et al., 2010c). The goal of dehulling is to completely remove the hull from pulse seeds while minimizing the production of powder, breaks, and in certain pulses-split seeds (Wang et al., 2005). According to Wood & Malcolmson (2011), the dehulling process without splitting is performed on pulses whose cotyledons are very tightly held together, such as lentil (*Lens culinaris).* In this case, a higher ratio of whole dehulled lentil seeds to split seeds is desired as whole dehulled lentil seeds are the most valuable fraction (Vandenberg & Bruce, 2008). Red lentils account for the majority of world lentil production and trade and nearly 90% of red lentils are consumed as cooked split or whole seeds where the seed coat has been removed by dehulling. Since most red lentils are dehulled before consumption, dehulling efficiency (yield of dehulled seeds) of red lentil is very important (Ross et al., 2010c; Vandenberg & Bruce, 2008; Wang, 2005). The rice industry cites breakage of rice kernels during milling/dehulling as one of its main problems (Iguaz et al., 2006). A large amount of work has been performed in the area of understanding the effect of drying temperature on rice milling quality (Cao et al., 2004; Cnossen et al., 2003; Cnossen & Siebenmorgen, 2000; Iguaz et al., 2006). The concept of the glass transition has been used to explain rice kernel fissure formation during drying and subsequent breakage during milling of rice (Siebenmorgen et al., 2004). The effect of drying temperature on the handling quality of whole green lentil seeds in terms of seed breakage upon handling has been investigated (Tang et al., 1990), however little work has been done on the effect of drying temperature on the milling quality of red lentils. A better understanding of pulse seed breakage and splitting could be achieved by studying dehulling from a polymer science approach. Therefore, the hypothesis used to explain rice breakage was adopted to explain breakage and splitting in red lentils upon dehulling. Overall, this chapter focuses on: 1) the state of knowledge of the glass transition temperature in food systems; 2) the importance of pulses and how they are processed; 3) the state of knowledge of water uptake in pulse seeds; 4) defining the role of the T_g with respect to water uptake in pulse seeds; and 5) defining the role of the T_g with respect to dehulling of red lentils.

2. The state of knowledge of the glass transition temperature in food systems

2.1. Definition of glass transition

The glass transition theory from polymer science has been studied from a food science perspective since the pioneering work of Slade and Levine in the 1980's as many phenomena related to food processing and stability can be systematically explained by the concept of glass transition (Abbas et al., 2010; Campanella et al., 2002; Kumagai & Kumaga, 2009; Lemeste et al., 2002; Rhaman, 2006; Roos, 2010). A material typically forms an amorphous glass if crystallisation is inhibited by steric hindrance and kinetic constraints (Norton, 1998). The glass transition is defined as a change in the state of an amorphous material from a solid/glassy-like to a liquid/rubbery-like state or vice versa (Figure 1) as the change of state (i.e. glass transition) exhibited by amorphous materials is a reversible transformation (Roos, 2010). The temperature range at which materials pass from a solid/glassy-like and liquid/rubbery-like structure or vice versa is considered the glass transition temperature (T_g).

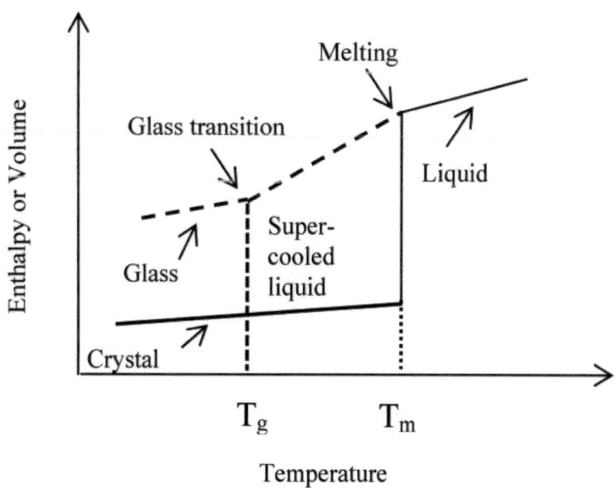

Figure 1. Enthalpy or volume of various states of materials as affected by temperature (Adapted from Debenedetti & Stillinger, 2001; Roos, 2010). T_g is glass transition temperature; T_m is melting temperature.

2.2. Theories of glass transition

The glass transition does not exhibit latent heat and no temperature can be defined where the both the solid/glassy-like and liquid/rubbery-like states coexist (Roos, 2010). Both the solid/glassy and the liquid/rubbery states of amorphous materials, separated by the glass transition, are non-equilibrium states. Thus, materials in the glassy state are not completely stable and are described as existing in a metastable solid state (Roos, 2010; Slade & Levine 1991). Rates of changes of amorphous materials are time-dependent and controlled by the ability of molecules within the material to respond to changes in their surroundings. Therefore, difficulties exist in understanding the properties of the non-equilibrium state of amorphous materials as it does not exhibit a characteristic order of molecular arrangement (Roos, 2010). Observations of the changes in thermodynamic properties (volume, enthalpy, entropy) and the kinetic nature of the glass formation have led to the development of several theories to explain the nature of the glass transition (Roos, 2010; Sperling 2006), which include: thermodynamic or entropic theories, free-volume theory, and kinetic theory. Thermodynamic or entropic theories state that the glass transition is a second order phase transition which is based on observed changes in the thermal expansion coefficient and heat capacity values that occur over the glass transition (Norton, 1998; Roos, 2010). These theories were devised by Gibbs (1956) and Gibbs & DiMarzio (1958) based on the work of Flory (1953) and it was suggested the glass transition occurs when the relaxation time of the segmental motion of polymer chains approaches that of the experimental time scale (Roos, 1995). As shown in Figure 1, the enthalpy of a material changes differently with temperature in the glassy and rubbery-like states, indicating that the glass transition is associated with a change in heat capacity, however, no single temperature in glass transition measurements can be identified for the change in heat capacity (Roos, 2010). The thermodynamic approach has been criticized as it assumes the system is at equilibrium, yet Gibbs & DiMarzio (1958) did note that although observed glass transitions are time-dependent, the real thermodynamic change in state occurs at infinitely long times (Roos, 2010). The free volume theory is rooted in the idea that if individual molecules are considered to be spheres in the glassy state, the unoccupied or free volume is reduced. For molecules to change their location or degree of motion, they must be able to move into the free volume (Roos, 2010). Creating a condition, such as increasing temperature or increasing moisture content, where the system temperature becomes higher than T_g, allows the system to transition from a glassy state into the rubbery state. This provides an increase in the free volume, allowing an increase in both rotational and translational molecular mobility. The fact that kinetically dependent processes such as viscosity or volume expansion (Norton, 1998) determine the free volume of a polymer system has led to the development of kinetic theories for explaining the nature of the glass transition. The kinetic theory of the glass transition considers the time dependent characteristics of the glass transition and time-dependent molecular relaxations that take place over the glass transition temperature range (Roos, 1995).

2.3. Measurement of glass transition

The glass transition has been typically determined by studying the changes in the thermal or rheological properties of a system, either as a function of sample temperature or composition, such as moisture content (Roos, 2010), as material properties including modulus, viscosity, volume, thermal expansion, and dielectric properties, exhibit a discontinuity around the glass transition. The glass transition associated with changes in rheological and dielectric material properties is attributed to mechanical and dielectric α-relaxations, respectively. Therefore, glass transition and relaxations of amorphous materials may be measured with thermal, dielectric, mechanical, and spectroscopic techniques. Using thermal methods, such as differential scanning calorimetry (DSC), the glass transition appears as a change in enthalpy and volume in the measurement of thermodynamic properties such as heat capacity, whereas the appearance of translational mobility of molecules around the glass transition results in a frequency-dependent α-relaxation manifested in changes in mechanical properties or permittivity, which are typically measured by mechanical and dielectric techniques, respectively, such as dynamic mechanical (thermal) analysis (DMA/DMTA) and dielectric thermal analysis (DEA/DETA), respectively. Spectroscopic techniques such as, infra-red and Fourier transform infra-red (IR/FTIR), Raman Electron Spin Resonance (ESR) and various NMR spectroscopy, have been used to provide information on chemical bonding and molecular mobility (Roos, 2010).

2.4. Effect of glass transition on molecular mobility

As noted, amorphous food materials are in a non-equilibrium state, which can be greatly affected during processing and storage conditions (Abbas et al., 2010; Campanella et al., 2002; Rhaman, 2006; Roos, 1991; Roos, 2010). The state of an amorphous material depends on its composition, temperature, and time. As temperature, relative humidity or moisture content increases, amorphous materials transform from the glassy state to the rubbery state, which reflects changes in molecular mobility and in mechanical and dielectric properties (Roos, 2010; Slade & Levine 1991). The movement of the matrix molecules in a system is greatly reduced in the glassy state compared to the rubbery state. Although, mobility of small molecules does occur in glassy biomolecular systems via vibration and short-range rotational motions (Sperling, 2006), the enhanced stability of food systems in the glassy state have been attributed to reduced molecular mobility below T_g (Norton, 1998). The control of moisture content or water activity and temperature of glassy foods is of great importance as the concept of the glass transition helps to explain changes which occur during food processing and storage. Typical changes in amorphous food materials above the T_g include stickiness, caking, collapse, and crystallization, textural changes such as softening and hardening, as well as chemical changes, such as enzymatic reactions and oxidation (Abbas et al., 2010; Li, 2010; Roos, 2010,). Furthermore textural changes, which are manifested as changes in mechanical properties, occurring over the glass transition, are characterized as a function of temperature, moisture content or water activity (Peleg, 1993; Roos, 2010). A state diagram, which reflects the relationship between T_g and moisture content (or water activity) and temperature, is a valuable tool to manipulate both T_g and material behavior under various storage and processing conditions (Roos 1993). Explicitly, a state diagram can be used to explain physical state changes of foods as a function of moisture content during water removal processes such as drying, baking, extrusion, evaporation, freezing, or water uptake processes such as agglomeration or tempering and flaking (Abbas et al., 2010; Campanella et al., 2002; Roos, 2010). Therefore application of the T_g as a processing parameter in the food industry is immense and is of relevance to the pulse processing industry. Hydration represents a significant requirement for processing of pulse products. Optimization of the dehulling/milling quality of pulses requires definition of the relationship between drying conditions and T_g.

2.5. Effect of water on the glass transition

As noted, water is a key factor affecting the glass transition. The molecular weight of water is significantly lower than most food components, which lowers the local viscosity and enhances molecular motion (Ferry, 1980), thereby adding free volume to the system. Uptake of water by an amorphous solid system will cause "plasticization" and will result in a decrease in the glass transition temperature (Li, 2010). The process of water uptake involves two main processes: 1) adsorption, which is the interaction of water with surface solids; and 2) absorption, in which water penetrates the bulk solid structure (Li, 2010). Van der Waals interactions and chemical adsorption by chemical bonding are the two kinds of forces involved in adsorption. Water molecules first adsorb onto the surfaces of dry material to form a monolayer, which is subjected to both surface binding and diffusional forces. As more water molecules adhere to the surface, a water multilayer forms and diffusional forces exceed the binding forces. As such, water is absorbed into the bulk structure by pores and capillary spaces (Barbosa-Canovas, 1996; Li, 2010). The glass transition process occurs when water uptake changes from surface adsorption to bulk absorption (Li, 2010). Work noting the importance of the glass transition and water uptake was reported in Oksanen & Zografi (1990) in which water vapor sorption isotherms of poly (vinylpyrrolidone) at various temperatures along with the measurement of T_g as a function of water content were analysed. It was suggested that sufficient water uptake (moisture content), which was designated by the upward inflection of the isotherm, was needed to be attained to cause T_g to be less than the experimental/environmental temperature and cause the polymer to transition into a rubbery state. At a higher temperature, less water was required to plasticise the sample because of the higher molecular mobility due to the higher temperature. This work is in agreement with the concept of critical water activity (a_w) and moisture content (mc) set forth by Roos (1993) who noted that moisture content and a_w can be considered as factors depressing T_g to the environmental temperature, albeit whatever environment-i.e. processing conditions or storage conditions (Roos, 1993), thereby enhancing the molecular mobility of the system. Thus, critical a_w or mc at which the glass transition occurs is a key parameter in predicting the behaviour of amorphous materials. Moreover, there is an analogy between the temperature driven glass transition (T_g) and the solvent driven glass transition (a_g) in polymer science literature (Laschitsch et al. 1999; Leibler & Sekimota 1993; Vrentas & Vrentas 1991). At a constant temperature, a sorption curve of a polymeric glass may remain relatively flat until a certain solvent concentration is attained, after which the sorption curve displays a dramatic increase in solvent uptake (Laschitsch et al., 1999). This remarkable increase in solvent uptake can be considered to have occurred at the glass transition solvent activity (a_g) and consequently this increase in solvent uptake can be attributed to the plasticization of the polymer by the sorbed penetrant (Vrentas & Vrentas 1991). Explicitly, the free volume of a polymer is affected by temperature in the same way as it is affected by the amount of plasticizer (e.g. water) present. This is a key concept in developing a hypothesis using the idea of glass transition to explain water uptake in pulse seeds and it will be revisited in a following section.

3. Pulses and pulse products

3.1. Definition of pulses

The family Leguminosae consists of more than 18 000 species of plants and members of the family are often referred to as legumes or pulses, which are the second most important food source in the world after cereals (Tiwari et al., 2011). Pulses are defined as the dry, edible seeds of legume plants (Maskus, 2010), notably this definition excludes fresh (i.e. non-dried) green beans and green peas which are consumed and considered as vegetables along with a few oil-bearing seeds like groundnut (*Arachis hypogaea*) and soybean (*Glycine max*) which are grown primarily for edible oil extraction (Tiwari et al., 2011). The terms "legumes" and "pulses" are used interchangeably because all pulses are considered legumes but not all legumes are considered pulses (Tiwari et al., 2011). Geographic region will influence the types that of pulses that are grown. In Canada, the most commonly grown pulses include: field peas, beans, lentils and chickpeas. However, cowpea (black-eyed pea) is a recognized a key pulse crop in the Southern United States while mung bean is commonly used in China (Maskus, 2010). The Food Agriculure Organization (FAO) recognizes 11 primary pulses (Tiwari et al., 2011), which are presented in Table 1.

3.2. Pulses: Utilization and processing

It has long been known that pulse crops are a good source of protein, energy (carbohydrate), fibre and mircronutrients (vitamins and minerals) (Black et al., 1998a; 1998b) and interest in the utilization of pulses in the developed world is on the increase (Tiwari et al., 2011). Factors contributing to this include: their reported nutritional and health benefits, changes in consumer preferences, increasing demand for variety/balance, changes in demographics (age, racial diversity), rise in the incidence of food allergies and ongoing research on production and processing technologies (Boye et al., 2010). In fact, over the past four decades, world pulse production has increased by 49% from 40.8 Mt in 1961 to 60.9 Mt in 2008 (Watts, 2011). In 2008, dry beans accounted for one-third of global pulse production, followed by peas at 16% and chickpeas at 14%. Cowpeas, pigeon peas, broad beans (or faba bean, horse bean), lentils, vetches and lupins accounted for the remaining third of production (Watts, 2011). Despite the increase global pulse production and the general trend toward the inclusion of health promoting foods into the diet, pulses are still considered to be an underutilized food source in Europe and the USA (Abu-Ghannam & Gowen, 2011).

Although, researchers have studied the development of whole seed products and use of pulse flours as ingredients in products conventionally formulated with non-pulse flour (Abu-Ghannam & Gowen, 2011; Maskus, 2010) and efforts in this area are increasing.

Pulse Class	**Common Name (*Scientific name*)**
Dry beans	Kidney bean, haricot bean, pinto bean, navy bean, black bean (*Phaseolus vulgaris*); Lima bean, butter bean (*Phaseolus lunatus*); Azuki bean, adzuki bean (*Vigna anularis*); Mung bean, black bean, golden gram, green gram (*Vigna radiata*); Black gram, urad (*Vigna mungo*); Scarlett runner bean (*Phaseolus coccineus*); Ricebean (*Vigna umbellata*); Moth bean (*Vigna acontifolius*); Tepary bean (*Phaseolus acutifolius*)
Dry broad beans	Horse bean (*Vicia faba equina*); Broad bean (*Vicia faba*); Field bean (*Vicia faba*)
Dry peas	Garden pea (*Pisum sativum* var. *sativum*); Protein pea (*Pisum sativum* var. *arvense*)
Chickpea	Garbanzo, Bengal gram (*Cicer arietinum*)
Dry cowpea	Black-eyed pea, black-eye bean (*Vigna unguiculata*)
Pigeon pea	Arhar/Toor, cajan pea, Congo bean (*Cajanus cajan*)
Lentil	Green lentil, red lentil (*Lens culinaris*)
Bambara groundnut	Earth pea (*Vigna subterranea*)
Vetch	Common vetch (*Vicia sativa*)
Lupins	Lupins (*Lupinus* spp.)
Minor pulses	Lablab, hyacinth bean (*Lablab purpureus*); Jack bean (*Canavalia ensiformis*); Sword bean (*Canavalia gladiata*); Winged bean (*Psophocarpus teragonolobus*); Velvet bean, cowitch (*Mucuna pruriens* var. *utilis*); Yam bean (*Pachyrrizus erosus*)

Table 1. Common Pulses (Adapted from Tiwari et al., 2011)

Some common forms of pulse based foods include: dry pulses (including whole, split, and/or dehulled); canned pulses; spouted pulses; fermented legumes, such as wadi and dhokla, which are chickpea based fermented products, and although technically not pulses but legumes, fermented soy bean products such as natto and chungkookjang which are short term fermented products and tauchu, miso, doenjang, and kochujang, which undergo long term fermentation. Value added pulse products such as micronized/infra-red heat treated pulses, quick-cook dehydrated pulses, extruded pulse products, roasted pulse seeds have been developed (Abu-Ghannam & Gowen, 2011; Bellido et al., 2006; Maskus, 2010). Foods conventionally prepared with non-pulse flours have been formulated with pulses that have been milled into flours such as: pasta, noodles, tortillas, batters, breads, extruded snacks, flours, fried snacks, infant food, and other baked goods (Maskus, 2010). However an important aspect of all of these pulses products is the necessity for the inclusion of water to the whole pulse seed at some point during processing (An et al., 2010; Bellido et al., 2006). Commercial processing of dry peas and beans usually involves soaking the seeds overnight (12-16 h) in water at ambient temperature to encourage maximum water uptake (Thanos et al., 1998).

Although water is a relatively inexpensive ingredient, time and costs associated with transporting hydrated seeds (typically canned) are relatively expensive factors. Quick-cook dehydrated pulses also require hydrothermal treatment (i.e. cooking in water) (Abu-Ghannam & Gowen, 2011). Literature has indicated that tempering (raising the moisture content of the seeds by addition of small amounts of water to a predetermined moisture content is required pre-treatment step to produce an acceptable micronized whole pulse seed product (Bellido et al., 2006). Tempering has also been noted as an essential step in the processing of roasted pulse snacks (Abu-Ghannam & Gowen, 2011). Furthermore, tempering of pulse seeds is an essential step in the milling and dehulling process as it improves yield of dehulled product material (i.e. dehulling efficiency), which is an important quality characteristic for pulse breeders, processors, and exporters, as it ultimately dictates whether a dehulling operation is economically feasible (Wood & Malcolmson, 2011), Therefore, adequate hydration of whole pulse seeds is important to both the scientific and industrial communities for developing whole seed products and products formulated with pulse flours.

4. The state of knowledge of water uptake in pulse seeds

4.1. Variations in water uptake behavior in pulse seeds

The water uptake behavior of seeds from the Leguminoseae family, which includes pulse seeds, has been studied extensively due to variation in imbibition patterns commonly observed in seeds belonging to this family (Ragaswamy et al., 1985; Ross et al., 2008). Seeds that demonstrate a long period (i.e. lag) before appreciable water uptake can be considered poorly hydrating seeds while seed that demonstrate a rapid uptake in water (i.e. no lag) can be considered well hydrating seeds. For example, seeds like peas (yellow, green and Marrowfat), navy beans, kidney beans, chickpeas, lentils, and most soybeans have been noted to possess rapid water uptake and do not possess a lag period prior to imbibition (An et al., 2010; Abu-Ghannam & McKenna, 1997; Hsu, 1983a; Hsu, 1983b; Liu et al., 2005; Meyer et al., 2007; Ross et al., 2008; Seyhan-Gurtas et al., 2001). However some pulses like pinto beans, black beans, dried green beans, red beans possess a significant lag period prior to the initiation of water uptake (Liu et al., 2005; Ross et al., 2008). Figure 2 shows a graph representing the water uptake behavior of well and poorly hydrating seeds.

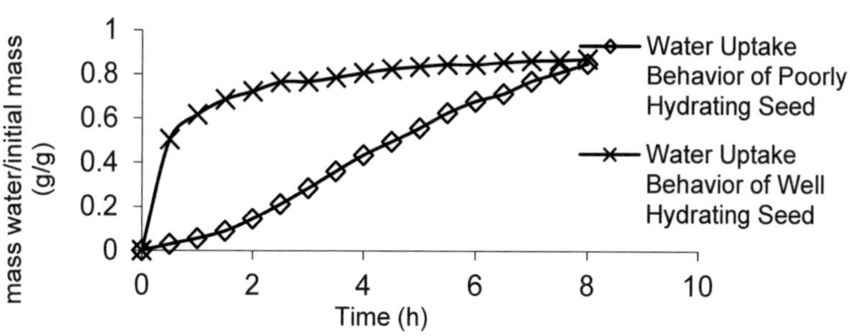

Figure 2. Water uptake behavior of well hydrating and poorly hydrating seeds

It should be noted that the terms hard or stone seeds appear in food science and botanical literature. Hard or stone seeds are seeds that are impervious to water and remain hard even after cooking (Argel & Parton, 1999). From a botanical perspective, hard seeds are defined as seeds which will not germinate, of which water imbibition is an integral step, even if subjected to conditions ideal for germination (DeSouza & Marcos-Filho, 2001). From a biological perspective hardseededness occurs as a long term seed survival mechanism. The presence of hard seeds in agricultural seed lots are detrimental as they contribute to uneven seedling emergence which may reduce yields and delay harvest. Hardseededness from a crop production perspective can be broken via thermal, chemical, and physical treatments (Argel and Paton, 1999; DeSouza and Marcos-Filho, 2001). Moreover, hard seededness is undesirable for the food processing industry due to its negative effects on product quality (Ma et al., 2005). Therefore water uptake in legume seeds is an issue that is important to both the scientific and industrial communities and much research has been devoted to understanding the cause and ultimate control of hard seededness and shortening the lag time of poorly hydrating seeds (Arechavaleta-Medina & Snyder, 1981; Marcbach and Mayer, 1974; Thanos et al., 1998; Zeng et al., 2005). A key point is that while both poorly hydrating seeds and hard/stone seeds exhibit a barrier to water uptake and do not take up water for hours, days or even longer-until dormancy is broken in the case of hard seeds, despite this long, variable lag before the start of imbibition, once water uptake begins the rate increases and the final amount of water absorbed is comparable to that of well hydrating seeds (Arechavaleta-Medina & Snyder, 1981). Therefore, the time needed before appreciable water uptake upon is a key factor in defining water uptake behavior. The long time required to soak pulses/legumes is one of the negative attributes associated with processing pulses/legumes. Understanding the cause of lag time is essential for effectively processing pulse seeds and ensuring opportunities for development of new pulse products and knowledge can be gained from examining the literature studying characteristics of well hydrating seeds, poorly hydrating seeds and hard/stone seeds.

4.2. Factors affecting water uptake

There is agreement in the literature indicating that the seed coat of a seed is the principle factor which determines water uptake behavior (Arechavaleta-Medina and Synder, 1981; DeSouza & Marcos-Filho, 2001; King & Ashton 1985; Ma et al., 2004; Meyer et al., 2007; Ross et al., 2008; Zeng et al., 2005) in pulses/legumes. With the aim of understanding differences in seed permeability, the physical, morphological and chemical characteristics of the seed coat of many pulses/legumes seeds has been extensively studied. A review of the literature shows that seed permeability, with respect to seed coat structure, has been studied from a food science and a plant anatomy perspective (Ross et al., 2008). Work in food science has accounted for physical differences for variation in water uptake behavior of seeds while work in botany has investigated differences between the morphology and anatomy to understand differences in imbibition patterns of seeds. Seed coat thickness has been noted in the literature as a factor affecting water uptake. Seeds with thicker seed coats have been shown in the literature to have slower water uptake rates (King & Ashton, 1985; Sefa-Dedeh

& Stanley, 1979). However it has been shown that certain varieties of cowpeas possessing thicker seed coats than other cowpea varieties exhibit faster water uptake than cowpeas with thinner seed coats (Sefa-Dedeh & Stanley, 1979). Water uptake behavior cannot solely be linked to seed coat thickness. Seed size has also been linked to water uptake behavior (Arechavaleta-Medina and Synder, 1981; Hsu et al., 1983a; Seyhan-Gurtas et al., 2001). As water uptake is determined via mass uptake such that a smaller seed would exhibit a proportionally greater increase in water mass uptake than a larger seed if both were taking up water at the same rate. Therefore, a difference in size could account for different water uptake rates but it would not offer an explanation for different starting times of water uptake (i.e. lag times) (Arechavaleta-Medina & Synder, 1981. Porosity of the seed coat has been implicated as a factor affecting water uptake behavior. Soybean seeds with porous seeds coats have been noted to be permeable while seeds with non-porous seeds coats are typically impermeable (DeSouza & Marcos-Filho, 2001). It is noted that while soybeans are not termed pulse seeds due to their high oil content, they are legume seeds and discussion of their seed coat properties with respect to water uptake is relevant and applicable to the seed coats of pulse seeds. In the discipline of botany the morphology and anatomy of a seed coat have been studied with the aim of understanding water uptake in seeds. Two opinions regarding the role of the seed coat on water uptake have been noted. One is that the seed coat has specialized regions for water loss and uptake such as the hilum, micropyle, lens, and raphe (DeSouza & Marcos-Filho, 2001) while the second opinion is that the whole seed coat is involved in water exchange (Zeng et al., 2005). Work by Ma et al. (2004) and Zeng et al. (2005) stated that the whole seed coat should be regarded as an integrated system responsible for water absorbing properties as the palisade layer, which develops from the outer epidermis of the seed coat, plays a significant role in water uptake. Moreover, the cuticle of the palisade layer has been implicated as the key factor that determines the permeability of the soybean seed coat (Ma et al., 2004). Impermeability of seeds has been attributed to differences in the seed coat structures such as contracted palisade cells and a thick cuticle (Rangaswamy & Nandakumar, 1985).

Additionally, there are numerous reports in the literature citing chemical differences in the seed coats of seeds that imbibe water rapidly and those that exhibit delayed water uptake (Arechavaleta-Medina & Snyder, 1981; Marbach & Mayer, 1974; Marbach & Mayer, 1975; Rangaswamy et al., 1985; Reyes-Moreno et al., 1994) however, no universal explanation has been reported. Phenolics substances may affect water uptake (Marbach & Mayer, 1974; Marbach & Mayer, 1975). Wild type pea species with naturally impermeable seed coats possessed high phenolic content while cultivated pea species had a low phenolic content (Marbach & Mayer, 1974). Phenolics, particularly tannins, have been reported to be responsible for reduced water uptake in bean seeds (Sievwright & Shipe, 1986). The seed coats of impermeable soybean seeds have been shown to contain high amounts of lignin (a complex polphenolic molecule) (DeSouza & Marcos-Filho, 2001). The work of McDougall et al. (1996) indicated that soybean genotypes with high lignin content in the seed coat tend to have low seed coat permeability, while genotypes with low lignin content tended to have high seed coat permeability. However, the work of Mullin and Xu (2001) which studied the composition of soy bean seed coat and water uptake indicated that lignin content did not influence seed permeability. Instead, their work demonstrated a relationship between hemicellulose, primarily xylan, content and poor water uptake. However, it has been shown that pulse seeds with inherently fast water uptake without a lag time contained lower levels of total phenolics and of these phenolics, the majority were non-tannin phenolics, while pulse seeds with inherently slow water uptake possessing a lag time contained higher total phenolics content and of these phenolics, the majority were tannins. (Ross et al., 2010a). Furthermore, this work (Ross et al., 2010a) showed that pulse seeds that were processed with a simple hydrothermal treatment to improve water uptake experienced a decrease in total phenolics content. Future work studying the influence seed coat chemistry on water uptake should include: 1) a more specific approach with regards to determination of seed coat phenolics, 2) concurrent determination of xylan content, and 3) determination of the monomeric phenolic composition of lignin.

Research in botany and plant science has linked loss of lipids in the seed coat to create permeable seeds (Zeng et al., 2005) however, food science research has not focused on relating the lipid chemistry of the seed coat and water uptake. As literature from a botanical perspective has implicated the cuticle of the palisade layer as the key factor that determines the permeability of the seed coat (Ma et al., 2004; Rangaswamy & Nandakumar, 1985; Zeng et al., 2005), the chemical properties of the cuticle have subsequently been characterized in this respect. The cuticular layer of the soybean has been implicated as a key factor in water uptake behavior, noting it was the site of the moisture barrier (Arechavaleta-Medina & Snyder, 1981). The cuticle is hydrophobic and consists of an insoluble, polymeric and structural component and a complex mixture of lipids including cutin, suberin, and waxes with different fatty acid compositions. Cutin, an insoluble lipophilic matrix, constitutes the framework of the cuticle (Casado & Heredia, 2001). Specifically, cutin is a high molecular weight polyester composed of various inter-esterified C16 and C18 hydroxyalkanoic acids. Structural and physicochemical studies on cutin have shown that cutin exhibits an amorphous structure. Rapidly hydrating seeds have been shown to possess seed coats containing cuticular waxes with a plasticized structure and altered hydrophobicity (Egerton-Warburton, 1998). It has been documented that well hydrating soybean seeds possessed seed coats with a cuticle lacking mid-chain hydroxylated fatty acids while the cuticle layer of seed coats from impermeable seeds contained a disproportionately high amount of hydroxylated fatty acids (Shao et al., 2007).

Zeng et al. (2005) hypothesized that the process of creating permeable legume seeds during growing conditions includes both physical and chemical changes in the lipids of the cuticle layer of seed coat. It was speculated that heat provided from the environment during growing conditions likely causes the polymeric structure of the lipids to change through a weakening of the hydrophobic interactions, rendering lipids vulnerable to degradations and as such, the presence of heat and water likely causes thermal degradation and hydrolysis of the lipids in the seed coat to free fatty acids (Zeng et al., 2005). It has been reported that the rate of water uptake of whole bean seeds could be increased by subjecting the whole bean seeds to a simple hydrothermal treatment while the application of dry heat did not significantly improve water uptake (Ross et al., 2008) which is in agreement with the work of Zeng et al. (2005). Hydrolysis of lipids in the seed coat to fatty acids, which requires water, likely contributed to the improved water uptake observed in seeds treated hydrothermally compared the hydration behavior seen in the unprocessed seeds and seeds subjected to dry heat only. Chemical changes in the seed coat as affected by hydrothermal processing in relation to water uptake were addressed along with identification of chemical differences between bean varieties with different water uptake profiles (i.e. bean varieties with a lag versus bean varieties with no lag). Chemical properties of the unprocessed seed coat from navy bean seeds (Galley variety-no lag) and pinto bean seeds (AC Ole variety-lag) were examined to explain differences in their rates of water uptake. Seeds that readily imbibed water presented lower levels of fatty acids. The seed coats of unprocessed pinto beans (AC Ole) exhibited a fatty acid content of 1.39 mg/g seed coat while the seed coats from unprocessed navy beans (Galley) presented 0.42 mg/g seed coat. Processed AC Ole seed coats displayed fatty acid levels up to 2.9 times lower than the unprocessed pinto bean seed coats. This work demonstrated a link between lower lipid content with enhanced water permeability. However it should be noted that the fatty acids identified and quantified in Ross et al. (2010a) were obtained from hydrolysis of the triglycerides present in the seed coat. Thus the amount of identified fatty acids present in the seed coat decreased and this was associated with a decrease in total lipids. Thus, although the amount of the identified fatty acids derived from the triglyceride lipids present in the seed coat decreased, it is possible that the amount of free fatty acids present in the seed coat increased. Future work in this area should include quantification of free fatty acids present in seed coats from seeds with different water uptake profiles.

Importantly, it has been stated that permeability of the seed coat may be affected by the mechanical properties of the cuticle (Ma et al., 2004; Zeng et al., 2005). The mechanical properties of a material are influenced by the polymers that are present in the material (i.e. their chemical composition, molecular arrangement, and interaction of the molecules) along with temperature and moisture content (Hoseney, 1994). As noted, the glass transition temperature (T_g) is a polymer science concept that has been applied to food science research for studying the material properties of biopolymers (Brent et al., 1997; Perdon et al., 2000). The stability and mechanical properties of a material vary depending on whether the material is above or below the T_g (Lin et al., 1991; Roos, 1995). With respect to material stability, the phase transition behavior of non-leguminous plant cuticles has been studied in relation to water loss as a second order phase transition was noted to occur at a temperature that coincided with a remarkable increase in the water permeability behavior of the plant cuticles (Casado & Heredia, 2001; Eckl & Gruler, 1980). The seed coat cuticle has been described as being functionally analogous to the cuticular layers in other organs of plants, such as leaves (Egerton-Warburton, 1998). This seems to imply that the T_g of a seed coat might influence water uptake behavior.

4.3. Relating the glass transition and water uptake behavior

Except for the work of Ross et al. (2008), to our knowledge, there have been no reports in the literature explicitly linking the glass transition behavior of seed coats and water uptake.

However, there is a large volume of literature indicating that dormant seeds reside in the glassy state (Tolstoguzov, 2000; Williams, 1995) as there is much agreement between the glass transition temperatures of seeds as a function of water content and their storage stability as a function of storage temperature and water content (Roos, 1995). The two main objectives of (Ross et al., 2008) were to measure the T_g of the seed coat of different beans and peas and relate T_g with water uptake behavior, and to utilize T_g as a processing parameter and subject the seeds to processing conditions above and below the T_g with the aim of altering the physico-chemical properties of the seed coat in order to modify the water uptake behavior. Adapted from Ross et al. (2008), Table 2 shows that the seed coats of the pea (Mozart yellow variety and Stratus green variety) and navy bean (Morden and Galley varieties) samples, which are described as well hydrating pulses not exhibiting a lag, have a T_g near room temperature (20-34°C) and the glass transition temperature range covers a relatively narrow 10°C range around the T_g. The beans AC Ole (pinto), AC Pintoba (pinto), and CDC Jet (black), seeds which as described as poorly hydrating demonstrating a lag, exhibit a glass transition temperature relatively higher than room temperature ranging from 37 to 81°C. Figure 3 shows the water uptake behavior of unprocessed/unmodified peas (Mozart and Stratus) and beans (Galley, Morden, AC Ole, AC Pintoba, and CDC Jet). The consequences of different glass transition temperatures on the water uptake behavior of seeds may also be inferred from Figure 3. The pulses with relatively fast water uptake (i.e. no lag before substantial water uptake) all have T_g values relatively close to room temperature and a narrow T_g range, while the relatively slowly hydrating samples (i.e. an appreciable lag before water uptake begins) have T_g values greater than room temperature and a broader T_g range. It was noted that the seed coats from samples with relatively faster hydration are thinner than the seed coats of the samples with slower hydration (Table 2). However, as previously discussed water uptake behavior is not solely linked to seed coat thickness.

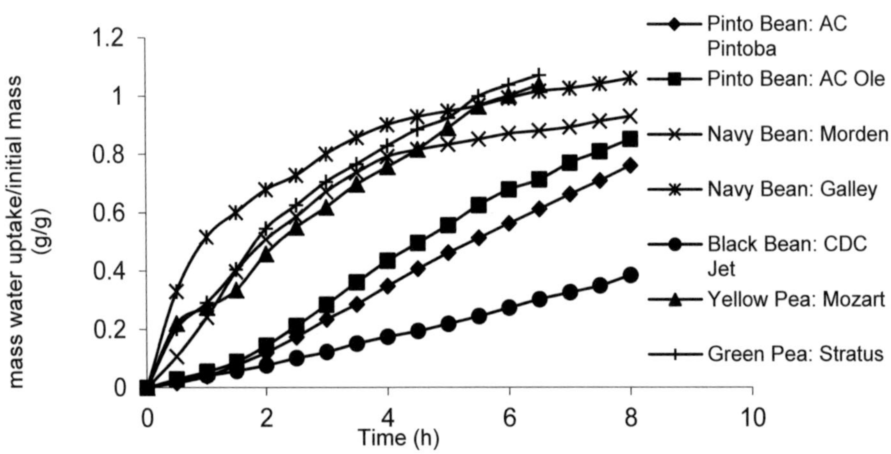

Figure 3. Water uptake behavior of native/unprocessed whole seed peas and beans (From Ross et al., 2008)

Pulse	Seed Coat Thickness (mm)	Seed Coat Moisture Content (%)	T_g Range (°C)	T_g midpoint (°C)
Mozart (yellow pea)	0.11	8.3	19-23	21.4
Stratus (green pea)	0.11	8.6	33-37	33.3
AC Pintoba (pinto bean)	0.23	9.7	43-65	45.2=1.1
AC Ole (pinto bean)	0.18	9.6	33-66	51.6
Processed* AC Ole (pinto bean):		9.4	33-47	37.7
Galley (navy bean)	0.11	9.7	24-48	26.2
Morden (navy bean)	0.11	9.2	20-30	22.2
CDC Jet (black bean)	0.19	9.3	22-57	40.2

*Beans were processed via tempering to 16% moisture content and subjected to cyclic heating (60 °C) and cooling

Table 2. Moisture content, T_g range and T_g midpoint measurements for the seed coats of selected pulses at ambient environment (Adapted from Ross et al., 2008)

Research studying the water uptake behavior of different lentils, chickpeas and beans explicitly indicated that the mechanism responsible for the initial observed lag phase in water uptake was temperature sensitive (Seyhan-Gurtas et al., 2001). The effect of temperature on lessening the lag time observed in the water uptake of soybeans has also been reported (Arechavaleta-Medina & Snyder, 1981). Moreover, a marked second order transition has been noted to occur at a temperature that coincides with a remarkable increase in the water permeability for non-leguminous plant cuticles (Matas et al. 2004; Matas et al. 2005; Schreiber & Schonherr, 1990). The glass transition determines the rheological and mechanical properties of the biopolyester cutin and in turn determines mass transfer between the environment and plant cell (Matas et al. 2004; Matas et al. 2005). It follows that a similar situation may exist regarding the water uptake behavior of seeds. The T_g of a seed coat might influence water uptake (Ross et al., 2008).

4.3.1. *Modifying water uptake behavior through processing*

In an attempt to modify the water uptake behavior of the bean samples exhibiting a lag prior to water uptake, a variety of processing regimes of cyclic heating (45, 60, and 85 °C) and cooling to achieve a seed coat temperature of 7°C and static heating (45, 60, 85°C) on non-tempered and tempered (13, 16, 24% moisture content) seeds has been investigated (Ross et al., 2008). It was shown that all of the processing regimes performed on the non-tempered beans in their native state did not produce significantly improved water uptake behavior compared against the control/unprocessed seeds. AC Ole and AC Pintoba seeds that were tempered to 16% moisture content and subjected to a 45, 60, and 85 °C cyclic heating and cooling treatment exhibited a significant increase in their water uptake behavior compared against the control seeds (i.e. reduction in lag time water uptake). CDC Jet seeds that were tempered to 16% and subjected to cyclic heating at 85 °C and cooling showed significantly better water uptake than control seeds. The results seemed to indicate that both water and heat are required to induce the changes necessary to improve water uptake (Ross et al., 2008). Also, there were no significant differences between the water uptake behavior of the 16% tempered AC Ole and the water uptake behavior of the seeds that were heated at 45, 60, and 85 °C. The 16% tempered AC Pintoba seeds subjected to 45 and 60 °C cyclic heating and cooling showed significantly better water uptake than the 16% tempered seeds subjected to 85 °C cyclic heating and cooling. This implied that the least aggressive and energy intensive treatment at 45 °C may be employed to elicit improved water uptake behavior. Interestingly, the T_g of the seed coat of AC Ole (51.6 °C) was higher than the T_g of the seed coat of AC Pintoba (45.2 °C). These results seem to indicate that the upper limit of the processing temperature which causes improved water uptake behavior may be influenced by the T_g of the seed coat. Figure 4 illustrates the improvement that hydrothermal heat treatment has on the water uptake behavior of pulse seeds possessing a lag (Ross et al., 2008). The effect of a cooling step was also investigated in this work (Ross et al., 2008). The 16% tempered AC Ole and AC Pintoba seeds subjected to cyclic heating and cooling did not show significantly better water uptake compared against the 16% tempered seeds subjected to static heating. A cooling step was not required to cause improved water uptake behavior. Although, the presence of small cracks in the seed coat has been attributed to improved water uptake in seeds as literature has implicated the expansion and contraction experienced by the seed coat during growing conditions as a cause of stress gradients which ultimately creates cracks in the seed coat (Ma et al., 2004; Zeng et al., 2005).

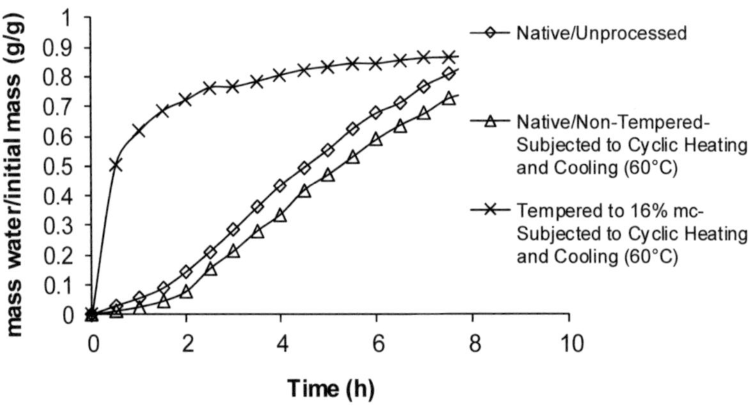

Figure 4. Water Uptake Behavior of Unprocessed and Whole Seeds (AC Ole, pinto bean) processed at 60°C with various conditions (Adapted from Ross et al., 2008).

The glass transition temperature of the seed coat from AC Ole seeds that were tempered at 16% and subjected to cyclic heating at 60 °C and cooling to achieve a seed coat temperature of 7 °C was examined to determine whether physico-chemical changes of the seed coat were induced via processing and altered its T_g (Ross et al., 2008). The measured T_g for the seed coat of the 16% tempered 60 °C cyclic heated AC Ole seeds was 37.7 ± 1.45 °C and the T_g range was 33-47 °C (Table 2). The measured T_g for the seed coat of unprocessed Ole was 51.6 ± 0.2 °C and the T_g range was 33-66 °C (Table 2). It appears that the T_g range is less broad for the seed coat from the 16% tempered 60 °C cyclic heated seeds and the T_g shifted to a lower temperature. The moisture content of the seed coat of the 16% tempered 60 °C cyclic heated AC Ole seeds was 9.4% (Table 2) and the moisture content of the seed coat of the unprocessed AC Ole seeds was 9.6% (Table 2) and therefore moisture content was not a factor in the decreased T_g observed in the seed coat of the 16% tempered 60 °C cyclic heated AC Ole seeds. It appears that subjecting seed to water and heat alters the physico-chemical properties of the seeds coats (Ross et al., 2008).

5. Linking the glass transition with water uptake in pulses: a polymer science approach

5.1. Modeling water uptake in pulse/legume seeds

Water uptake by legume seeds has been discussed extensively in the literature. Many attempts have been made to shorten the required soaking time and other efforts have been focused on defining and predicting water absorption during soaking as a function of time and temperature (Abu-Ghannam & McKenna 1997; Hung et al 1993; Sopade & Obekpa 1990). Abu-Ghannam & McKenna (1997) applied Peleg's (1988) two parameter non-exponential equation to model water absorption during the soaking of red kidney beans of both blanched and unblanched beans at 20, 30, 40 and 60°C. They indicated that Peleg's (1988) two parameter non-exponential equation described the hydration process of blanched beans more adequately than the unblanched beans as the unblanched beans exhibited a significant lag before water uptake began.

Also, there have been attempts to model water uptake and modes of water transfer in legumes using the laws of diffusion, however these models typically fail to describe water uptake data for pulses exhibiting a lag (Seyhan-Gurtas et al., 2001). Soaking at higher temperatures typically reduces/eliminates the lag observed in water uptake data which indicates a temperature sensitivity of the mechanism for the lag. Also, the importance of temperature on lag time and water uptake behavior of California white beans has been documented (Kon, 1979). A mathematical model based on Fick's diffusion, noting the influence of temperature on the concentration-dependent diffusivity of water into soybeans has been reported (Hsu 1983a; Hsu, 1983b]. Water uptake by dry beans was successfully modelled using diffusion theory along with a seed coat wetting theory (Liu et al., 2005). These works from food science literature (Hsu, 1983a; Hsu, 1983b; Liu et al., 2005) were key in proposing the effect of variable surface concentration at the seed coat and lag time in water uptake of legumes. Polymer science research has implicated (a) the degree of movement of the molecules at the polymer surface, and (b) surface concentration conditions, as being important variables affecting solvent/penetrant absorption (Fujita, 1961). Evidence has been provided showing that in rubbery polymers (i.e. polymers above their glass transition) a saturated surface concentration is attained instantaneously at the polymer surface and that many of the non-Fickian anomalies, such as sigmoid absorption curves, may be ascribed to the slow establishment of a saturated surface concentration condition at the surface of the sample below the glass transition (Richman & Long, 1960). Above the glass transition temperature, solvent uptake by a polymer can be modelled with the following exponential equation (Kanamaru & Hirata 1969):

$$\frac{H(t)}{H_{eq}} = 1 - \exp(-kt) \quad (1)$$

Where: H(t) is the mass of the penetrant sorbed at a given time (t), H_{eq} is the saturated amount of penetrant sorbed at equilibrium, k is the rate constant.

The water uptake profile generated by Eq. 1 has been shown to effectively model water uptake in seeds that do not possess a significant lag time (Meyer et al 2007). Alternatively, Nakano (1994a; 1994b) used an empirical equation to model water uptake in wood that exhibited sigmoid-type water uptake (i.e. possesses a lag time). The sigmoid-type water uptake behavior was attributed to a variable surface concentration during the initiation of water uptake and the importance of surface concentration conditions during water uptake was noted. The empirical equation used by Nakano (1994; 1994b) is similar in mathematical form to the empirical equation used by Peleg (1994a; 1994b) to model the mechanical changes in biopolymers at and around their glass transition. Peleg (1994a) indicated that many materials at certain conditions (i.e. temperature and moisture content) undergo considerable physical changes as a result of passing through the glass transition region. The glass transition affects not only mechanical properties of the material but many other physical properties especially those governed by internal molecular mobility. Near the glass transition, the relationship between the mechanical property and temperature (at constant moisture content) or that between the mechanical property and moisture content (at a constant temperature) has a characteristic sigmoidal shape. The mathematical form of the equation used by Peleg (1994a; 1994b) to provide a description of the mechanical behavior of biological materials around their glass transition was borrowed from Fermi's distribution function. The relationship between the mechanical property and temperature (T) at constant moisture content described by Peleg (1994a; 1994b) was given by:

$$Y(T) = \frac{Y_S}{1 + \exp\left[\frac{(T - T_C)}{a}\right]} \quad (2)$$

Also, the relationship between mechanical property and moisture content at a constant temperature described by Peleg (1994a; 1994b) was given by:

$$Y(M) = \frac{Y_S}{1 + \exp\left[\frac{(M - M_C)}{a}\right]} \quad (3)$$

Where: Y(T) and Y(M), in the above equations are the values of any mechanical property at the corresponding temperature (T) or moisture content (M). Y_s is the value of the mechanical property in the glassy or unplasticized state, and T_c and M_c are the characteristic temperature or moisture content, respectively. The parameters T_c and M_c occur at the inflection point on the curve and it occurs where there is a 50% reduction in the mechanical property. The empirical constant a, possesses the same units as the corresponding independent variable.

By tying these concepts together, Peleg's (1994a; 1994b) model of mechanical changes in biomaterials at and around their glass transition, was used to model the water uptake behavior of seeds that possess a lag time and an exponential equation of the form given by Kanamaru & Hirata (1969) was used to model the water uptake behavior of seeds that do not possess a lag time (Ross et al., 2010b).

5.1.1. Use of Peleg's mechanical model for characterizing water uptake in seeds exhibiting a lag phase

Figures 3 and 4 display water uptake data for the seeds demonstrating a lag in their water uptake profile; the pinto beans (AC Ole) that were in their native (unprocessed) state, and those that were processed with cyclic heating (60 °C) and cooling in a non-tempered (moisture content unaltered) state. This data was used in Peleg's (1994a; 1994b) equation, which describes mechanical changes in biopolymers at and around their glass transition, to model the water uptake. In doing so, the original water uptake data was expressed as a fraction of potential water sorption in order to present the data in a manner similar to Peleg's (1994b) report in which the data starts at an initially high value and falls as the independent variable increases (Ross et al., 2010b). This type of treatment is acceptable based on the symmetry of the curves. The equations are given as follows:

$$m_g(t) = \frac{m(t) - m_0}{m_0} \quad (4)$$

Where: $m_g(t)$ is the mass of water sorbed at time (t), m(t) is the mass at time (t), and m_0 is the original mass at time zero.

$$F_h(t) = \frac{m_g(t)}{m_f} \quad (5)$$

$$F_u(t) = 1 - \left[\frac{m_g(t)}{m_f}\right] \quad (6)$$

Where: $F_h(t)$ is the fraction of water sorbed at time (t), $F_u(t)$ is the fraction of potential water sorbed at time (t), and m_f is the total mass of water sorbed at the equilibrium time point.

It was proposed that the amount of water sorbed could be considered the dependent variable, and soaking time at a constant temperature could be considered the independent variable. Their relationship was then described using a modified version of Eqs. 2 or 3. The modified version is given in Eq. 7, and it was explicitly noted that the independent variable is soaking time whereas the independent variable in Eqs. 2 and 3 is temperature or moisture content, respectively (Ross et al., 2010b).

$$F_{uP}(t) = \frac{F_S}{1+\exp\left(\frac{(t-t_C}{a}\right)} \tag{7}$$

Where: $F_{uP}(t)$ is the fraction of potential water sorption at time (t), F_s is the magnitude of this parameter in the glassy or unplasticized state (i.e. during the lag), t_c is the characteristic time which occurs at the inflection point on the curve and it occurs where there is a 50% reduction in the fraction of the potential water sorption, a is the empirical constant with the same units as time.

F_s=1 in the above equation because the value was taken at t=0 (Ross et al., 2010b). Experimental water uptake data provided in Liu et al. (2005) for green beans soaked at 20 and 50 °C was also analysed with Eq. 7 (Ross et al., 2010) as these beans presented a lag in water uptake profiles. Figure 5 shows the experimental data and values generated with Eq. 7 for the native/unprocessed AC Ole pinto beans.

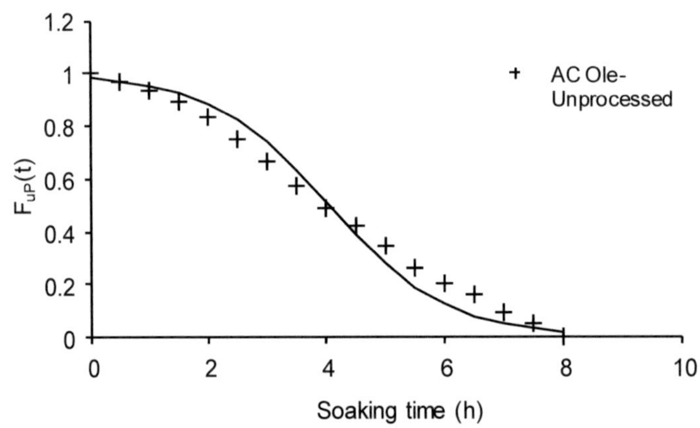

Figure 5. Fraction of potential water sorption as affected by soaking time fit into Peleg's mechanical model: Unprocessed AC Ole beans. Symbols indicate data points, line indicates prediction from model. (Ross et al., 2010b)

5.1.2. Use of an exponential model for characterizing water uptake in seeds exhibiting no lag phase

Figures 3 and 4 also display water uptake data for the seeds that did not demonstrate a lag in their water uptake profile; the peas and AC Ole pinto beans tempered to 16% moisture content and treated with cyclic heating at 60 °C and cooling. This data was analysed with a modified version of the exponential equation presented by Kanamaru & Hirata (1969) (i.e. Eq. 1). The data generated from the modified version of Eq. 1 were consequently expressed in a manner similar to the data obtained from Eq. 7 as the fraction of potential water sorption (F_{uK}) at time (t) (Ross et al., 2010b). This type of treatment is acceptable based on the symmetry of the curves. The equations were given as follows:

$$F_{uK}(t) = 1 - \left(\frac{m_g(t)}{m_{eq}} \right) \quad \text{(a)}$$
$$F_{uK}(t) = 1 - \left[1 - \exp(-kt) \right] \quad \text{(b)} \qquad (8)$$

Where: $m_g(t)$ is the mass of the water sorbed at a given time (t), m_{eq} is the amount of water sorbed at equilibrium or saturation, k is the rate constant.

Figure 6 shows the experimental data and values generated with Eq. 8b for the beans tempered and treated with cyclic heating at 60 °C. Table 3 provides a summary of the equations and constants used to model the data for all of the seed types along with the duration of lag time noted from respective water uptake curves, for the unprocessed AC Ole seeds, untempered processed AC Ole seeds, green beans soaked at 20°C and green beans soaked at 50°C from the work of Lui et al. (2005) and Ross et al. (2010b). The values obtained with Peleg's mechanical model (Eq. 7) for all of the samples exhibiting a lag agree well with the experimental data: the R^2 values corresponding to Eq. 7 are all greater than 0.993.

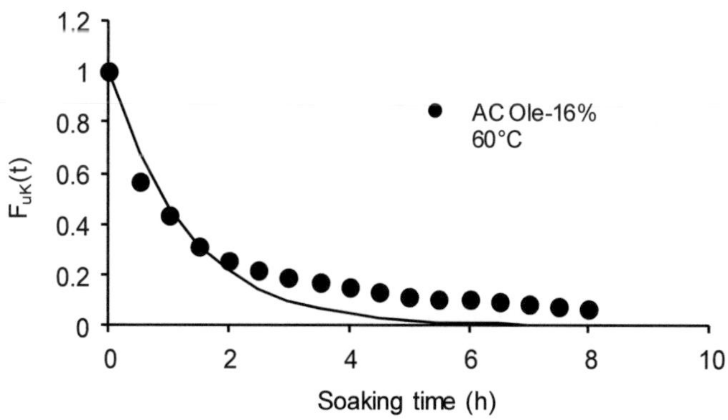

Figure 6. Fraction of potential water sorption as affected by soaking time fit into Kanamaru & Hirata's sorption model: 16% tempered, 60°C cyclic heated and cooled AC Ole beans. Symbols indicate data points, line indicates prediction from model. (Ross et al., 2010b)

Material	Lag Time (h)	Equation	Peleg constants		R^2
			t_c	A	
AC Ole: Unprocessed	2	7	4.2	0.854	0.993
AC Ole: Non-tempered, 60°C heat treatment	2	7	4	1.049	0.997
Green Bean: soaking water 20°C (Lui et al., 2005)	5	7	7.2	1.46	0.997
Green Bean: soaking water 50°C (Lui et al., 2005)	0.5	7	1.8	0.429	0.998
			Kanamaru & Hirata constant		
			K		
AC Ole: tempered to moisture content 16%, 60°C heat treatment	na	8b	0.769		0.989
Pea: Mozart	na	8b	0.276		0.992
Pea: Stratus	na	8b	0.302		0.998

na=not applicable

Table 3. Parameters used to describe water uptake behaviour (Adapted from Ross et al., 2010b)

The values generated by the exponential equation (8b) were in good agreement with the experimental values of the unprocessed peas and beans tempered and treated with cyclic heating at 60 °C and cooling. The lowest R^2 value corresponding to Eq 8b. is 0.989. Therefore, fitting the experimental water uptake data with these equations provided evidence that the glass transition temperature of the seed coat is an important factor in the mechanism of water uptake in seeds. The seeds that possessed a lag time were well represented by Eq. 7, which was based on Peleg's (1994a; 1994b) model describing changes near the glass transition, while seeds that did not possess a lag time were well characterized by Eq. 8a, which was based on a model describing water uptake in polymer above the glass transition. Using this approach, the implication of the glass transition in the mechanism of water uptake in seeds was demonstrated.

5.2. Hypothesis of the mechanism describing water uptake in pulse seeds

As the implication of the glass transition in the mechanism of water uptake provided in section 5.1 mainly depends upon the experimental data fitting models, a hypothesis for the mechanism of water uptake for seeds possessing seed coats with a glass transition above ambient soaking conditions has been developed based on an analogy between a temperature driven glass transition (T_g) and a solvent driven glass transition (a_g) presented in polymer science literature (Laschitsch et al. 1999; Leibler & Sekimota 1993; Vrentas & Vrentas 1991). At a constant temperature, a sorption curve of a polymer below the glass transition remains relatively flat (i.e. possesses a lag period) until a certain solvent concentration is attained, after which the sorption curve displays a steep increase in solvent uptake. This steep increase in solvent uptake can be considered to have occurred at the glass transition solvent activity (a_g) and consequently this increase in solvent uptake can be attributed to the plasticization of the polymer by the sorbed penetrant (Li & Lee, 2006; Vrentas & Vrentas 1991). For polymers in the glassy state (i.e. below the glass transition), the length of time required to reach a saturated surface concentration, a state where water uptake is exponential, is affected by solvent content of the polymer and temperature (Fujita, 1961). If the seed coat of legume seeds has a glass transition below soaking temperature, it can be considered a biopolymer in the glassy state, thus the length of time required to reach saturated surface concentration would be affected by the temperature of the soaking water. Higher soaking temperatures would allow for more molecular movement and therefore promote faster attainment of saturated surface conditions, consequently reducing the lag time. This result was observed with the green bean data of Lui et al. (2005).

Furthermore, the thermal behavior of the seed coat of the unprocessed AC Ole bean seed was examined with differential scanning calorimetry (Ross et al., 2010b) to further test this hypothesis. The seed coat was placed in distilled water under isothermal conditions at 25 °C with the aim of imitating soaking conditions. The heat flow was measured as a function of time and subsequently the rate of heat flow change was examined as a function of soaking time. The DSC data showed a sharp change in the amount of energy needed for soaking per unit time up to the two hour time point. Around the two hour time point, the change in the energy rate slows down and by four hours of soaking the energy rate seems to plateau. This result indicated an energy barrier to soaking exists in the first two hours of soaking, which does agree with the lag time observed in the water uptake data for the unprocessed AC Ole bean seed (Table 3). The gradual energy rate change from two hours to four hours corresponds to the concave up portion of the water uptake curve and the four hour time point after which the energy rate change becomes constant corresponds with the inflection point of the water uptake curve and may correspond to the time required to reach surface concentration saturation. The rate of energy change in the seed coats upon soaking is affected by the amount of water sorbed by the seed coat at a constant temperature. The time point where there is a substantial variation in the rate of energy change corresponds to the seed coat adsorbing the necessary amount of water (i.e. solvent) required to cause the seed coat to have a glass transition at room temperature. This time point corresponds to the lag time. After the seed coat moves through the glass transition temperature, the energy barrier to water uptake is lessened. These results are in agreement with the work of Gunnells et al. (1994) where it was shown that the glass transition temperature of wood when measured with DSC corresponded to the temperature where there was a maximum in the first derivative of heat flow (i.e. rate of energy change) after a steep increase.

The hypothesis for the mechanism of water uptake in seeds possessing seed coats with a glass transition near ambient soaking conditions was based on work reported in the field of polymer science. Polymers at their glass transition have a small relaxation time which allows for a sudden, almost instantaneous, increase to a saturated surface concentration upon exposure to solvent (Fujita, 1961; Richman & Long, 1960) and therefore rapid initial solvent uptake is achieved. Thus, seeds that possess seed coats with a glass transition temperature near ambient temperature reach a saturated surface concentration rapidly upon immersion in water and exhibit exponential water uptake behavior with no significant lag time. Consequently, water uptake in seeds that possess seed coats with a glass transition temperature near ambient temperature and were successfully modelled with the exponential equation provided by Kanamaru & Hirata (1969) used for describing solvent uptake by polymers above the glass transition.

6. A polymer science approach to dehulling of pulses: The role of the glass transition

6.1. Importance of optimizing red lentil dehulling efficiency

This section provides original experimental evidence determined by the authors which supports the role of the glass transition in defining the dehulling quality of red lentils and is presented as such. Red lentils account for the majority of world lentil production and trade (Agblor, 2006). With the exception of North America and Australia, most lentils are consumed in the region of production. Thus, the export market for red lentils is of paramount importance to North American and Australian producers. More than 90% of red lentils produced are consumed as dehulled split or dehulled whole seeds (Vandenberg & Bruce, 2008). Since most red lentils are dehulled before consumption, milling or dehulling efficiency of red lentil is very important to consumer acceptability. For this work, dehulling efficiency (DE) as indicated by Wood & Malcolmson (2011), was defined as the sum of dehulled whole seed (DW) and dehulled split seed (DS) relative to total initial mass of seeds. Dehulling efficiency is an important quality characteristic for lentil breeders, processors, and exporters as it ultimately dictates whether a dehulling operation is economically feasible (Wang, 2005). Upon milling/dehulling, the various fractions are obtained: 1) powder, 2) breaks/broken cotyledons, 3) hull, undehulled whole seeds, 4) dehulled split seeds, and 5) dehulled whole seeds (Wang, 2005). The goal of milling is to completely dehull the red lentils while minimizing the production of powder, breaks and splits. Based on the above definition theoretical DE is about 92% (complete hull removal, no undehulled whole seeds, no breaks, and no powder), yet millers strive to achieve 85% DE as a DE >80% is necessary to achieve an economically viable process (Vandenberg & Bruce, 2008; Wang, 2005). Also a higher ratio of whole dehulled seeds to split seeds is desired as whole dehulled seeds are the most valuble fraction (Vandenberg & Bruce, 2008).

The problem that Canadian red lentil producers face is that the Western Canadian climate is very different from global competitors. Canadian grown red lentils have a higher moisture content compared to non-Canadian grown red lentils, and as a result dehulling is difficult (Vandenberg & Bruce, 2008). Red lentils are typically harvested at 16-18% moisture content and to ensure safe storage red lentils are dried to 13% moisture content or less (CGC, 2008; McVicar, 2006). The use of aeration fans to reduce moisture has been recommended (McVicar (2006) yet supplemental heat drying may be necessary and it has also been recommended that air temperatures should not exceed 45 °C (Saskatchewan Government, 2007). The Canadian grain grading system (CGC, 2008) indicates that broken seeds must be less than 2% and 3.5% in Canada No. 1 and Canada No. 2 grades, respectively. The effect of drying temperature on the handling quality of whole seeds in terms of seed breakage upon handling has been investigated (Tang et al., 1990), however little work has been done on the effect of drying temperature on the milling quality of red lentils. Alternatively, a large amount of work has been performed in the area of understanding the effect of drying temperature on rice milling quality (Cao et al., 2004; Cnossen & Siebenmorgen, 2000; Cnossen et al., 2003; Iguaz et al., 2006). The rice industry cites breakage of rice kernels during milling as one of its main problems (Iguaz et al., 2006). Rice must be dried to 12-13% moisture content for safe storage. During drying moisture content gradients are created within the rice kernels that induce stresses that can cause the rice kernels to fissure/crack (Cao et al., 2004). Fissured/cracked rice kernels usually break during milling which results in poor cooking quality and a low value product (Cao et al., 2004). The ultimate goal of the rice industry is to maximize head rice yield (HRY), which is defined as the weight in percentage of rough rice that remains as head rice (rice kernels are at least ¾ of the original kernel length) after milling (Cnossen & Siebenmorgen, 2000). As analogous situation exists in red lentil dehulling where upon milling maximizing the whole dehulled seed fraction and minimizing broken cotyledons and split seeds is desired.

The concept of the glass transition (T_g) has been used to explain rice kernel fissure/crack formation during drying and subsequent breakage during milling (Siebenmorgen et al., 2004). At temperatures below T_g, grain seeds exist in a glassy solid state, starch granules are compact, water is relatively immobile and diffusion of moisture inside a grain seed is slow while at temperatures above T_g, starch exists in a rubbery state, free volume is increased, the water associated with starch has greater mobility, and the diffusion of moisture is enhanced (Siebenmorgen et al., 2004). During rice drying, the temperature will increase and moisture will diffuse from the grain seed and a temperature and moisture content gradient will develop from the surface to the center of the grain seed (Cnosssen et al., 2003). The temperature gradient disappears rapidly but the moisture content gradient remains to play an important role during and after drying. During the equilibrating (sometimes termed tempering) stage, following drying, moisture migrates from the center (higher moisture region) to the surface of the grain seed (lower moisture region) and consequently the moisture content gradient decreases. If the equlibrating air temperature is below the T_g of the grain seed, the seed will cool and go through a glass transition and become glassy as the grain seed temperature decreases. If a sufficient moisture content gradient exists when the tempering/equilibrating environment is one that produces a change of state of the starch (transitioning from the rubbery to the glassy state), the different sections of the grain seed (surface, midpoint, centre), resulting from the moisture content gradient, pass through the T_g at different moisture content values, which is depicted by situation B of Figure 7 (Cnossen & Siebenmorgen, 2000; Cnossen et al., 2003). Due to large differences in kernel properties between the rubbery and the glassy states, specifically the thermal expansion coefficients (Perdon et al., 2000; Cnossen et al., 2003), differential stresses within the grain seed will likely cause fissuring/cracking (Siebenmorgen et al., 2004). If the tempering/equilibrating air temperature is above T_g, then the moisture content will equilibrate between the different sections of the kernel (surface, midpoint, centre) and the moisture content gradient will cease to exist. Upon subsequent exposure to ambient temperature, the kernel will pass through the glass transition at a common moisture content, stresses will be minimized and therefore fissure/cracking will be minimized. This is illustrated in situation A of Figure 7. The hypothesis used to explain rice breakage was adopted to explain breakage and splitting in red lentils. The objectives of the research linking the glass transition with dehulling quality were to: 1) determine the glass transition temperature of two varieties of red lentils, Impact and Redberry as affected by moisture content; and 2) use knowledge of the glass transition temperature to examine the effect of drying temperature on dehulling efficiency in terms of breakage and the ratio of dehulled whole seeds to dehulled split seeds.

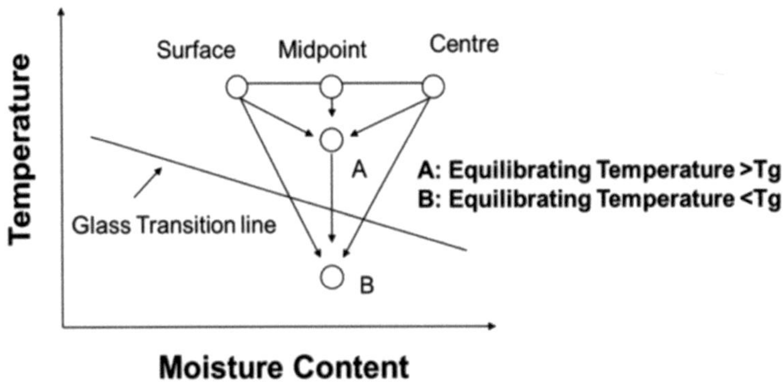

Figure 7. Paths followed by the surface, mid-point between the surface and the center, and the center of the seed for an equilibration temperature above or below the glass transition (T_g). (Adapted from Cnossen et al., 2003)

6.2. Experimental work

6.2.1. Samples

Two varieties of red lentils in commercial production (CDC Impact and CDC Redberry) were chosen for this study. Both lentil varieties (2007 crop) were grown in Saskatchewan and transported to the University of Manitoba in polyethylene bags in January 2008. The Impact and Redberry varieties were purchased as cleaned seeds and thus were relatively free of foreign materials upon arrival and did not require further cleaning prior to use.

6.2.2. DSC Experiments

6.2.2.1 DSC Sample preparation

Seed coats were removed from red lentil cotyledons by dehulling, which is described in detail in a following section. The red lentil cotyledon material was ground with a coffee grinder (Persona, ON) to reduce particle size and sieved to pass an 825 μm sieve. Ground samples were sub-sampled and placed in four different relative humidity environments to alter the moisture content of the samples. The saturated salt solutions used were lithium chloride (LiCl), potassium acetate, (KAc), magnesium nitrate ($MgNO_3$), and sodium chloride (NaCl) (Sigma Chemicals, St. Louis, MO) and they produced the following RH environments respectively, 11.3, 22.5, 52.9, and 75.3% at 25 °C. The samples were allowed to sit in the various relative humidity environments until a constant mass was attained. The moisture content of the ground samples were determined by drying duplicate samples for 24 h in a drying oven (Precision Thelco Laboratory Oven, Thermo Fisher Scientific, Waltham, MA) set at 130 °C.

6.2.2.2. DSC conditions

The T_g of the ground red lentil cotyledons were measured with differential scanning calorimetry (DSC). A DSC, (DSC7, Perkin Elmer, Norwalk, CT) equipped with a thermal analyzer controller (TAC7/DX, Perkin Elmer, Norwalk, CT) and Pyris software v.8 (Perkin Elmer, Shelton, CT) were used to obtain the DSC thermograms. Samples were heated isothermally at 5 °C for 3 minutes prior to temperature ramping. A temperature ramp of 10 and 30 °C/min and a temperature range of 5 to 170 °C were employed. The DSC was calibrated with indium (melting point =156.6 °C and enthalpy = 28 J/g). At least 25 mg of ground sample was accurately weighed into sample pans. The sample pans were sealed and an empty sample pan was used for the reference. The T_g of each thermogram was determined by identifying the transition corresponding to a mid-point in slope change in the heat capacity of the sample. T_g measurements were performed in duplicate.

6.2.3. Drying experiments: Preparation of lentils dried at 40 and 80 ºC

For the drying experiments, the methods of Ross et al. (2010c) were followed. Briefly, Impact and Redberry variety red lentils were subjected to near ambient drying at 40 °C and high temperature drying at 80 °C. Since the samples in their native/untreated state had a moisture content of 7.4% (Impact) and 9.2% (Redberry), the samples were tempered prior to drying to create an initial moisture content (mc) of 13% (wb). The tempered samples were left for at least 48 h at room temperature to absorb and evenly distribute moisture throughout the individual seeds. These samples were then rewetted to raise the mc to 18% (wb) to simulate moisture content at harvest. A starting common moisture content of 13% was chosen to ensure a moisture content increase of 5% was imposed on all samples. Five hundred grams of lentils were placed in aluminium pans (29.8 x 21.6 x 3.2 cm) creating a 15 mm thin layer. The 80 °C dried samples were placed in a drying oven (Precision Thelco Laboratory Oven, Thermo Fisher Scientific, Waltham, MA) set at 80 °C for 2h to lower the moisture content of the samples to 11%. The 40 °C dried samples were placed in an environmental chamber (model IH 400, Yamato, Tokyo, Japan) set at 40 °C for 6.75 h to lower the moisture content of the samples to 11%. After drying, the samples were sub-sampled. A portion of both the 40 and 80 °C dried samples were placed in sealable polyethylene plastic bags. The bags were sealed and allowed to equilibrate at 25 °C for 16 h. A portion of both the 40 and 80 °C dried samples were placed in air-tight glass jars and placed in a drying oven set a 72 °C for 1.5 h. After heating at 72 °C for 1.5 h, the samples were placed sealable polyethylene plastic bags. The bags were sealed and allowed to equilibrate at 25 °C for 16 h. After equilibrating at 25 °C for 16 h the samples were prepared for dehulling.

6.2.4. Dehulling conditions

In order to prevent seed size from being a variable in the dehulling tests all seeds were passed through a series of sieves (Carter Day International Inc., Minneapolis, MN) with round hole openings of 5.56, 4.76, 4.37, 3.97, and 3.37 mm diameter and separated into fractions to obtain red lentils with uniform size. For the Redberry seeds that had not been treated (i.e. native seed samples), the most abundant fraction (90%) belonged to the size range of 4.37 to 5.56 mm. The Impact seeds were smaller and showed more size heterogeneity; the most abundant fraction (78%) for the native/untreated samples belonged in the size range of 3.97 to 4.76 mm. Seeds fitting in these ranges after pre-milling treatment were used in the dehulling evaluation tests. This practice was in accordance with the work of Erksine et al. (1991) and Ross et al. (2010c).

A grain testing mill (TM05C, Satake Engineering Co., Hiroshima, Japan), fitted with a 36 mesh abrasive wheel was used for the dehulling tests. The dehulling experiments were performed using the methods of Ross et al. (2010c) and dehulling variables such as abrasive milling speed, milling time and milling moisture content were chosen in accordance with the work of Ross et al. (2010c) and Wang (2005) to achieve optimal dehulling efficiency (DE). The Impact and Redberry red lentil samples were milled in 30 g batches using an abrasive milling speed of 1100 RPM for 40 s. The samples were tempered to 12.9 % moisture content prior to milling/dehulling. It should be noted that the moisture level at milling/dehulling was independent of the moisture content that the samples were tempered to simulate moisture content at harvest. Tempering samples prior to dehulling was considered a milling/dehulling pre-treatment. The Impact and Redberry red lentils were moisturized to 12.9% moisture content by adding the amount of water needed to achieve the desired sample moisture content. Tap water was added to the sample using a graduated cylinder, and then the plastic bag was sealed and shaken for at least 60 s, ensuring that the lentils were evenly coated with water. For up to two hours after the initial addition of moisture, the bags were periodically shaken for 60 s at 30 min intervals. The sealed samples were left for 48 h at room temperature to absorb and evenly distribute moisture throughout the individual seeds. Upon completion of the 48 h tempering period the samples were milled/dehulled. After dehulling, the milled lentils were screened on a US standard No. 20 mesh sieve (850 μm) to collect the powder. The milled seeds remaining on top of the 850 μm sieves were separated into whole seeds, split seeds, broken seeds and hulls using the following method. The seeds were sent through a husk aspiration unit (S.K. Engineering and Allied Works, Bahraich, India) to remove the hulls. The seeds were then separated into split and whole fractions by sieving over a No. 4.5 (1.79 x 12.7 mm) slotted sieve (Carter Day International Inc., Minneapolis, MN). The split seeds were screened on a US standard No. 8 sieve (2.36 mm) to separate out the broken seeds. The whole seeds were separated by hand into their respective hulled and dehulled classes. It is noted that the split seeds were visually inspected for contaminating attached hull, yet none to negligible amounts were detected. All fractions were weighed and then expressed as a proportion of the total original sample weight. Dehulling efficiency (DE) was defined as the sum of percent dehulled whole seed (DW) and percent dehulled split seed (DS) relative to total initial mass of seeds and was calculated as:

$$DE(\%) = \left(\frac{M_{DW} + M_{DS}}{M_T} \right) * 100 \tag{9}$$

Where: M_{DW} is mass of dehulled whole seeds, M_{DS} is the mass of dehulled split seeds, M_T is the total initial mass of seeds.

6.2.5. Statistical analysis

Statistical analysis was conducted using SAS Institute Inc. Software, version 9.1 (SAS Institute, 2001). Data were subjected to analysis of variance (ANOVA) with replication using the SAS PROC GLM procedure to generate Least square (LS) means. Significance was accepted at $p \leq 0.05$.

6.3. The role of the glass transition in defining the dehulling quality of red lentils

6.3.1. Determination of glass transition temperature (Tg) of red lentils

Table 4 shows the results for the T_g of the ground red lentils samples as a function of moisture content. The data presented in Table 4 are the midpoint T_g values. These results indicate a clear dependence of T_g and moisture content, as sample moisture content increased, the T_g decreased. This is in agreement with the work of Cao et al. (2004) and Perdon et al. 2000, which indicated that the second order transition associated with the glass transition temperature was correlated with moisture content. It should be noted that there was no significant difference between the moisture content of the samples stored in the relative humidity environment of the LiCl saturated salt solution and the relative humidity environment of the KAc salt solution. This result was unexpected as these saturated salt solutions provide different relative humidity environments 11.3 and 22.5% at 25 °C, respectively. Although sufficient amounts of these salts were added to water to create a saturated solution, it could be that one of the containers failed to properly seal. Nevertheless, samples with at least three different moisture contents were tested and a clear relationship between T_g and moisture content was observed. Table 4 also shows the effect of heating rate on measured T_g. Generally, as the heating rate is increased there was an observed increase in T_g, which in agreement with the work of Bruning & Samer (1992).

Table 5 gives coefficients for the linear regression equation relating glass transition temperature (T_g) with respect to moisture content of Redberry and Impact red lentils at two different heating rates. Using respective correlation coefficients in the equation relating moisture content and T_g, it was determined that Redberry red lentils with a moisture content of 11% would possess a T_g of 66 and 69 °C with a 10 and 30 °C/min heating rate, respectively while Impact red lentils with a moisture content of 11% would have a T_g of 65 and 67 °C with a 10 and 30 °C/min heating rate, respectively. This equation is given as:

$$Tg = \beta_1 \times MC + \beta_0 \tag{10}$$

Where: T_g is the glass transition temperature (°C); MC is % moisture content; β_1 is the slope of the regression line; β_0 is the y-intercept of the regression line.

Relative Humidity Environment	Moisture Content (%)	Heat Rate (10°C/min) T_g (°C)		Heat Rate (30°C/min) T_g (°C)	
		Redberry	Impact	Redberry	Impact
LiCl	6.1a	75.1a	74.8a	81.3a	79.1a
KAC	6.5a	71.7a	76.0a	78.9a	79.9a
MgNO3	10.1b	66.7b	66.7b	74.5b	66.2b
NaCl	14.0c	62.2c	58.4c	61.3c	60.5c

Different letters within a column indicate significant differences between the means (n=2)

Table 4. Effect of Moisture Content and Heating Rate on T_g

Variety	Heat Rate (10°C/min)			Heat Rate (30°C/min)		
	B_0 (°C)	B_1	R^2	B_0 (°C)	B_1	R^2
Impact	89.24	-2.21	0.988	94.8	-2.54	0.954
Redberry	82.68	-1.49	0.949	95.8	-2.38	0.956

MC is % moisture content; β_1 is the slope of the regression line; β_0 is the y-intercept of the regression line; R^2 is correlation coefficient

Table 5. The Linear Relationship Between Glass Transition Temperature and Moisture Content for Red Lentils

The T_g values obtained for the ground red lentil cotyledons were higher than the T_g values obtained by Perdon et al. (2000) and Cao et al. (2004) for rice at comparable moisture contents. Possible reasons for this result may be due to the fact that ground red lentils were used instead of individual grain sections. It is possible that ground samples have different heat transfer properties than individual grain sections. Also, a cereal starch (rice) is different in composition than a legume starch (lentil). These different starches possess different granule organization and different degrees of branching which will affect T_g (Parker & Ring, 2001). Therefore it is not unreasonable to accept different T_g values at a common moisture content for the different starch types. In all, for red lentils, the linear regression equations provided in Table 5 can be used to predict T_g from moisture content.

6.3.2. *Effect of drying & equilibration temperature on dehulling quality*

Table 6 shows the effect of drying temperature and equilibration temperature on the dehulling quality of Impact and Redberry red lentils. For the Impact red lentils, drying at a temperature above T_g (80 °C) and equilibrating at a temperature below Tg (25 °C) did result in significantly more breaks, more split cotyledons and a significantly higher ratio of split seed to whole dehulled seeds than any other condition. Although drying below T_g (40 °C) and equilibrating below T_g (25 °C) did result in significantly less breaks and split cotyledons than drying above T_g (80 °C) and tempering below T_g (25 °C), there were still significantly more breaks and split cotyledons observed compared to the reference (non-dried) Impact seeds. The Impact samples dried at either 40 or 80 °C and equilibrated above T_g (72 °C) resulted in a significant reduction in broken and split cotyledons. Within the Impact seeds equilibrated above T_g, the samples that were dried at 40 °C possessed significantly less splits compared to the Impact seed samples dried at 80 °C upon milling. Also, the ratio of split to dehulled whole seeds and dehulling efficiency (DE) of the Impact samples, dried at either 40 or 80 °C and equilibrated above T_g (72 °C) was not significantly different than the Impact samples that were not subjected to any artificial drying. For the Redberry red lentils, drying at either 40 or 80 °C and equilibrating at a temperature below T_g (25 °C) did result in significantly more split cotyledons and a significantly higher ratio of split seed to whole dehulled seeds than the condition of drying below or above T_g (40 or 80 °C) and equilibrating above T_g (72 °C). Within the Redberry seed samples equilibrated above T_g (72 °C) or below T_g (40 °C), the Redberry samples that were dried at 40 °C possessed significantly less splits compared to the sampled dried at 80 °C upon milling. For the Redberry seeds, there was no observed effect on the amount of breaks or dehulling efficiency (DE) at any drying and equilibrating condition. Therefore the results observed for Redberry showed that equilibrating above T_g (72 °C) resulted in a significant reduction in split cotyledons. Additionally, the ratio of split to dehulled whole seeds for the Redberry samples that were equilibrated above T_g (72 °C) was not significantly different than the Redberry samples that were not subjected to any artificial drying (i.e. the reference samples).

For both red lentil varieties, in the situation where the samples were dried below T_g (40 °C) and equilibrated at a temperature below T_g (25 °C), there should be no change of phase of starch and therefore no or limited stress gradient should result as will result as the surface, midpoint and center of the grain reach a common moisture content. Breakage and splitting upon dehulling should be minimized. Although the results of this work seemed to indicate drying below and equilibrating below T_g will negatively impact dehulling quality as significantly higher values of split seeds were obtained for this condition compared to values observed for samples equilibrated above T_g. Possibly the drying temperature used in these experiments were too close to the measured T_g values and possibly overlapped onset T_g values. However, seeds dried at the higher temperature condition (80 °C) and equilibrated below T_g (25 °C) showed even higher levels of split seeds than those dried at the lower temperature condition (40 °C) and equilibrated below T_g (25 °C), which helps to support this explanation. Also, research in rice drying has shown that the use of higher drying temperature promote rice fissuring and breakage upon milling. When higher drying temperatures are employed, the rate of moisture removal increases which promotes the development stress gradients within the seed and thereby promotes fissuring and breakage (Iguaz et al., 2006). When the samples are dried below T_g and equilibrated at a temperature above T_g, the starch will experience a change of state, yet the surface, midpoint, and center of the seed will reach a common moisture content above T_g and stress gradients will also be minimized. This situation would also lead to less breakage and splitting upon dehulling, which as observed in this work. When the samples are dried above T_g and equilibrated at a temperature above T_g, the starch will be in the rubbery state and the seed will reach a common moisture content above T_g and stress gradients will also be minimized. This situation would also lead to less breakage and splitting upon dehulling, which was observed in this work. However, samples dried above T_g (80 °C) and equilibrated above T_g (72 °C) presented more splits than corresponding samples dried below T_g (25 °C) and equilibrated above T_g (72°C).

Samples	Drying Temp (ºC)	Equilib Temp (°C)	MC (%)	Powder (%)	Breaks (%)	Hull (%)	UDW (%)	DW (%)	DS (%)	DE (%)	DS:DW Ratio
Impact	40	25	11.1	1.58a	0.73a	8.0s	12.1a	22.2a	55.1a	77.1a	2.5a
Impact	80	25	11.2	1.47a	1.26b	7.6a	10.0a	15.5b	61.2b	76.7a	3.9b
Impact	40	72	11.2	1.75a	0.75a	7.3a	3.4b	63.1c	23.5c	86.6b	0.37c
Impact	80	72	11.2	1.60a	0.90a	7.2a	4.6b	53.0d	32.5d	85.5b	0.62c
Impact: Reference	Na	na	7.4 "as is"	1.94a	0c	7.4a	2.7b	63.4c	23.5c	86.9b	0.37c
Redberry	40	25	11.3	1.89a	0.23a	8.6a	3.4a	50.1a	35.0a	85.2a	0.7a
Redberry	80	25	11.2	1.92a	0.18a	8.6a	2.95a	42.1b	42.8b	84.9a	1.1b
Redberry	40	72	11.2	1.62a	0.27a	9.2b	1.15a	77.7c	10.8c	88.5a	0.14cd
Redberry	80	72	11.1	1.90a	0.22a	8.3a	0.73a	71.1d	17.3d	88.4a	0.24c
Redberry: Reference	Na	na	9.2 "as is"	2.06a	0b	8.2a	0b	59.1e	28.5e	87.6a	0.49ac

MC= moisture content; Powder=powder yielded from dehulling; Breaks=broken seed yielded from dehulling; Hull=hull yielded upon dehulling; UDW=Undehulled whole seeds yielded from dehulling; DW=Dehulled whole seeds yielded from dehulling; DS=dehulled split seeds yielded from dehuling; DE=Dehulling Efficiency; na=not applicable
Different letters within a column and variety indicate significant differences between the means (n=2)

Table 6. Effect of Drying Temperature and Equilibration Temperature on Dehulling Quality

In summary of this experimental work, moisture content has a significant effect on the thermal properties of Redberry and Impact variety red lentils. The glass transition temperature, T_g, increased with decreasing moisture content. Single broad transitions were observed from the DSC thermograms of Redberry and Impact variety red lentils at different moisture contents. The linear equations calculated to predict T_g from moisture content proved useful in understanding the mechanism of red lentil breakage and splitting from drying and milling. With knowledge of the T_g, the experimental procedure tested the T_g drying hypothesis by drying and equilibrating the red lentils at temperatures above and below their T_g. Results for both varieties of red lentils showed that equilibration of the dried lentils at temperatures above T_g caused a decrease in splitting, implying that stress gradients were minimized using this treatment. Also, equilibrating dried samples at a temperature above T_g caused a remarkable increase in the ratio of dehulled whole lentils to split lentils. The Impact variety red lentils also showed significantly less breaks and a higher dehulling efficiency when equilibrated above T_g for either drying condition. Therefore, the results that were obtained in this work are supported by the rice drying hypothesis put forth by Cnossen & Siebenmorgen (2000) and indicated that the splitting of cotyledons upon dehulling was more affected by drying temperature and equilibration temperature than breakage.

7. Conclusion

To the best of the authors' knowledge, no other work has explicitly linked the glass transition of the seed coat with water uptake behaviour. Implication of the glass transition of the seed coat as a key factor influencing water uptake is important as delayed water uptake behavior has served as the main impediment to the processing of legumes and the creation of value added whole seed legume products. A detailed chemical analysis of components in the seed coat affecting the glass transition temperature is required in future work as it would allow for focussed breeding efforts to reduce the chemical components that cause high seed coat glass transition temperatures thereby improving the processability of some legume seeds. The concept of the glass transition has been used to explain rice kernel fissure formation during drying and subsequent breakage during milling of rice. The hypothesis used to explain rice breakage was adopted to explain breakage and splitting in red lentils upon dehulling. Red lentil varieties, Impact and Redberry, showed an increase in the amount of split seeds when dried above or below T_g and equilibrated below T_g. The red lentils that were dried above or below T_g and equilibrated at temperatures above T_g showed a significant decrease in splitting. The Impact variety also showed significant decrease in seed breakage and increased dehulling efficiency. Furthermore, equilibrating dried samples above T_g caused a remarkable increase in the ratio of dehulled whole seeds to split seeds. Future work must include investigating: a minimum equilibration temperature, a minimum equilibration time, a maximum moisture content removal per drying step, a very gentle drying condition (21 °C at 50% RH) and the effects on dehulling quality. However, the work at present does implicate the T_g as having an effect on breakage and splitting of red lentils. Overall, the challenges in pulse processing can be addressed by a following a polymer science approach.

Author details

Kelly A. Ross*
Agriculture & Agri-Food, Canada

Susan D. Arntfield
Dept. of Food Science, University of Manitoba Canada

Stefan Cenkowski
Dept. Biosystems Engineering, University of Manitoba, Canada

8. References

Abbas, K.A.; Lasekan, O. & Khalil, S.K. (2010). The Significance of Glass Transition Temperature in Processing of Selected Fried Food Products: A Review. *Modern Applied Science*, Vol.4, No.5, pp. 3-21.

Abu-Ghannam, N. & Gowen, A. (2011). Pulse-based food products. In: *Pulse Foods: Processing, Quality and Nutraceutical Applications*, B. Tiwari, A. Gowen, B. McKenna (Eds.), 249-278, Academic Press

Abu-Ghannam, N. & McKenna, B.; (1997). The Application of Peleg's Equation to Model Water Absorption During the Soaking of Red Kidney Beans (*Phaseolus vulgaris* L.). *Journal of Food Engineering*, Vol.32, pp. 391-401.

Agblor, K. (2006). The Cropportunity Strategy for Red Lentils-Spot Light on Research. PulsePoint. June, pp.25-27: Saskatchewan Pulse Growers.

An, D.; Arntfield, S.D.; Beta, T. & Cenkowski, S.; (2010). Hydration Properties of Different Varieties of Canadian Field Peas (Pisum sativum) form Different Locations. *Food Research International*, Vol.43, No.2, pp. 520-525.

Arechavaleta-Medina, F. & Snyder, H.E. (1981). Water Imbibition by Normal and Hard Soybeans. *Journal of the American Oil Chemical Society*, Vol.58, pp.976-979.

Argel, P.J. & Parton, C.J. (1999). Overcoming Legume Hardseededness. In: *Forage Seed Production: Tropical and Sub-tropical Species*, D.S. Loch, J.E. Ferguson (Eds.), 247-267, CAB International, Wallingford.

Barbosa-Cánovas G.V. & Vega-Mercado, H. (1996). Physical, Chemical, and Microbiological Characteristics of Dehydrated Foods. In: *Dehydration of Foods*, G.V. Barbosa-Canovas & H. Vega-Mercado (Eds.), 29-99, International Thompson Publishing, New York, USA.

Bellido, G.; Arntfield, S.D.; Cenkowski, S. & Scanlon, M.G. (2006). Effects of micronization pretreatments on the physicochemicalproperties of navy and black beans (Phaseolus vulgaris L.) *Lebensm. Wiss. Technol.* Vol.39, pp. 779–787.

Black, R.G.; Singh, U. & Meares, C. (1998a). Effect of Genotype and Pre-Treatment of Field Peas (*Pisum Sativum*) on their Dehulling and Cooking Quality. *Journal of the Science of Food and Agriculture*, Vol.77, pp. 251-258.

Black, R.G.; Brouwer, J.B.; Meares, C. & Iyer, L. (1998b). Variation in Physico-Chemical Properties of Field Peas (*Pisum sativum*). *Food Research International*, Vol.31, pp. 81-86.

* Corresponding Author

Boye, J.; Zare, F. & Pletch, A. (2010). Pulse Proteins: Processing, Characterization, Functional Properties and Applications in Food and Feed Food Research International, Vol.43, pp. 414–431.

Brent, J.L.; Mulvaney, S.J.; Cohen, C. & Bartsch, J.A. (1997). Thermomechanical Glass Transition of Extruded Cereal Melts. Journal of Cereal Science, Vol. 26, pp. 301-312.

Bruning, R. & Samer, K. (1992). Glass Transition on Long Time Scales. *Physics Review B* Vol.46, No.18, pp. 11318-11322.

Campanella, O.H.; Li, P.X.; Ross, K.A. & Okos, M.R. (2002). The Role of Rheology in Extrusion. In: *Engineering and Food for the 21st Century*, J. Welti-Chanes, G.V. Barbosa-Canovas, J.M. Aguilera (Eds.), 393-413, Technomic Publishing.

Canadian Grain Commission (2008). Official Grain Grading Guide – Lentils (Chapter 18) Canadian Grain Commission ISSN 1704-5118.

Cao, W.; Nishiyama, Y. & Koide, S. (2004). Physicochemical, Mechanical, and Thermal Properties of Brown Rice with Various Moisture Contents. *International Journal of Food Science and Technology*, Vol.39 pp. 899-906.

Casado, C.G. & Heredia, A. (2001). Specific Heat Determination of Plant Barrier Lipophillic Components: Biological Implications. *Biochimica et Biophysica Acta* Vol.1511, pp. 291-296.

Cnossen, A.G.; Jimenez, M.J. & Siebenmorgen, T.J. (2003). Rice Fissuring Response to High Drying and High Tempering Temperatures. *Journal of Food Engineering*, Vol.59, pp. 61-69.

Cnossen, A.G. & Siebenmorgen. TJ. (2000). The Glass Transition Temperature Concept in Rice Drying and Tempering Effect on Milling Quality. *Transactions of the ASAE*, Vol. 23, pp. 1661-1667.

DeSouza, F.H., & Marcos-Filho, J. (2001). The seed coat as a modulator in seed-environment relationships in Fabaceae. *Revta Brasil Botany*, Vol.24, No.4, pp. 365-375.

Debendedetti, P.G. & Stillinger, F.H. (2001). Supercooled liquids and the glass transition. Nature, Vol49, pp.259-267.

Ferry, J.D. (1980). Viscoelastic Properties of Polymers. 3rd ed. John Wiley & Sons, New York. 641 p.

Eckl, K. & Gruler, H. (1980). Phase Transitions in Plant Cuticles. *Planta*, Vol. 1150, pp. 102 113.

Egerton-Warburton, L.M. (1998). A Smoke Induced Alteration of the Sub-Testa Cuticle in Seed of the Post Fire Recruiter, *Emmenanthe penduliflora* Benth. (Hydrophyllaceae). *Journal of Experimental Botany*, Vol.49, No.325, pp. 1317-1327.

Flory, P.J. (1953). Principles of Polymer Chemistry. Cornell University Press, Ithaca, New York. 672 p.

Fujita, H. (1961). Diffusion in Polymer-Diluent Systems. *Algebra Universalis*, Vol.3, No.1, pp. 1-47.

Gibbs, J.H. (1956). Nature of the Glass Transition in Polymers. *Journal of Chemical Physics*, Vol.25, pp. 185-186.

Gibbs, J.H. & DiMarzio, E.A. (1958). Nature of the Glass Transition and the Glassy State. *Journal of Chemical Physics, Vol.*,28, pp. 373-383.

Gunnells, D.W.; Gardner, D.J. & Wolcott, M.P. (1994). Temperature Dependence of Wood Surface Energy. *Wood and Fibre Science*, Vol.26, No.4, pp. 447-455.

Hoseney, R.C. Glass Transition and its Role in Cereals. 1994. In: *Principles of Cereal Science and Technology*, R.C. Hoseney (Ed.), 307-320, American Association of Cereal Chemists, St. Paul, MN.

Hsu, K.H. (1983a). A Diffusion Model with a Concentration-Dependent Diffusion Coefficient for Describing Water Movement in Legumes During Soaking. *Journal of Food Science*, Vol.48, pp, 618-622,645.

Hsu, K.H. (1983b). Effect of Temperature on Water Diffusion in Soybean. *Journal of Food Science*, Vol.48, pp. 1364-1365.

Hung, T.V.; Liu, L.H.; Black, R.G. & Trewhella, M.A. (1993). Water Absorption in Chickpea (*C. arietinum*) and Field Pea (*Pisum sativum*) Cultivars Using the Peleg Model. *Journal of Food Science*, Vol.58, pp. 848-852.

Iguaz, A.; Rodriguez, M. & Virseda, P. (2006). Influence of handling and processing of rough rice on fissure and head rice yields. *Journal of Food Engineering*, Vol. 77, pp. 803-809.

Kanamaru, K,. & Hirata, M. (1969). The Two-Stage Sorption Process of Polymers Immersed in Water Viewed from the Process of the Lowering of ξ-Potential with Time. Colloid Polymer Science, Vol.230, No.1, pp. 206-221.

King, R.D. & Ashton, S.J. (1985). Effect of Seed Coat Thickness and Blanching on the Water Absorption by Soybeans. Journal of Food Technology, Vol.20, pp. 505-509.

Kon, S. (1979). Effect of Soaking Temperature on Cooking and Nutritional Quality of Beans. Journal of Food Science, Vol.44, pp. 1329-1334, 1340.

Kumagai, H. & Kumaga, H. (2009). Glass Transition of Foods and Water's Effect on It. Foods Food Ingredients Journal of Japan, Vol.214, No.2. pp. 1.

Laschitsch, A.; Bouchard, C.; Habicht, J.; Schimmel, M.; Ruhe, J. & Johannsmann, D. (1999). Thickness Dependence of the Solvent Induced Glass Transition in Polymer Brushes. *Macromolecules*, Vol.32, pp. 1244-1251.

Leibler, L. & Sekimota, K. (1993). On the Sorption of Gases and Liquids in Glassy Polymers. *Macromolecules*, Vol.26, pp. 6937-6939.

Lemeste, M.; Champion, D.; Roudaut, G.; Blond, G. & Simatos, D. (2002). Glass Transition and Food Technology: A Critical Appraisal. *Journal of Food Science*, Vol.67, No.7, pp. 2444-2458.

Li, J.X. & Lee, P.I. (2006). Effect of Sample Size on Case II Diffusion of Methanol in Poly(Methyl Methacylate) Beads. *Polymer*, Vol.47, pp. 7726-7730.

Li, Q. (2010). Investigating the Glassy to Rubbery Transition of Polydextrose and Corn Flakes Using Automatic Water Vapor Sorption Instruments, DSC, and Texture Analysis [Thesis]. Urbana, IL: University of Illinois at Urbana-Champaign.

Lin, L.S. ; Yuen, H.K. & Varner, J.E. (1991). Differential Scanning Calorimetry of Plant Cell Walls. Proceeding of the National Academy of Sciences, Vol.88, pp. 2241-2243.

Liu, C.K.; Lee, S.; Cheng, W.J.; Wu, C.J. & Lee, I.F. (2005). Water absorption in dried beans. *Journal of the Science of Food and Agriculture*, Vol.85, pp. 1001-1008.

Ma, F.; Cholewa, E.; Mohamed, T.; Peterson, C.A. & Gijzen, M. (2004). Cracks in the Palisade Cuticle of Soybean Seed Coats Correlate with their Permeability to Water. *Annals of Botany*, Vol.94, pp. 213-228.

Marbach, I. & Mayer, A.M. (1975). Changes in Catechol Oxidase and Permeability to Water in Seed Coats of *Pisum elatius* during Seed Development and Maturation. *Plant Physiology*, Vol.56, pp. 93-96.

Marbach I. & Mayer, A.M. (1974). Permeability of Seed Coats to Water as Related to Drying Conditions and Metabolism of Phenolics. *Plant Physiology*, Vol.54, pp. 817-820.

Maskus, H. (2010). Pulse Processing, Functionality and Application-Literature Review. Pulse Canada. pp. 1-146.

Matas, A.J.; Cuartero, J. & Heredia, A. (2004). Phase Transitions in the Biopolyester Cutin Isolated from Tomato Fruit Cuticles. *Thermochimica Acta*, Vol.409, pp. 165-168.

Matas, A.J.; Lopez-Casado, G.; Cuartero, J. & Heredia, A. (2005). Relative Humidity and Temperature Modify the Mechanical Properties of Isolated Tomato Fruit Cuticles. *American Journal of Botany*, Vol.92, No.3, pp. 462-468.

McVicar, R. (2006). Red Lentils: Handle with Care. PulsePoint. June, p.9-10.

McDougall, G.J.; Morrison, I.M.; Stewart, D. & Hillman, J.R. (1996). Plant Cell Walls as Dietary Fiber: Range, Structure, Processing and Function. *Journal of the Science of Food and Agriculture*, Vol.70, pp. 133-150.

Meyer, C.J.; Steudle, E. & Peterson, C.A. (2007). Patterns and Kinetics of Water Uptake by Soybean Seeds. Journal of Experimental Botany, Vol.58, pp. 717-732.

Mullin, W.J. & Xu, W. (2001). Study of Soybean Seed Coat Components and Their Relationship to Water Absorption. *Journal of Agricultural and Food Chemistry*, Vol.49, No.11, pp. 5331-5335.

Nakano, T. (1994a). Non-Steady State Water Adsorption of Wood. Part 1. A Formulation for Water Adsorption. *Wood Science Technology*, Vol. 28, pp. 359-363.

Nakano, T. (1994b). Non-Steady State Water Adsorption of Wood. Part 2. Validity of the Theoretical Equation of Water Adsorption. *Wood Science Technology*, Vol.28, pp. 450-456.

Norton, C. (1998). Texture and Hydration of Expanded Rice. [Thesis] Nottingham University.

Oksanen, C.A. & Zografi, G. (1990). The Relationship Between the Glassy Transition Temperature and Water Vapor Absorption by Poly(Vinylpyrrolidone). *Pharmaceutical Research*, Vol.7, No.6, pp. 654-7.

Parker, R. & Ring, S.G. (2001). Aspects of the Physical Chemistry of Starch. Journal of Cereal Science Vol.34, pp. 1-17.

Peleg, M. (1988). An Empirical Model for the Description of Moisture Sorption Curves. *Journal of Food Science*, Vol.53, pp. 1216-1217, 1219.

Peleg, M. (1993). Mapping The Stiffness-Temperature-Moisture Relationship of Solid Biomaterials at and Around Their Glass Transition. *Rheologica Acta*, Vol.32, No.6, pp. 575-580.

Peleg, M. (1994a). A Model of the Mechanical Changes in Biomaterials at and Around their Glass Transition. *Biotechnology Progress*, Vol.10, pp. 385-388.

Peleg, M. (1994b). Mathematical Characterization and Graphical Presentation of the Stiffness-Temperature-Moisture Relationship of Gliadin. *Biotechnology Progress*, Vol. 10, pp. 652-654.

Perdon, A.; Siebenmorgan, T.J. & Mauronmoustakos, A. (2000). Glassy State Transition and Rice Drying: Development of a Brown Rice State Diagram. *Cereal Chemistry*, Vol.77, No.6, pp. 708-713.

Rangaswamy, N.S. & Nandakumar, L. (1985). Correlative Studies on Seed Coat Structure, Chemical Composition, and Impermeability in the Legume *Rhynchosia minima. Botanical Gazette*, Vol. 146, No.4, pp. 501-509

Reyes-Moreno, C.; Carabez-Trejo, A.; Paredes-Lopez, O. & Ordorica-Falomir, C. (1994). Physicochemical and Structural Properties of Two Bean Varieties which Differ in Cooking Time and the HTC Characteristic. *Lebensm-Wiss. u-Technol.* Vol.27, pp. 331-336.

Rhaman, M.S. (2006). State Diagram of Foods: Its Potential use in Food Processing and Product Stability. *Trends in Food Science and Technology*, Vol. 17, pp. 129–141.

Richman, D. & Long, F.A. (1960). Measurement of Concentration Gradients for Diffusion of Vapors in Polymers. *Journal of the American Chemical Society*, Vol. 82, pp. 509-513.

Roos, Y. & Karel, M. (1991). Plasticizing Effect of Water on Thermal Behavior and Crystallization of Amorphous Food Models. *Journal of Food Science and Technology*, Vol. 56, pp. 38-43.

Roos, Y.H. (1993). Water Activity and Physical State Effects on Amorphous Food Stability. *Journal of Food Processing and Preservation*, Vol.16, No.6, pp. 433-47.

Roos, Y.H. (1995). Phase Transitions in Foods. Academic Press, San Diego, USA, 360 p.

Roos, Y.H. (2010) Glass Transition Temperature and its Relevance in Food Processing. *Annual Review of Food Science and Technology*, Vol.1, pp. 469-496.

Ross, K.A.; Arntfield, S.D.; Beta, T.; Cenkowski, S. & Fulcher, R.G. (2008). Understanding and Modifying Water Uptake Of Seed Coats Through Characterizing the Glass Transition. *International Journal of Food Properties*, Vol.11, No3, pp. 544-560.

Ross, K.A.; Zhang, L. & Arntfield, S.D. (2010a). Understanding Water Uptake from Changes Induced During Processing: Chemistry of Pinto and Navy Bean Seed Coats. *International Journal of Food Properties*, Vol.13, No.3, pp. 631-647.

Ross, K.A.; Arntfield, S.D.; Cenkowski, S. & Fulcher, R.G. (2010b). Relating Mechanical Changes at the Glass Transition with Water Absorption Behavior of Dry Legume Seeds. *International Journal of Food Engineering*, Vol.6, No.4, Article 13, pp. 1-22.

Ross, K.A.; Alejo-Lucas, D.; Malcolmson, L.J.; Arntfield, S.D. & Cenkowski, S. (2010c). Effect of Milling Treatments and Storage Conditions on the Dehulling Characteristics of Red Lentils. *International Journal of Postharvest Technology and. Innovation*, Vol.2, No.1, pp. 89-113.

Saskatchewan Government, 2007. Crops – Overview: Red Lentil. [Online]. Available at: http://www.agriculture.gov.sk.ca/Default.aspx?DN=a88f57f0-242b-40f6-8755-1fc6df4dfa14 [accessed on 31 March 2012]

Schreiber, L. & Schonherr, J. (1990). Phase Transitions and Thermal Expansion Coefficients of Plant Cuticles. The Effects of Temperature on Structure and Function. *Planta*. Vol. 182, pp. 186-193.

Sefa-Dedeh, S. & Stanley, D.W. (1979). The Relationship of Microstructure of Cowpeas to Water Absorption and Dehulling Properties. *Cereal Chemistry*, Vol.56, No.4, pp. 379-386.

Seyhan-Gurtas, F.; Ak, M.M. & Evranuz, E.O. (2001). Water Diffusion Coefficients of Selected Legumes Grown in Turkey as Affected by Temperature and Variety. *Turkish Journal of Agriculture and Forestry*, Vol.25, pp. 297-304.

Shao, S.; Meyer, C.J.; Ma, F.; Peterson, C.A. & Bernards, M.A. (2007). The Outermost Cuticle of Soybean Seeds: Chemical Composition and Function During Imbibition. *Journal of Experimental Botany*, Vol. 58, No.5, pp. 1071-1082.

Siebenmorgen, T.J.; Yang, W. & Sun, Z. (2004). Glass Transition Temperature of Rice Kernels Determined by Dynamic Mechanical Thermal Analysis. *Transactions of the ASAE*, Vol. 47, No.3, pp. 835-839.

Sievwright, C.A. & Shipe, W.F. (1986). Effect of Storage Conditions and Chemical Treatments on Firmness, *In Vitro* Protein Digestibility, Condensed Tannins, Phytic Acid And Divalent Cations of Cooked Black Beans (*Phaseolus vulgaris*). *Journal of Food Science*, Vol. 51, No.4, pp. 982-987.

Slade, L. & Levine, H. (1991). Beyond Water Activity: Recent Advances Based on an Alternative Approach to the Assessment of Food Quality and Safety. *Critical Reviews in Food Science and Nutrition*, Vol.30, pp. 115–360.

Sopade, P.A. & Obekpa, J.A. (1990). Modelling Water Absorption in Soybean, Cowpea and Peanuts at Three Temperatures Using Peleg's Equation. *Journal of Food Science*, Vol. 55, pp. 1084-1087.

Sperling L.H. (2006). Introduction to Physical Polymer Science. 3rd ed. Wiley, New York, 845 pp.

Tang, J.; Sokhansanj, S.; Slinkard, A.E. & Sosulki. F (1990). Quality of Artificially Dried Lentil. *Journal of Food Process Engineering*, Vol.13, pp. 229-238.

Thanos, A.J. (1998). Water Changes in Canned Dry Peas and Bean During Heat Processing. *International Journal of Food Science and Technology*, Vol. 33, pp. 539-545.

Tiwari, B.; Gowen, A. & McKenna, B. (2011). Introduction. In: *Pulse Foods: Processing, Quality and Nutraceutical Applications*, B. Tiwari, A. Gowen, B. McKenna (Eds.), 1-7, Academic Press.

Tolstoguzov, V.B. (2000). The Importance of the Glassy Biopolymer Components in Food. *Nahrung* Vol.44, No.2, pp. 76-84.

Vandenberg, B. & Bruce, J. (2008). Producing Better Quality Red Lentils. PulsePoint. March, p.31-33.

Vrentas, J.S. & Vrentas, C.M. (1991). Sorption in Glassy Polymers. *Macromolecules*. Vol.24, pp. 2402-2412.

Wang, N. (2005). Optimization of a Laboratory Dehulling Process for Lentil (*Lens culinaris*). *Cereal Chemistry*, Vol.82, pp. 671-676.

Watts P, (2011) Global pulse industry: state of production, consumption and trade; marketing challenges and opportunities. In: *Pulse Foods: Processing, Quality and Nutraceutical Applications*, B. Tiwari, A. Gowen, B. McKenna (Eds.), 437-464, Academic Press.

Williams, R.J. (1994). Methods for Determination of Glass Transitions in Seeds. *Annals of Botany*, Vol.74, pp. 525-530.

Wood JA, Malcolmson LJ (2011) Pulse Milling Technologies. In: *Pulse Foods: Processing, Quality and Nutraceutical Applications*, B. Tiwari, A. Gowen, B. McKenna (Eds.), 193-221, Academic Press.

Zeng, L.W.; Cocks, P.S.; Kailis, S.G. & Kuo, J. (2005). The Role of Fractures and Lipids in the Seed Coat in the Loss of Hardseededness of Six Mediterranean Legume Species. *Journal of Agricultural Science*, Vol.143, pp. 43-55.

Chapter 4

Polymer Characterization with the Atomic Force Microscope

U. Maver, T. Maver, Z. Peršin, M. Mozetič,
A. Vesel, M. Gaberšček and K. Stana-Kleinschek

Additional information is available at the end of the chapter

http://dx.doi.org/10.5772/51060

1. Introduction

1.1. Atomic force microscopy

Atomic force microscopy is a powerful characterization tool for polymer science, capable of revealing surface structures with superior spatial resolution [1]. The universal character of repulsive forces between the tip and the sample, which are employed for surface analysis in AFM, enables examination of even single polymer molecules without disturbance of their integrity [2]. Being initially developed as the analogue of scanning tunneling microscopy (STM) for the high-resolution profiling of non-conducting surfaces, AFM has developed into a multifunctional technique suitable for characterization of topography, adhesion, mechanical, and other properties on scales from tens of microns to nanometers [3].

1.2. The technique

A schematic representation of the basic AFM setup is shown in Figure 1. Using atomic force microscopy (AFM), a tip attached to a flexible cantilever will move across the sample surface to measure the surface morphology on the atomic scale. The forces between the tip and the sample are measured during scanning, by monitoring the deflection of the cantilever [1]. This force is a function of tip sample separation and the material properties of the tip and the sample. Further interactions arising between the tip and the sample can be used to investigate other characteristics of the sample, the tip, or the medium in-between [4].

1.2.1. Force between the sample and the tip

To understand the mechanisms behind the interacting components in multi-component formulations, we have to take into account all the contributing forces. This is especially important if a quantitative analysis of the interaction is required, like in the case of interactions between polymers and biological macromolecules [6]. The forces between the tip and the substrate have short- and long-range contributions. When measurements are performed, it is crucial that we can separate the contributions of various forces and eliminate the undesired ones. This ensures the measurement of desired sample properties only and makes further quantitative analysis possible [7]. In vacuum, chemical forces of very short range (less than 1 nm), electrostatic, magnetic and Van der Waals forces can be determined, while in air forces with longer range, which can be up to 100 nm, cover them, making the measurements mostly qualitative [8]. At room conditions water moisture can condense on the tip, which is a source of capillary force. Capillary forces are relatively big and can cover the contributions of other forces; therefore they have to be avoided if possible. The latter is possible by measuring in special, water free conditions, like in a N_2 or Ar atmosphere or in liquid environments.

To represent forces on the atomic level, different potentials corresponding to changes of potential energy at various particle positions, are used. Known empirical models used to illustrate chemical bonds are the Lennard-Jones and Morse potential [9]. These models quite satisfactory fit the force regime curve shown in Figure 2, which represents the course of tip-sample interaction.

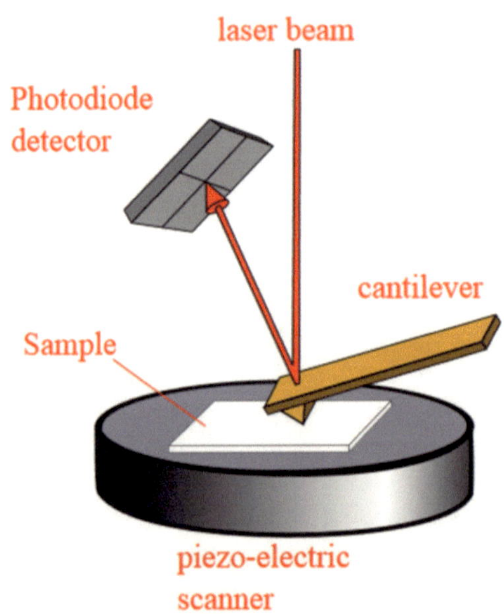

Figure 1. Schematical representation of the AFM. The image was reproduced with permission of C. Roduit [5].

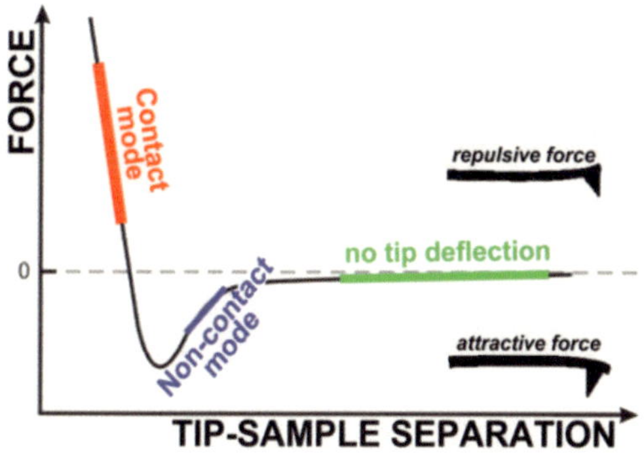

Figure 2. Force regimes governing the AFM measurement.

1.2.2. *AFM modes for polymer examination*

Many different variations of the basic AFM setup have been developed through the years of its use. Although most of them are applicable to all types of samples, not all yield the same amount and quality results. Proper use of these versatile measurement variations enables one to study and understand processes even at the fundamental, namely molecular level [10]. Considering various different samples, several modes have been developed and adapted to cope with the demand of field specific research [11]. In the scope of the next few paragraphs only some of the most popular will be presented.

1.2.3. *Contact mode*

Contact mode was the first developed mode of atomic force microscopy. In this mode, the tip is moving across the surface and deflects according to its profile (Figure 3). Two types of contact mode measurements are known, the constant force and the constant height mode. In the constant force type, a feedback loop is used to move the sample or the tip up and down and keep its deflection constant. The value of z-movement is equal to the height changes of the sample's surface. The result of such measurement is the information about the surface topography. Since the tip is in constant contact with the surface, significant friction forces, which can destroy or sweep soft samples like polymers or biological macromolecules on the surface, appear [12].

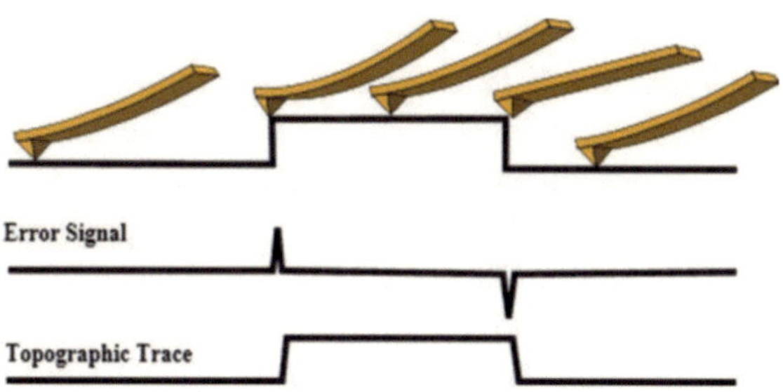

Figure 3. Schematic representation of the contact mode. The image was reproduced with permission by C. Roduit [5].

The other type of contact mode AFM measurement is based on the constant height, while the forces are changing. In this case, the cantilever deflection is measured directly and the deflection force on the tip is used to calculate the distance from the surface. Since no feedback loop is required for this type of measurement, it is appropriate for quick scans of samples with small height differences (if height differences are big, the tip will very likely crash into the surface, by which it gets destroyed or damages the samples' surface). With this type of measurements atomic resolution was achieved at low temperatures and in high vacuum. Such measurements are often used for quick examination of fast changes in biological structures [13].

1.2.4. Noncontact mode

In noncontact mode, the sample's surface is investigated using big spring constant cantilevers. The tip attached to the cantilever is hovering very close to the surface (at a distance of approximately 5-10 nm), but never gets into contact with it, hence the name noncontact mode (Figure 4). A major advantage of this mode is negligible friction forces, making this mode capable for measurements of biological and polymeric samples without alteration of their surface. The biggest drawbacks of this mode are low lateral and z-resolution when compared to the contact mode. Recently it was used for characterization of single polymer chains [14].

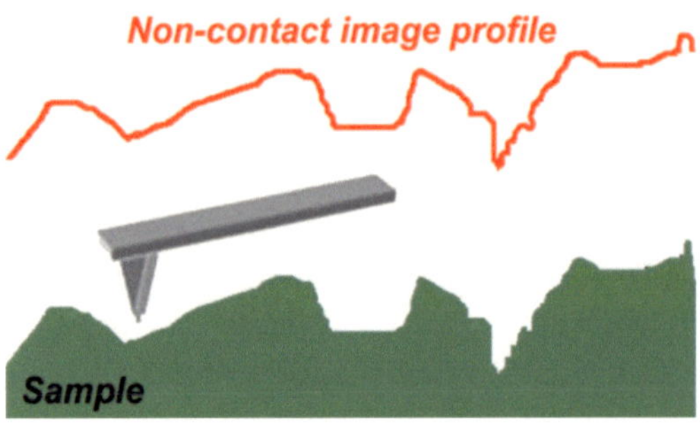

Figure 4. Schematical depiction of the non-contact AFM mode.

1.2.5.Amplitude. modulation mode or dynamic force mode

This mode is often called the *intermittent-contact* or *tapping mode* and it eliminates major weaknesses of the noncontact mode (such as the low lateral and z-resolution). Instead of hovering above the sample, the cantilever vibrates above the surface and moves through the force gradient above the surface, during which it might momentarily touch the surface [15]. Due to interactions of the AFM tip with the sample surface, the amplitude of vibrations decreases and a phase shift occurs (Figure 5). We can choose either of these parameters (amplitude or phase shift) and keep it constant through the feedback loop by moving either the sample or the tip in z-direction. This gives us information about the surface topography similar to the contact mode. To measure in the amplitude modulation mode we need much stiffer cantilevers, which exhibit the smallest possible damping factors (this factor is commonly referred to as the Q-factor) [16]. Amplitude modulation mode is the most often used AFM mode due to its high resolution, almost non-destructive nature of the imaging and its applicability in air and also in liquid conditions [17].

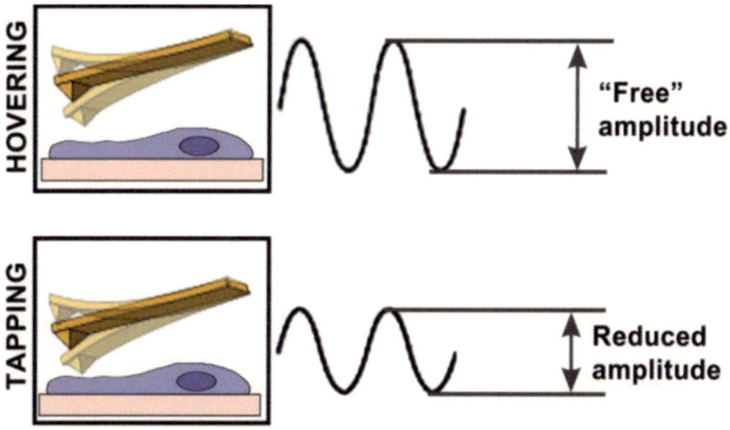

Figure 5. Schematical representation of the amplitude modulation mode. Parts of the image were reproduced with permission by C. Roduit [5].

1.2.6. Force spectroscopy

Force spectroscopy has proved to be one of the most promising techniques using AFM. In an AFM experiment, a tip is attached to a flexible cantilever, which is moved across the sample surface. During this procedure, the surface morphology is measured with a nanometer resolution. Upon contact with the sample surface, the tip experiences a force, which is monitored as a change in the deflection of the cantilever [18]. This force is a function of tip sample separation and the material properties of the tip and the sample and can be used to investigate other characteristics of the sample, the tip, or the medium in-between [4]. The procedure of an AFM force measurement is schematically depicted in Figure 6 and goes as follows: the tip attached to a cantilever spring is moved towards the sample in a normal direction, during this movement the vertical position of the tip and the deflection of the cantilever are recorded and converted to force-versus-distance curves, briefly called force curves [1].

In the early nineties only skilled and specialized physicists were able to interpret the complex behavior, which occurs after an AFM tip gets close to a specific sample surface. But these days many more researchers try to explore these measurements to better understand mechanisms behind more and more phenomena. In addition to evaluation of interaction forces between the tip and model surfaces, AFM can also produce two-dimensional chemical affinity maps by modifying the cantilever tip with specific molecules [19]. In such a way, it is possible to characterize differently responding regions on the material's surface, resulting in a better understanding and, consequently, application of the examined materials [20]. In this way even quantitative data can be gathered, which can be used to identify the forces involved in specific biological systems [21].

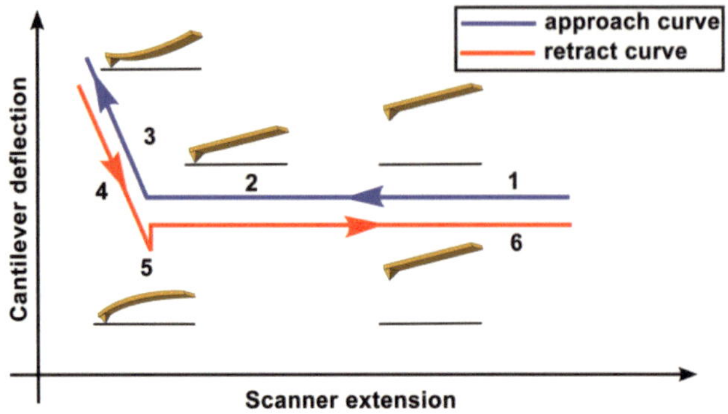

Figure 6. A typical force curve. When approaching the surface, the cantilever is in an equilibrium position (1) and the curve is flat. As the tip approaches the surface (2), the cantilever is pushed up to the surface – being deflected upwards, which is seen as a sharp increase in the measured force (3). Once the tip starts retracting, the deflection starts to decrease and passes its equilibrium position at (4). As we start moving away from the surface the tip snaps in due to interaction with the surface, and the cantilever is deflected downwards (5). Once the tip-sample interactions are terminated due to increased distance, the tip snaps out, and returns to its equilibrium position (6). The image was reproduced with permission by C. Roduit [5].

Mapping chemical functional groups and examining their interactions with different materials is of significant importance for problems ranging from lubrication and adhesion, to the recognition of biological systems and pharmacy [22]. Changing environmental conditions during the measurement has also been extensively used to monitor changes in the interactions between different functional groups and surfaces to simulate the material behavior upon exposure to a real environment [23].

2. Tip functionalization

At the moment, one of the most promising AFM related techniques for polymer examination is surely the chemical force microscopy (CFM) [2, 24]. CFM enables the measurement of interactions appearing between polymer molecules or polymers, and different surfaces [23]. This additional information allows the prediction of final material characteristics based on the examined polymers, even before their finalization. Quantitative assessment of the involved forces and their extent makes it easier to choose the correct polymers for achieving desired interactions between the materials used in several different interest fields (adhesion, adsorption, repelling etc.). Multilayer polymeric materials are lately also the first choice materials for the preparation of modern wound dressings. When sticking together layers of different polymeric origins, their interaction gains importance regarding the behavior of the final product. A CFM experiment has to be conducted with specially designed tips, which for themselves act as chemical sensors. Success of such measurements is impossible without proper tips, so choosing the right ones is crucial in this regard. Many commercial ones are available at the moment, but only some exhibit characteristics that allow for a simple and repeatable functionalization. Whilst the functionalization of tips may seem quite easy during the first iteration, it quickly becomes clear, that a lot of chemical skills are needed to bind the right species to the right place in the desired amount [19]. Additionally a lot of statistical evaluation is needed in order to prove and evaluate the success of any attachment [25].

2.1. Polymers and AFM

Polymers have found their way into all fields of science and industry over the last decades. Their potential applications range from binders in batteries [26] to composite materials in drug delivery [27]. Whilst the range of possible combinations between different monomers is endless, polymers found or based on natural polymers have recently become the subject of thorough research, once again [28]. Synthetic changes to their native structure make them even more appealing; especially cellulose derivatives exhibit a lot of potential for satisfying most industrial needs [29].

Within the field of polymer sciences, AFM has been used to quantify the entropic elasticity of single polymer chains[30], the elastic moduli of nanowires [31], single polymer chain elongation [32], molecular stiffness of hyperbranched macromolecules [33], friction of single polymers on surfaces [34], influence of temperature on the stability of single chain conformation [35], and surface glass transition temperature [36]. It has also been used to perform stretching experiments on single carboxy-mehtylated amylase [37], and to differentiate between sugar isomers [38].

2.2. AFM measurements in polymer science

Recent progress in the understanding of the underlying mechanisms during AFM force measurements enabled thorough research of the interaction between different polymer molecules and the materials, with which these get in contact upon use. Such knowledge is of utter importance in the development stages of polymeric materials, because they allow prediction of materials behavior during use. The use of controlled environments during measurement enables the simulation of the exact conditions one desires, while the measurement in liquids allows measurements in even simulated physiological conditions, which is especially desired in the testing stages of drug delivery systems.

Without proper experiment design, quantitative measurements using AFM are not possible. In this light several preparation steps have to be included in the planning phase of an experiment (Figure 7). On the following pages, we will explain them a little further and expand them with our own results and experiences.

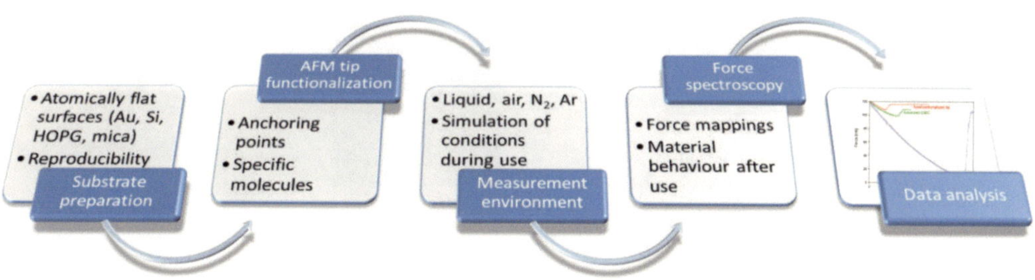

Figure 7. Schematical depiction of the necessary steps for successful AFM measurement.

2.2.1. Substrate preparation for AFM measurements

Chemical force microscopy (CFM) was derived from AFM for the examination of interactions between different materials and even molecules by exploiting their chemical characteristics [39]. Quantitative assessment of such interactions can be used for identification purposes, for determination of compatibility between different materials to be put into one single final product, and to predict interactions with the target site in drug delivery systems [40].

CFM is best used with a defined experimental setup, comprising model surfaces, a controlled environment during measurements and materials of high purity. When using a high resolution technique like AFM, we have to be very careful not to confuse the information about the desired species with the substrate characteristics [41]. That is why atomically flat surfaces, apart from mica, which is commonly used for much longer, were introduced a couple of years ago, when researchers realized that not all of the data they gathered corresponded to actual species' properties, but were in fact more related to the substrates' characteristics [42]. Atomically flat surfaces are free of surface roughness and proper choice of an inert material for their preparation makes it possible to gather reliable high resolution data after desired sample attachment [43].

During our research we had to find the best possible technique to prepare such surfaces on a daily basis. Therefore we upgraded and combined different previous methods into one highly efficient preparation procedure, which enabled us to progress much faster in our experiments. A detailed explanation of this method can be found elsewhere [44], while a brief description is depicted in Figure 8 and goes as follows. Prior to any preparation steps, all used laboratory accessories were cleaned in a multi-step procedure, combining different chemicals, to assure extreme cleanliness. In the next step, high-grade mica was coated with gold of high purity. A two stage heating/annealing step was introduced afterwards, which yielded atomically flat gold terraces of sizes in the range from a couple hundred nm to 2 microns.

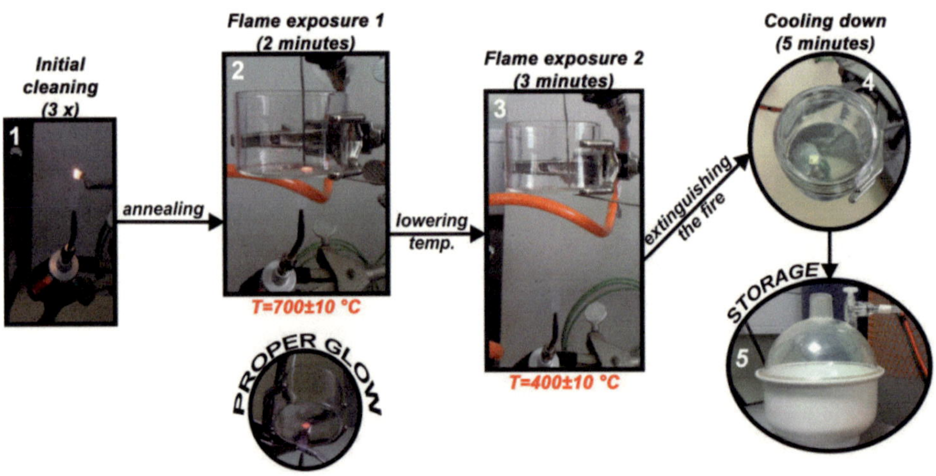

Figure 8. Scheme of the annealing procedure with corresponding photographs. The initial cleaning step comprises three passes of the gold coated mica piece through the hydrogen flame [44].

The value of such substrates cannot be evaluated without their inclusion into sample preparation. In our case, we tested them by preparing a sample with attached carbon nanotubes. If their morphology has to be evaluated, we have to use flat surfaces, which do not temper their actual properties, measured on the nanoscale. In our study, the substrates and samples were evaluated using two different types of microscopy, namely the scanning electron microscopy (SEM) and AFM. Figure 9 shows the improvement from not annealed to annealed surface with attached test molecules.

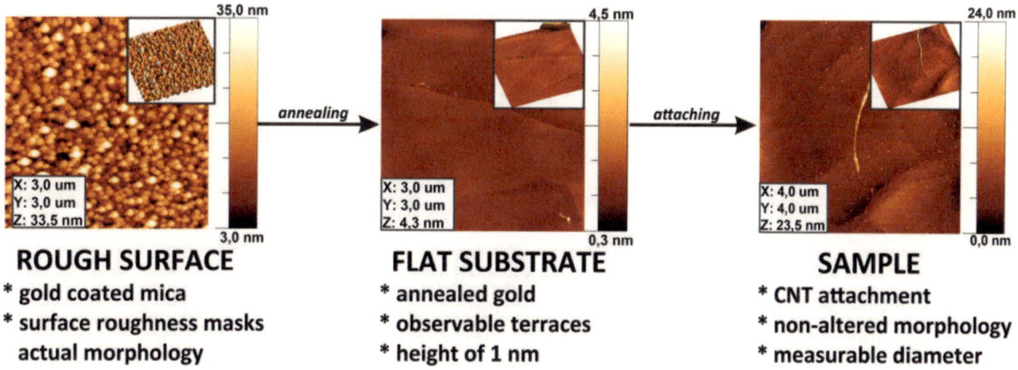

Figure 9. Progress from non-annealed gold-coated mica to the actual sample preparation and examination.

2.2.2. AFM tip functionalization for chemical sensing

Specific interaction mappings and identification of mechanisms behind processes require tip functionalization, which turns AFM cantilevers into chemical sensors. Depending on the degree of surface coverage with newly added functional species on the tip surface, one can measure even single molecule interactions. Such high resolution is desired, when interactions between biomolecules are tested, especially when novel drug targets are being investigated. Tip functionalization depends on the tip composition and on the desired functional groups, which in turn serve as anchoring points for more specific chemical sensing. Especially useful is CFM during the development of multilayered materials, which are commonly exploited in wound dressing preparation. The heterogeneity of the commonly used materials for this cause renders the preparation of their surfaces to stick together upon application on the wound, a tough job. CFM is capable of delivering such information in vitro. In our group, we introduced several ways of tip functionalization, which enable CFM measurements.

There are different types of commercially available AFM tips. We mostly use tips from two different groups. In the first are silicon-based tips, which can be functionalized in two step procedures. The first step involves the introduction of functional groups, which can serve as non-specific chemical sensors on their own. The second step adds specificity to them by binding desired molecules to these anchoring points, which serve as efficient sensors of desired species. In the second group are tips, which are coated with different coatings, which enable superior measuring capabilities, but on the other hand require different chemical means to transform them into chemical sensors. From this group, we use gold coated AFM tips the most, because of their relatively simple functionalization options, whilst bi-functional molecules, bearing on one end thiol moieties, which are known to stick to gold and on the other the desired species. Schemes of both mentioned preparation procedures are shown in Figure 10.

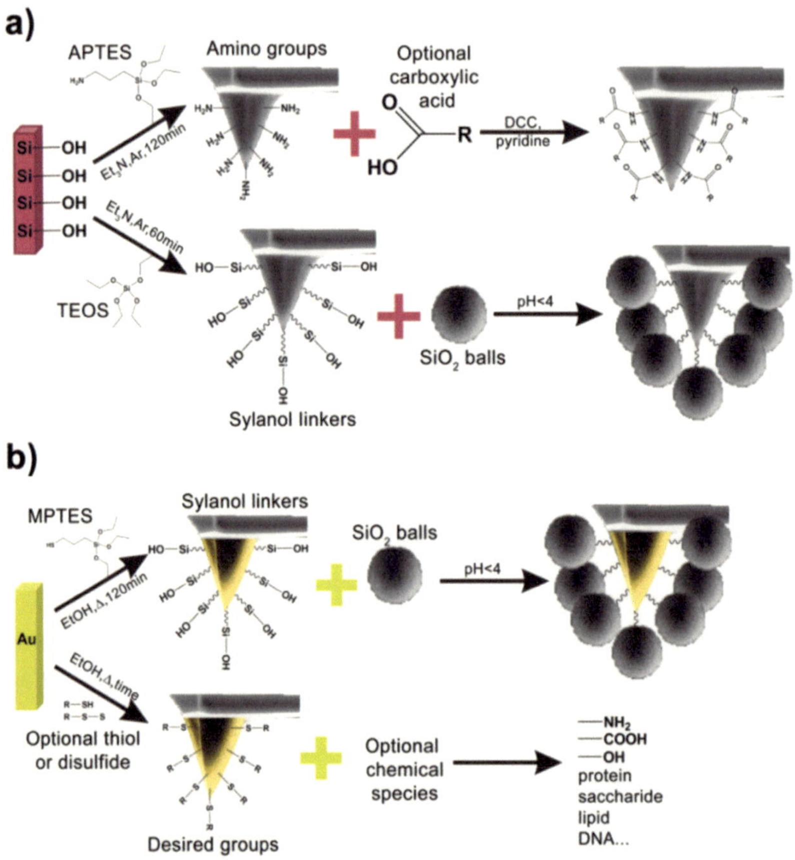

Figure 10. Schematical depiction of the AFM tips functionalization procedures: a) functionalization of silicon-based tips and b) functionalization of gold-coated tips [45].

Such custom made AFM tips serve as ideal chemical sensors for many different applications [23]. As mentioned before, functionalized AFM tips can be divided into two groups, differing by the extent of their specificity towards certain chemical species. In one of our studies, we prepared functionalized AFM tips with several different functional groups [45] and showed how differently they interact with a model surface. By this, we have proven that the functionalization actually resulted in different surface functional groups and how this successful functionalization can be confirmed by using AFM. Figure 11 shows some of the results of our measurements with corresponding SEM micrographs. A clear distinction between tips with different functionalizations can be observed.

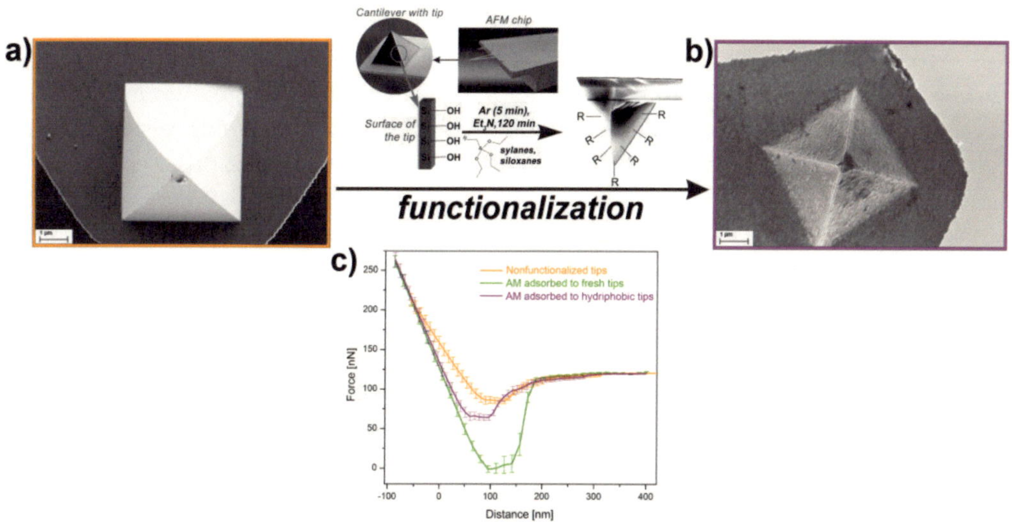

Figure 11. Scheme of the functionalization process with the corresponding results: a) SEM micrograph of nonfunctionalized AFM tips, b) SEM micrograph of an AFM tip after functionalization and c) different forces as measured with the non- and functionalized AFM tips [45].

Our results suggest that by employing some alterations to the known functionalization procedures, we are now able to attach different functional groups to the tip surface, thus providing numerous possibilities for the further attachment of a wide variety of different species. All used procedures resulted in mainly decorating the edge of tips, leaving the surroundings almost as clean as before the functionalization. In this way, there is no decrease in the response of the AFM feedback system and therefore no resolution is lost.

2.2.3. Relationship between the polymer exposure to a specific environment and its function

Miniaturization demands and the characteristics of specialized AFM measurements are the origin of an increasingly more and more important field of cantilever biosensing. This technique enables the determination of material and molecule behavior upon exposure to a desired environment in vitro, and by this contributes to a decrease of overall development costs for modern drug delivery systems with targeted capabilities. The main research fields, which gained the most from this technique over the past years, are pharmaceutical technology (measurements in simulated body fluids and in vitro detection of interactions between different components in complex formulations [3]), supramolecular chemistry (real time follow up of formation of self-assembled monolayers [46]), biochemistry (simulating the binding of drugs to their targets [6]), and microbiology (measurements of interactions between materials and bacteria [47]).

Our main interest in this field was the evaluation of materials performance after different exposure times in simulated physiological environments. As a consequence of the products we develop (mostly materials for use in preparation of advanced wound dressings), we tried to simplify the testing environments to simple physico-chemical parameters, which enable logical correlation with the results of AFM force spectroscopy [48]. Accessible in vitro testing of material response to environments similar to the ones during their use is of high importance for modern product design. Wound dressing development is not different. Several different polymer based materials are used in this field and combinations of them are often found in the most advanced products. Cellulose derivatives are by far the most spread materials for development of all kinds of plasters, bandages, gauzes etc. Because we are also focused on the development of different products made of cellulose derivatives, we tried to extend our understanding of their behavior in different environments, to better predict and more efficiently choose the right derivate for the desired purpose.

In light of the mentioned facts, we designed an experimental setup, which serves as the platform for such testing. To be reproducible, effective and to allow proper evaluation, it had to be simplified as far as possible. It consists of a model surface (atomically flat silicon wafer), two different polymer molecules (carboxy methyl cellulose and amylose) and solutions exhibiting different pHs and ionic strengths. The setup is schematically depicted in Figure 12.

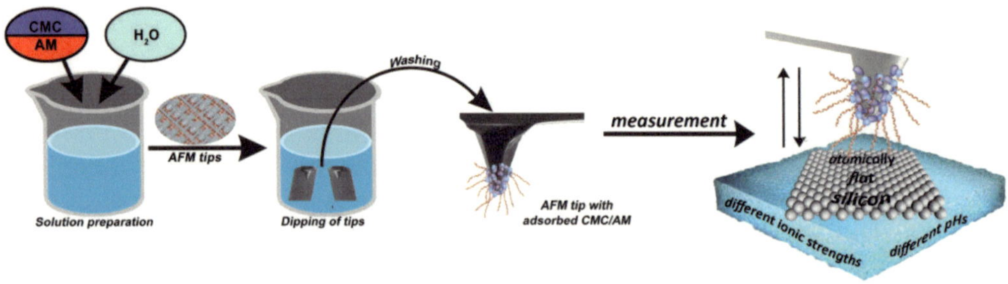

Figure 12. Scheme of the used procedure for evaluation of forces in different environments. This figure is partly reproduced from [48].

Force spectroscopy has proven to be a perfect method for assessing any interactions over a wide range of environmental conditions, especially in liquid media. The latter is especially important because capillary forces, if present (as in measurements in air), are capable of hiding smaller interaction contributions. Our research was focused on finding a reliable method for determination of environmental influences of polymer materials after exposure to a healing wound. During the healing process several physico-chemical parameters of the wound exudates change. While not all can be easily simulated, we tried to reproduce conditions, which are known to have a bigger implication on exposed materials, namely the pH and ionic strength. Both can induce structural changes in the polymeric chains, which in turn causes different behavior and material stability. By simplifying the setup to only two changing-parameters separately, it was possible to show that our proposed technique could serve as a good platform for assessing any changing wound-environment during healing. Some of our results are shown in Figure 13. A more detailed explanation of the measurement results can be found elsewhere [48].

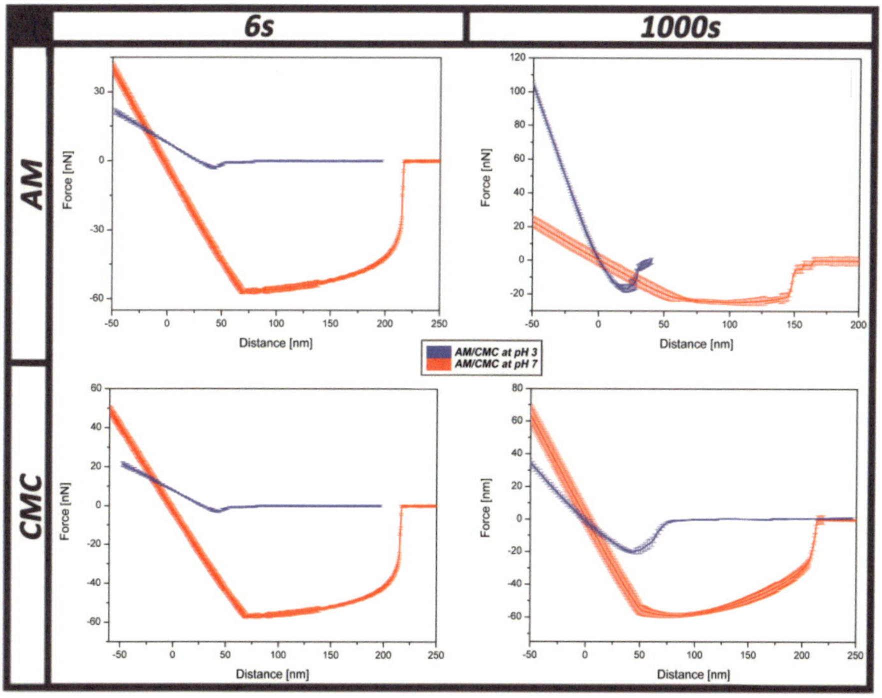

Figure 13. Force spectroscopy results for the measurements in two solutions with different pHs. TOP: retract force curves for amylose at two different measurement durations with two pHs, BOTTOM: retract force curves for carboxymethyl cellulose at two different measurement durations with two pHs. The results are reproduced from article [48].

2.2.4. Force spectroscopy as the source of quantitative information

One of the greatest contributions of AFM to scientific community in the last decade is its ability to probe interaction forces between different species (surfaces, molecules, functional groups) on a quantitative basis [49]. Many researchers know that quantitative interaction mappings between species, interacting in real systems, are the basis for the comprehension of their appearance. AFM force spectroscopy yielding information about single molecules interactions was used for several important discoveries. For example, Allison et al. measured forces between adenine coated AFM tips and thymine coated surfaces, which led to the development of a methodology to study the required forces for unraveling immunoglobulin [50]. Several other research groups used the same type of experiment (attachment of specific molecules to the AFM tip edge to probe the interaction with a desired surface) to gain interaction mappings, which they used as the basis for understanding of processes on the molecular scale [51].

The mechanisms behind appearing interactions between surfaces are of utter importance for many research areas, ranging from the development of polymers for protective films to preparation of implants for medical use [52]. Quantitative assessment of these mechanisms can be used in many ways. For example, it can act as the input data for sophisticated modeling of polymer behavior [53], it can lead to understanding of processes on the molecular level, by which novel drugs can be developed or pathological factors filtered out several stages earlier in the development of a disease [54] or it can be used as the input data for the design of novel drug delivery systems, by which the development of such gets cheaper and less time consuming [55].

In our case, we wanted to understand a process, involved in the working process of Li-ion based battery system. Such systems comprise several components, which are connected into a sort of net via polymers, which act as the binding material [56]. Although such systems do not comprise a lot of different components, are the present ones not easy to include into calculations either due to their complex molecular structure or due to the fact that their morphology is not the same throughout the whole material. Therefore we had to develop a novel methodology of data assessment and analysis, which enables us to get more insight into the ongoing reactions during the preparation of this material [57]. The latter serves as the basis for the prediction of the loss of initial characteristics during prolonged use (the durability). By this we were able to show what binding occurs in the material, and how to correlate such data with the choice of binding material. The developed methodology is shown in Figure 14.

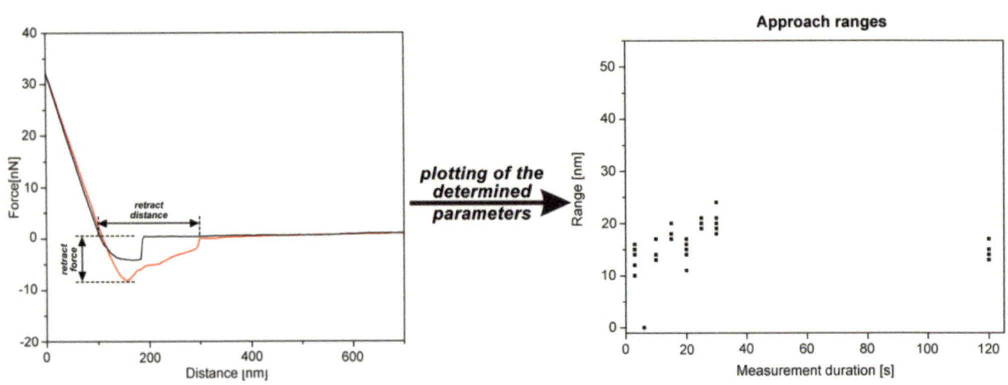

Figure 14. LEFT: typical force curve (black – approach curve, red – retract curve); RIGHT: typical plot of the extracted parameters as a function of measurement duration (approach labels were removed to increase plainness of the scheme). Reproduced with permission of the Royal Society of Chemistry from [57].

Upon introducing measurements with different durations and the final extraction of four parameters form the force curves, we were able to first define both borderline scenarios, namely the case, where a covalent bond occurs and the other, where the bond type is reversible. The next experiment was carried out at conditions, which are known to be present during the material preparation. After the comparison of this set of measured data with the previously taken ones, we found a remarkable similarity for three of the four extracted parameters. Due to the fact that the similarity was highly pronounced and due to the fact that other publications suggest the same, we are certain that the bond type in the examined material between the used binder molecules and the silicon particles is covalent. Some of the results are depicted in Figure 15, while a more detailed version can be found in our article [57].

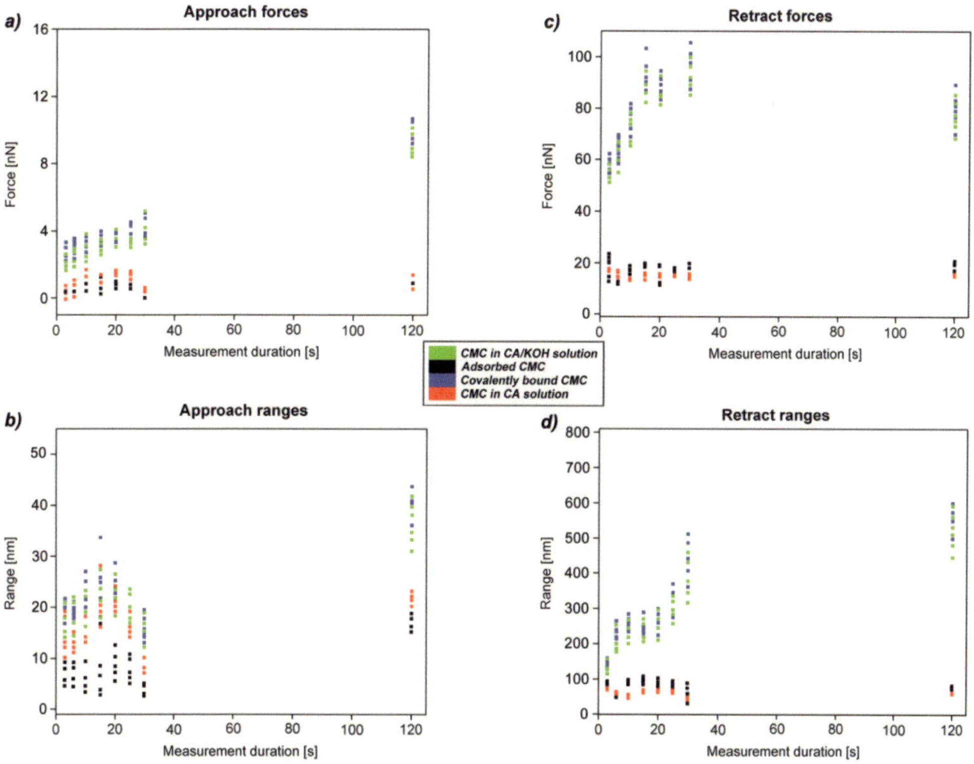

Figure 15. Plots of extracted data (determination procedure shown in Fig. 2.) for the measurements performed in water. The plots respectively show the results for four chosen parameters of each force curve taken for four different functionalization types: a) approach forces, b) approach ranges, c) retract forces and d) retract ranges vs. measurement duration. Reproduced with permission of the Royal Society of Chemistry from [57].

Such an approach is certainly not limited for the present study, but is a very good method for all other samples, where either by theory or experiment, no unambiguous data can be obtained. Additionally it can be used also for more complex molecules, where direct measurements cannot result in quantitative data or bond type confirmation. The latter is especially important in testing of polymeric materials for medical use, where the bond type between material and tissue is of high importance for the actual outcome of the healing process.

3. Further research

Probably the most advanced study with an AFM is the examination of single molecule behavior in its natural environment. AFM was proven as the perfect tool for identification and characterization of single polymer chains [35]. Our future efforts will be in conducting measurements on single polymer chains, compare them with results of other mechanical methods and finally try to correlate both sets of results with the final polymer material characteristics. If successful, we will be able to design and predict several novel materials, with a far greener and cheaper approach, which is the result of a drastically reduced number of needed experiments for desired material preparation. Our goal is to define methods, which enable effective correlation of easy obtainable laboratory data with final products characteristics even in the development stages.

4. Conclusion

The present chapter introduces some basic concepts of AFM measurements on polymers and explains the most used modes for their examination. Our own results are added at sections, where our knowledge represents good ground knowledge for other researchers to examine their own materials. The chapter is divided into sections, which follow the steps, needed for a thorough, and more importantly a correct analysis. Nearing the end of the chapter the complexity increases, which climaxes in the future research section, where our efforts lie at the moment.

Author details

U. Maver*
Centre of Excellence for Polymer Materials and Technologies, Ljubljana, Slovenia
National Institute of Chemistry, Ljubljana, Slovenia

T. Maver, Z. Peršin and K. Stana-Kleinschek
Centre of Excellence for Polymer Materials and Technologies, Ljubljana, Slovenia
University of Maribor, Faculty of Mechanical Engineering,
Laboratory for Characterisation and Processing of Polymers, Maribor, Slovenia

M. Mozetič
Institut "Jožef Stefan", Ljubljana, Slovenia

A. Vesel
Centre of Excellence for Polymer Materials and Technologies, Ljubljana, Slovenia

M. Gaberšček
National Institute of Chemistry, Ljubljana, Slovenia

Acknowledgement

The authors acknowledge the financial support from the Ministry of Education, Science, Culture and Sport of the Republic of Slovenia through the contract No. 3211-10-000057 (Centre of Excellence for Polymer Materials and Technologies).

5. References

[1] Cohen SH, Bray MT, Lightbody ML. Atomic force microscopy/scanning tunneling microscopy. New York: Plenum Press; 1994. x, 453 p. p.

[2] Kocun M, Grandbois M, Cuccia LA. Single molecule atomic force microscopy and force spectroscopy of chitosan. Colloids and Surfaces B: Biointerfaces. 2011;82(2):470-6.

[3] Sitterberg J, Ozcetin A, Ehrhardt C, Bakowsky U. Utilising atomic force microscopy for the characterisation of nanoscale drug delivery systems. Eur J Pharm Biopharm. 2010;74(1):2-13.

[4] Butt H-J, Cappella B, Kappl M. Force measurements with the atomic force microscope: Technique, interpretation and applications. Surface Science Reports. 2005;59(1-6):1-152.

[5] Roduit C. "AFM figures", www.freesbi.ch,. Creative Commons Attribution. 2010.

[6] La R, Arnsdorf MF. Multidimensional atomic force microscopy for drug discovery: A versatile tool for defining targets, designing therapeutics and monitoring their efficacy. Life Sci. 2010;86(15-16):545-62.

[7] Uchihashi T, Higgins MJ, Yasuda S, Jarvis SP, Akita S, Nakayama Y, et al. Quantitative force measurements in liquid using frequency modulation atomic force microscopy. Appl Phys Lett. 2004;85(16):3575-7.

[8] van Noort SJT, Willemsen OH, van der Werf KO, de Grooth BG, Greve J. Mapping electrostatic forces using higher harmonics tapping mode atomic force microscopy in liquid. Langmuir. 1999;15(21):7101-7.

[9] Israelachvili JN. Intermolecular And Surface Forces: Academic Press; 2010.

[10] Cohen SH, Lightbody ML, Foundation for Advances in Medicine and Science. Atomic force microscopy/scanning tunneling microscopy 3. New York: Kluwer Academic/Plenum Publishers; 1999. viii, 210 p. p.

[11] Braga PC, Ricci D. Atomic force microscopy : biomedical methods and applications. Totowa, N.J.: Humana Press; 2004. xiv, 394 p. p.

[12] Alsteens D, Dupres V, Dague E, Verbelen C, André G, Francius G, et al. Imaging Chemical Groups and Molecular Recognition Sites on Live Cells Using AFM. In: Bhushan B, Fuchs H, editors. Applied Scanning Probe Methods XIII: Springer Berlin Heidelberg; 2009. p. 33-48.

[13] Ando T, Uchihashi T, Kodera N, Yamamoto D, Miyagi A, Taniguchi M, et al. High-speed AFM and nano-visualization of biomolecular processes. Pflug Arch Eur J Phy. 2008;456(1):211-25.

[14] Goddard JM, Barish JA. Topographical and Chemical Characterization of Polymer Surfaces Modified by Physical and Chemical Processes. J Appl Polym Sci. 2011;120(5):2863-71.

[15] Parrat D, Sommer F, Solleti JM, Due TM. Imaging Modes in Atomic-Force Microscopy. J Trace Microprobe T. 1995;13(3):343-52.

[16] Basdogan C, Varol A, Gunev I, Orun B. Numerical simulation of nano scanning in intermittent-contact mode AFM under Q control. Nanotechnology. 2008;19(7).

[17] Nnebe I, Schneider JW. Characterization of distance-dependent damping in tapping-mode atomic force microscopy force measurements in liquid. Langmuir. 2004;20(8):3195-201.

[18] Binnig G, Quate CF, Gerber C. Atomic Force Microscope. Physical Review Letters. 1986;56(9):930.

[19] Ebner A, Wildling L, Zhu R, Rankl C, Haselgrübler T, Hinterdorfer P, et al. Functionalization of Probe Tips and Supports for Single-Molecule Recognition Force Microscopy. In: Samorì P, editor. STM and AFM Studies on (Bio)molecular Systems: Unravelling the Nanoworld: Springer Berlin / Heidelberg; 2008. p. 29-76.
[20] Kienberger F, Pastushenko VP, Kada G, Puntheeranurak T, Chtcheglova L, Riethmueller C, et al., editors. Improving the contrast of topographical AFM images by a simple averaging filter. 2006.
[21] Selvin PR, Ha T. Single-molecule techniques : a laboratory manual. Cold Spring Harbor, N.Y.: Cold Spring Harbor Laboratory Press; 2008. vii, 507 p. p.
[22] Frisbie CD, Rozsnyai LF, Noy A, Wrighton MS, Lieber CM. Functional-Group Imaging By Chemical Force Microscopy. Science. 1994;265(5181):2071-4.
[23] Kienberger F, Ebner A, Gruber HJ, Hinterdorfer P. Molecular recognition imaging and force spectroscopy of single biomolecules. Accounts Chem Res. 2006;39(1):29-36.
[24] Ito T, Ibrahim S, Grabowska I. Chemical-force microscopy for materials characterization. TrAC Trends in Analytical Chemistry. 2010;29(3):225-33.
[25] Hinterdorfer P, Gruber HJ, Kienberger F, Kada G, Riener C, Borken C, et al. Surface attachment of ligands and receptors for molecular recognition force microscopy. Colloid Surface B. 2002;23(2-3):115-23.
[26] Kaneko M, Nakayama M, Wakihara M. Lithium-ion conduction in elastomeric binder in Li-ion batteries. Journal of Solid State Electrochemistry. 2007;11(8):1071-6.
[27] Satarkar NS, Biswal D, Hilt JZ. Hydrogel nanocomposites: a review of applications as remote controlled biomaterials. Soft Matter. 2010;6(11):2364-71.
[28] Huang S, Fu X. Naturally derived materials-based cell and drug delivery systems in skin regeneration. J Control Release. 2010;142(2):149-59.
[29] Liu ZH, Jiao YP, Wang YF, Zhou CR, Zhang ZY. Polysaccharides-based nanoparticles as drug delivery systems. Advanced Drug Delivery Reviews. 2008;60(15):1650-62.
[30] Ortiz C, Hadziioannou G. Entropic elasticity of single polymer chains of poly(methacrylic acid) measured by atomic force microscopy. Macromolecules. 1999;32(3):780-7.
[31] Shanmugham S, Jeong JW, Alkhateeb A, Aston DE. Polymer nanowire elastic moduli measured with digital pulsed force mode AFM. Langmuir. 2005;21(22):10214-8.
[32] Bemis JE, Akhremitchev BB, Walker GC. Single polymer chain elongation by atomic force microscopy. Langmuir. 1999;15(8):2799-805.
[33] Shulha H, Zhai XW, Tsukruk VV. Molecular stiffness of individual hyperbranched macromolecules at solid surfaces. Macromolecules. 2003;36(8):2825-31.
[34] Kuhner F, Erdmann M, Sonnenberg L, Serr A, Morfill J, Gaub HE. Friction of single polymers at surfaces. Langmuir. 2006;22(26):11180-6.
[35] Giannotti MI, Rinaudo M, Vancso GJ. Force spectroscopy of hyaluronan by atomic force microscopy: From hydrogen-bonded networks toward single-chain behavior. Biomacromolecules. 2007;8(9):2648-52.
[36] Bliznyuk VN, Assender HE, Briggs GAD. Surface Glass Transition Temperature of Amorphous Polymers. A New Insight with SFM. Macromolecules. 2002;35(17):6613-22.

[37] Lu Z, Nowak W, Lee G, Marszalek PE, Yang W. Elastic Properties of Single Amylose Chains in Water: A Quantum Mechanical and AFM Study. J Am Chem Soc. 2004;126(29):9033-41.
[38] Zhang Q, Marszalek PE. Identification of Sugar Isomers by Single-Molecule Force Spectroscopy. J Am Chem Soc. 2006;128(17):5596-7.
[39] Noy A, Frisbie CD, Rozsnyai LF, Wrighton MS, Lieber CM. Chemical Force Microscopy - Exploiting Chemically-Modified Tips To Quantify Adhesion, Friction, And Functional-Group Distributions In Molecular Assemblies. J Am Chem Soc. 1995;117(30):7943-51.
[40] Tumer YTA, Roberts CJ, Davies MC. Scanning probe microscopy in the field of drug delivery. Advanced Drug Delivery Reviews. 2007;59(14):1453-73.
[41] El Kirat K, Burton I, Dupres V, Dufrene YF. Sample preparation procedures for biological atomic force microscopy. J Microsc-Oxford. 2005;218:199-207.
[42] Hegner M, Wagner P, Semenza G. Ultralarge Atomically Flat Template-Stripped Au Surfaces For Scanning Probe Microscopy. Surf Sci. 1993;291(1-2):39-46.
[43] Stroh C, Wang H, Bash R, Ashcroft B, Nelson J, Gruber H, et al. Single-molecule recognition imaging-microscopy. P Natl Acad Sci USA. 2004;101(34):12503-7.
[44] Maver U, Planinšek O, Jamnik J, Hassanien AI, Gaberšček M. Preparation of atomically flat gold substrates for AFM measurements. Acta Chimica Slovenica. 2012;59(1):212-9.
[45] Maver T, Stana-Kleinschek K, Peršin Z, Maver U. Functionalization of AFM tips for use in force spectroscopy between polymers and model surfaces. Mater Tehnol. 2011;45(3):205-11.
[46] Kluge D, Abraham F, Schmidt S, Schmidt H-W, Fery A. Nanomechanical Properties of Supramolecular Self-Assembled Whiskers Determined by AFM Force Mapping. Langmuir. 2010;26(5):3020-3.
[47] Ivanova EP. Nanosacale structure and properties of microbial cell surfaces. Hauppauge, N.Y.: Nova Science Publishers; 2007. xiv, 269 p. p.
[48] Maver U, Maver T, Znidarsic A, Persin Z, Gaberscek M, Stana-Kleinschek K. Use of Afm Force Spectroscopy for Assessment of Polymer Response to Conditions Similar to the Wound, during Healing. Mater Tehnol. 2011;45(3):259-63.
[49] Roberson ED. Alzheimer's disease and frontotemporal dementia : methods and protocols. New York: Humana Press; 2011. x, 277 p. p.
[50] Allison DP, Hinterdorfer P, Han WH. Biomolecular force measurements and the atomic force microscope. Current Opinion in Biotechnology. 2002;13(1):47-51.
[51] Ebner A, Wildling L, Kamruzzahan ASM, Rankl C, Wruss J, Hahn CD, et al. A new, simple method for linking of antibodies to atomic force microscopy tips. Bioconjugate Chem. 2007;18(4):1176-84.
[52] Lehnert M, Gorbahn M, Rosin C, Klein M, Koper I, Al-Nawas B, et al. Adsorption and Conformation Behavior of Biotinylated Fibronectin on Streptavidin-Modified TiOX Surfaces Studied by SPR and AFM. Langmuir. 2011;27(12):7743-51.
[53] Junker JP, Rief M. Single-molecule force spectroscopy distinguishes target binding modes of calmodulin. Proceedings of the National Academy of Sciences. 2009.

[54] Stolz M, Gottardi R, Raiteri R, Miot S, Martin I, Imer R, et al. Early detection of aging cartilage and osteoarthritis in mice and patient samples using atomic force microscopy. Nature Nanotechnology. 2009;4(3):186-92.
[55] Lyon LA, South AB. Direct Observation of Microgel Erosion via in-Liquid Atomic Force Microscopy. Chem Mater. 2010;22(10):3300-6.
[56] Magasinski A, Zdyrko B, Kovalenko I, Hertzberg B, Burtovyy R, Huebner CF, et al. Toward Efficient Binders for Li-Ion Battery Si-Based Anodes: Polyacrylic Acid. Acs Appl Mater Inter. 2010;2(11):3004-10.
[57] Maver U, Žnidaršič A, Gaberšček M. An attempt to use atomic force microscopy for determination of bond type in lithium battery electrodes. J Mater Chem. 2011;21(12):4071-5.

Chapter 5

Absorption Kinetics of Phenol into Different Size Nanopores Present in Syndiotactic Polystyrene and Poly (p-methylstyrene)

Kenichi Furukawa and Takahiko Nakaoki

Additional information is available at the end of the chapter

http://dx.doi.org/10.5772/46142

1. Introduction

The stereoregularity of polyolefins is one of their important characteristics that regulate their physical properties. Since Natta et al. successfully synthesized isotactic polypropylene having high stereoregularity,[1] many investigations have been reported, not only from the viewpoint of stereoregular polymerization, but also with regard to the molecular structure based on the stereoregularity. Isotactic polystyrene (iPS) is one of the earliest stereoregular polymers synthesized using the Ziegler-Natta catalyst in 1955.[2] The stereoregular repeating unit allows it to form a crystal with 3/1 helical structure. Syndiotactic polystyrene (sPS), the counterpart of iPS, was first synthesized by Ishihara et al. in 1986.[3]

One of the most attractive features of this material is that there exists polymorphic structures. It has been reported that there are five crystalline modifications. The α [1-8] and β [9, 10] forms consist of a planar zigzag chain with period of 5.1 Å, whereas the γ [11-13], δ [14-17], and ε [18-20] forms adopt a helical structure with a trans-trans-gauche-gauche (ttgg) conformation, having a period of 7.7 Å. The crystal with an all-trans conformation is a thermally stabilized form induced above about 180 °C and its high melting temperature around 270 °C is expected to be useful as an engineering plastic. The ttgg conformation is induced by the presence of organic solvents such as aromatic and chlorinated compounds and crystals of both the δ- and ε-forms are characterized by the formation of complex structures containing solvent molecules. The difference between these crystalline forms is that δ-sPS is formed by casting from solution or exposing it to the vapor of organic compound, whereas ε-sPS is formed by treating γ-sPS in chloroform. The molecular dipoles of guest molecules in the crystal units are respectively perpendicular and parallel to the chain axis of δ- and ε-sPS. It is worth noting that a nanoporous structure in these crystals can be prepared by removing the incorporated solvent in acetone or supercritical carbon dioxide ($scCO_2$). [21-25]

Similar molecular structures have been reported for syndiotactic poly(*p*-methylstyrene) (sPPMS); the all-trans conformation (forms III, IV, and V) induced by annealing, and the ttgg conformation (forms I and II) induced by organic solvents. [26-28] In addition, three clathrate forms of α-, β-, and γ-classes which contain solvent in the crystal unit have been reported. The α and β class clathrates have a ttgg conformation with a period of 7.8 ± 0.1 Å. [29-32]The $t_6g_2t_2g_2$ conformation γ class clathrate has a period of 11.7 ± 0.1 Å. [33] A porous crystalline structure can be constructed in a crystal with the ttgg conformation by removing the solvent. However, this solvent-depleted crystal is unstable, being readily collapsed on removing the solvent.[34]

One of the most important features of sPS and sPPMS is their abilities to form nanoporous structures. The nanoporous structure of δ_e-sPS, in particular, is a very attractive morphology for the absorption of specific compounds. Many investigations have been carried out on the sorption and desorption kinetics of liquid organic compounds, such as 1,2-dichloroethane, and also gases such as CO_2. [35, 36] The absorption rate of solvent molecules was enhanced by a porous aerogel with **a** mesh structure, which was prepared by treating the sPS gel in $scCO_2$.[37-39] The relationship between the size of the solvent molecule and the cavity volume of δ_e-sPS was reported by Tsujita et al. [40-43] It was shown that the cavity volume of δ-sPS is reduced after extracting the solvent molecules.

Our research group reported that the pore dimension can be regulated by the size of the solvent molecule, being larger for solvents with a larger molar volume. Furthermore, the absorption of linear and branched alcohols with four and five carbons depended on the molecular shape. [44] The alcohols with a bulk substituent, such as a methyl group, were impossible to incorporate in the pore because of their bulk. In addition, the absorption process of ethanol and butanols for sPS and sPPMS was investigated from the viewpoint of the pore size in the crystal. [34] The diffusion of ethanol into the large cavity pores of sPPMS was faster than that of sPS. A bulky alcohol, such as *tert*-butanol, was not absorbed by sPS but did incorporate into sPPMS.

Our previous report concerned aliphatic alcohols.[44] In this report, the absorption of phenol in nanoporous crystals of sPS and sPPMS is investigated.

2. Experimental section

The sPS was supplied by Idemitsu Petrochemical Co. Ltd. The weight average molecular weight (M_w) and polydispersity (M_w/M_n) were 2.4×10^5 and 2.3, respectively. The sPPMS was prepared using tetramethylcyclopentadienyl titanium trichloride (Cp^*TiCl_3) and methylaluminoxane (MAO) as a polymerization catalyst. The weight average molecular weight (M_w) and polydispersity (M_w/M_n) were 5.8×10^5 and 1.6, respectively. Films of the two polymers were prepared as follows. A melt-quenched film was obtained by quenching in ice-water after melting at 280 °C. A cast-crystallized film was obtained by treating the film cast from *m*-xylene solution with $scCO_2$. As described in our previous paper, [34] the cast-crystallized sPS and sPPMS from m-xylene adopt the δ_e-form and form I, respectively. The treatment in $scCO_2$ was performed using a supercritical fluid apparatus (Jasco SCF-Get) under the mild conditions of 40 °C and 10 MPa. Sorption of phenol was carried out by soaking the sPS or sPPMS film a 5% aqueous solution of phenol.

Infrared spectra were measured on a JASCO FT-IR660 Plus over the standard wavenumber range of 400 - 4000 cm^{-1}. The film thickness was calibrated by the Lambert-Beer equation as follows.

$$A = \varepsilon \cdot c \cdot l \tag{1}$$

Here A, ε, c, and *l* denote the absorbance, absorption coefficient, molar concentration, and the path length, respectively. The following relationship between the absorbance and the film thickness was reported for sPS using the CH out-of-plane mode of the phenyl ring at 1028cm^{-1} as an internal standard.[44]

$$A_{1028} = 1.004 \times 10^{-2} \cdot l \tag{2}$$

In order to calibrate the film thickness for sPPMS, the 1022 cm^{-1} band, which does not depend on the crystalline form, was used. Figure 1 shows the absorbance of melt-quenched sPPMS at 1022 cm^{-1} against the film thickness, which was determined by a micrometer. The slope provided the following relationship.

$$A_{1032} = 1.010 \times 10^{-2} \cdot l \tag{3}$$

From these relationships, the film thickness of sPS and sPPMS were estimated, respectively.

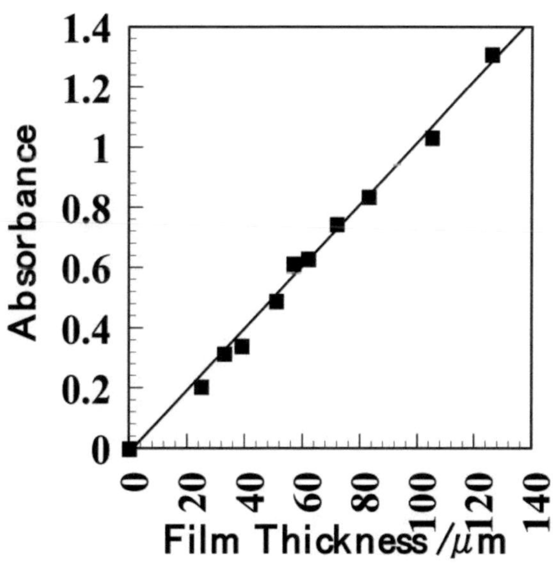

Figure 1. Dependence of the absorbance of the OH stretching mode for phenol on path length. Nakaoki et al.

The degree of crystallinity was estimated by a flotation method. Water and *tert*-butanol were used for sPS as a mixed solvent. In the case of sPPMS, solvents with large molar volume such as 3-ethyl-3-pentanol and dimethysulfoxide (DMSO) were used as a mixed solvent. We confirmed by infrared spectroscopy that these solvents were not absorbed in the polymer films. The density was determined by a Lipkin-type pyknometer at 25 °C. The measurements were carried out three times and the average value was taken to estimate the crystallinity. The following equation was used to determine the crystallinity (χ_C).

$$\chi_C = \frac{\rho - \rho_C}{\rho_a - \rho_C} \tag{4}$$

Here ρ_C and ρ_a denote the densities of crystalline and amorphous material, respectively and ρ is the observed density.

3. Results and discussion

3.1. Absorption of phenol into the pores in crystals of sPS and sPPMS

The sPS and sPPMS films were soaked in phenol solution for one week, and the infrared spectrum was remeasured. Figure 2 shows the infrared spectrum of the OH stretching mode of phenol. The absorption peak at 3540 cm^{-1} assignable to the H bond-free OH stretching mode was observed for the cast-crystallized sPS and sPPMS. This indicates that there is no interaction between phenol molecules, because each phenol molecule is isolated in a pore of the sPS and sPPMS crystal. Contrary to the cast-crystallized films, two peaks were observed for the melt-quenched films. In addition to the peak at 3540 cm^{-1}, a broad absorption peak around 3300 cm^{-1}, which is assignable to the OH stretching mode associated with a hydrogen bond was observed. This indicates that the phenol molecules exist in a non-crystalline region. The fact that the 3540 cm^{-1} band was also observed indicates that the crystallization proceeded to the δ-form, which then incorporated phenol into the crystal.

Figure 3 shows the infrared spectra of the phenyl out-of-plane mode of sPS and sPPMS, before and after soaking in phenol solution. The peaks characteristic of the ttgg conformation at 572 cm^{-1} and 566 cm^{-1} for sPS and sPPMS are well-known to be very sensitive to the conformation. These peaks increased in intensity for both cast-crystallized and melt-quenched films after soaking in phenol solution. This is direct evidence that the crystal with a ttgg conformation was induced by the presence of phenol.

The crystallinity of cast-crystallized δ_e-sPS and sPPMS was estimated by the flotation method. The calculated values are listed in Table 1. In general, the crystal takes a tightly packed structure so that the density of the crystal tends to be higher. However the densities of the cast-crystallized sPS and sPPMS films were lower than those of the amorphous polymer, because of the porous structure obtained after excluding solvent molecules. The crystallinity of the cast-crystallized sPPMS was as low as 4.7 %, which is in contrast to that

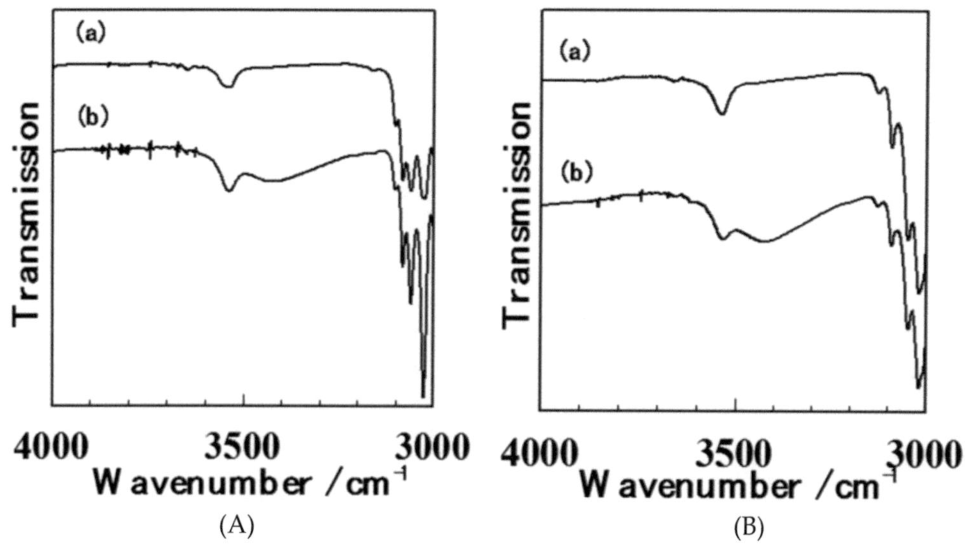

Figure 2. Infrared spectra for the cast-crystallized (a) and melt-quenched (b) sPS (A) and sPPMS (B) after soaking in a 5 wt% phenol aqueous solution for 24 h. The peaks at 3540 cm^{-1} and 3300 cm^{-1} are assignable to the OH stretching mode of phenol without, and with hydrogen bonding, respectively. Nakaoki et al.

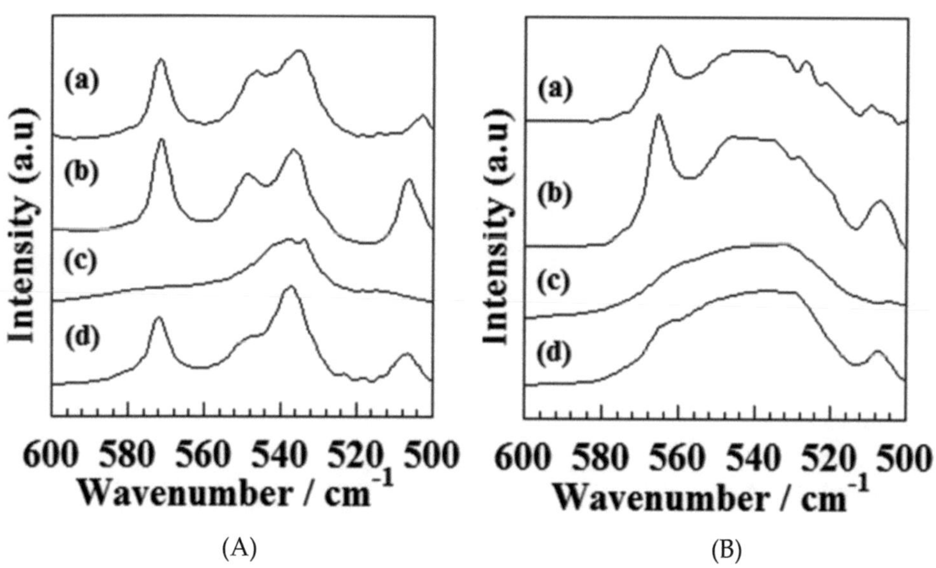

Figure 3. Infrared spectra of sPS (A) and sPPMS (B). The (a) and (b) spectra correspond to the cast-crystallized films before, and after, soaking in a 5 wt% phenol aqueous solution, respectively. The (c) and (d) spectra correspond to the melt-quenched films, before and after, soaking in a 5 wt% phenol aqueous solution, respectively. Nakaoki et al.

	ϱ (observed) g/cm³	ϱC g/cm³	ϱA g/cm³
sPS	1.003	0.977 a)	1.055 b)
sPPMS	1.010	0.815 c)	1.02 d)

a) Ref. 25, b) ref. 45, c) ref. 32 and, d) ref. 31.

Table 1. Densities of cast-crystallized sPS and sPPMS determined by the flotation method.

of sPS at 38 %. As described in our previous paper,[34] a complex structure including a solvent molecule is a stable form for sPPMS, and the crystal would collapse on excluding the solvent. Although the crystallinity of the cast-crystallized sPPMS was very low, a porous structure remained, even after removing the solvent because the hydrogen-bond free OH stretching mode was observed for the cast-crystallized sPPMS, as shown in Fig. 2(B). Since the phenol molecule is incorporated in the crystal unit, the flotation method cannot be used to determine the crystallinity because the phenol comes out from the film. In order to estimate the crystallinity after soaking in phenol solution, the infrared spectrum was measured. As described above, the 572 cm^{-1} and 566 cm^{-1} bands for sPS and sPPMS are very sensitive indicators of the formation of the ttgg conformation in during crystallization. Therefore the relationship of the crystallinity with the absorbance at 572 cm^{-1} for sPS (A_{572}) and 566 cm^{-1} for sPPMS (A_{566}) was estimated as follows.

$$\chi_C(sPS) = 1.66 \times 10^{-2} \cdot A_{572} \tag{5}$$

$$\chi_C(sPPMS) = 8.14 \times 10^{-2} \cdot A_{566} \tag{6}$$

The increase in crystallinity after soaking in phenol solution was estimated using these equations and is shown in Table 2. The crystallinity changed from 38 to 52 % for sPS and from 4.7 % to 6.5 % for sPPMS after soaking in phenol solution. In our previous paper concerning the absorption of aliphatic alcohols in sPS and sPPMS,[34] there was no trace of further crystallization. Phenol is basically a poor solvent for these polymers, but it was established that the crystallization can be promoted by soaking in the phenol solution.

	Crystallinity / %	
	before	after
sPS/*m*-xylene cast	38	52
sPPMS/*m*-xylene cast	4.7	6.5
Melt-quenched sPS	0	7.2
Melt-quenched sPPMS	0	1.3

Table 2. Crystallinity before and after soaking in a 5 % phenol aqueous solution.

3.2. Diffusion coefficient of phenol in the crystal pores

Figure 4 plots the absorbance of the OH stretching mode of phenol at 3540 cm^{-1} depending on the time soaked in the phenol solution. The time dependence of the band at 3540 cm^{-1}

corresponds to the diffusion of phenol in the pores of crystal. The absorption was slow and took as long as a few days. The diffusion coefficient (D) can be estimated from the following Fickian equation.

$$\frac{c}{c_0} = \left(\frac{4}{d}\right) \cdot \sqrt{\frac{Dt}{\pi}} \tag{7}$$

Here d is the film thickness, and c and c_0 denote the concentration at time t and the equilibrium state, respectively. The concentration can be replaced by the absorbance using the Lambert-Beer equation (1)

$$\frac{I}{I_0} = \left(\frac{4}{d}\right) \cdot \sqrt{\frac{Dt}{\pi}} \tag{8}$$

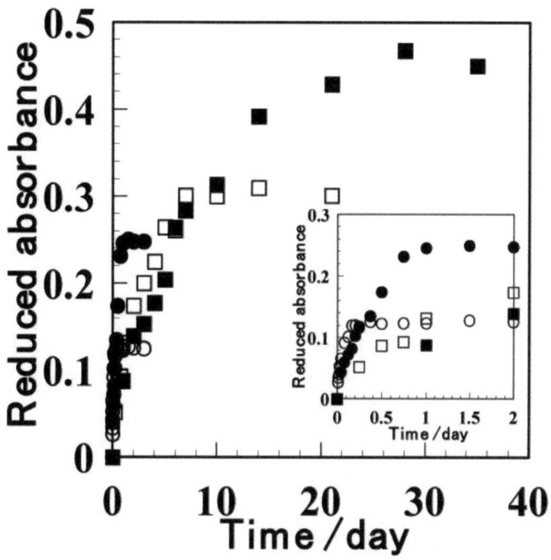

Figure 4. Absorbance of the OH stretching mode of phenol free from hydrogen bonding depending on the time soaked in a 5 wt% phenol aqueous solution. The ▪ and • notations indicate the cast crystallized and melt-quenched sPS, while □ and ○ represent the cast-crystallized and melt-quenched sPS, respectively. Nakaoki et al.

Here, I_0 and I denote the absorbance at the equilibrium state and the observed absorbance, respectively. Figure 5 plots the normalized intensity I/I_0 against $t^{1/2}$. The diffusion coefficient was estimated from the initial slope and is listed in Table 3. The diffusion coefficient of the cast-crystallized sPPMS was about 4 times larger than that of the cast-crystallized sPS. This can be explained by the large pores present in the sPPMS crystal because of the existence of the methyl group attached to the phenyl group. As for the melt-quenched films, the diffusion was much faster than that of the cast-crystallized films. This is because there is no hindrance, such as transformation of a crystal, to the movement of a phenol molecule. In addition, the non-crystalline chain has a large free volume so that the solvent molecules tended cause a transformation in the film with ease. When the diffusion of phenol was compared to that of ethanol [33], the diffusion coefficient of phenol is roughly two orders smaller than that of ethanol. This can be interpreted as due to the different molecular size between phenol and ethanol.

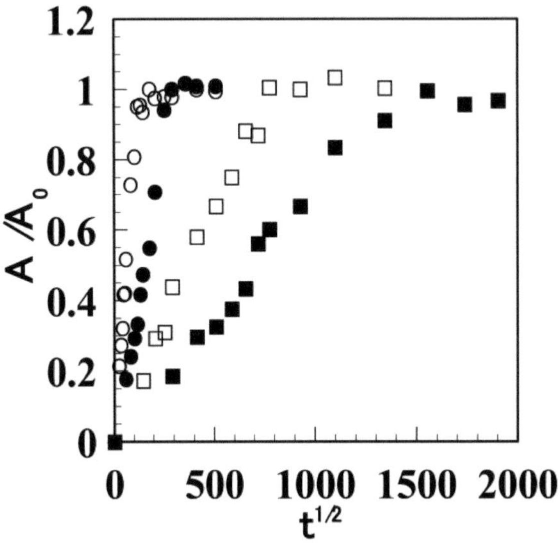

Figure 5. Normalized absorbance of the OH stretching mode of phenol as a function of $t^{1/2}$. The ■ and ● notations indicate the cast-crystallized and melt-quenched sPS and □ and ○ represent the cast-crystallized and melt-quenched sPS, respectively. Nakaoki et al.

	Diffusion coefficient / $\times 10^{-12}$ cm^2s^{-1}	Absorbance at the equilibrium state	Concentration / $\times$ 10^{-4} mol cm^{-3}	Number of phenol in the pore
sPS/*m*-xylene cast	2.3	0.47	4.37	1.2
sPPMS/*m*-xylene cast	8.2	0.30	2.80	2.0
Melt-quenched sPS	55	0.25	2.12	1.2
Melt-quenched sPPMS	320	0.13	1.23	4.4

Table 3. Diffusion coefficient of phenol in the pore of sPS and sPPMS, and IR absorbance and concentration of phenol at the equilibrium state. The number of phenol molecules in the pores was calculated from the concentration of phenol at the equilibrium state.

It is clear from Fig. 4 that the absorbance became constant after a few tens of days. The absorbance of the 3540 cm^{-1} band at the equilibrium state is listed in Table 3. The equilibrium state for the cast-crystallized sPS provided the highest absorbance of these films. This indicates that the incorporation of phenol molecules in the crystal was the largest, corresponding to the high crystallinity of the cast-crystallized δ_e-sPS. The number of phenol molecules in the pores of crystals can be estimated by the absorbance at the equilibrium state. According to Lambert-Beer's law, the absorbance is in proportional to the concentration. In order to estimate the concentration of phenol in the pores of sPS and sPPMS, the absorption coefficient of the hydrogen bond-free OH stretching mode is required. However, normal phenol is hydrogen bonded. So, in order to obtain the required coefficient, each phenol molecule must be isolated from the others. The same problem occurred in the work described in our previous paper which reported the incorporation of ethanol molecules into the pores of cast-crystallized sPS and sPPMS.[34] In that case, the absorption coefficient of the hydrogen bond-free OH stretching mode of ethanol was estimated by diluting the ethanol in toluene. Following this method, toluene was used here as a diluent in order to cancel the interaction between phenol molecules. Figure 6 shows infrared spectra of 5% phenol in aqueous solution, the cast-crystallized sPS film after soaking in phenol solution and a 1 % phenol in toluene solution. The OH stretching mode in the aqueous solution was a broad peak due to hydrogen bonding around 3300 cm^{-1}. However the OH stretching mode diluted by toluene provided a sharp peak at 3540 cm^{-1}, which is identical with that in the pore of cast-crystallized δ_e-sPS. This shows that the phenol is isolated from others by toluene molecules. Therefore the absorption coefficient of the hydrogen bond-free OH stretching mode of phenol was estimated by that of the phenol/toluene solution.

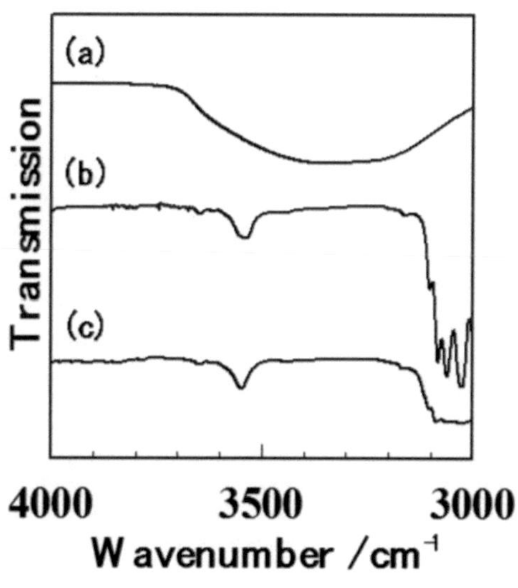

Figure 6. Infrared spectra of the OH stretching mode of phenol: (a) a 5 wt% phenol aqueous solution, (b) cast-crystallized sPS after soaking in 5 wt% phenol aqueous solution, and (c) a 1 v/v% phenol in toluene solution. Nakaoki et al.

Figure 7 plots the absorbance of the OH stretching mode of phenol in the 1 % phenol in toluene solution as a function of path length. The absorption coefficient was estimated to be 2.44×10^{-3}. This value was applied to estimate the molar concentration of phenol in the pore of cast-crystallized δ_e-sPS and sPPMS. The molar concentration of phenol at equilibrium state is listed in Table 3. The concentration of phenol in the sPS/m-xylene cast film was the largest of the four films. Since the concentration was estimated from the 3540 cm^{-1} hydrogen bonding-free band, these solvent molecules were absorbed in the pores of crystal. If the crystallinity is known, the number of phenol molecules in the pores of the crystals can be calculated. So the next section will deal the quantification of phenol molecules in the pores.

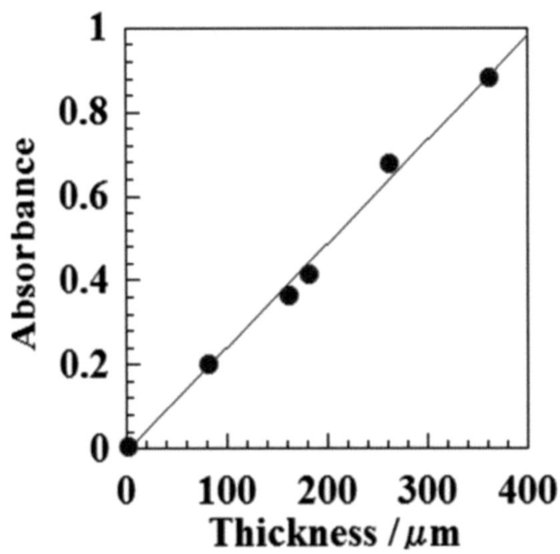

Figure 7. Absorbance of the OH stretching mode of 1 v/v% phenol in toluene as a function of path length. The absorption coefficient was estimated to be 2.44×10^{-3} cm^2/mol. Nakaoki et al.

3.3. Number of phenol molecules in the pore of sPS

The number of phenol molecules in the pores (n_{phe}) was estimated from the concentration of phenol at the equilibrium state and the crystallinity. The sPS and sPPMS have one pore per four monomer units. As shown in our previous report,[34] the number of phenol molecules trapped in one pore can be estimated by the following equation.

$$n_{phe} = \frac{400c_{phe}}{c_{poly} \times \chi_c} \tag{9}$$

Here c_{poly} and c_{phe} denote the molar concentrations of polymer and phenol in the pores, respectively. These values were estimated using Lambert-Beer's equation by converting from the IR absorbance at the equilibrium state. Then the number of phenol molecules incorporated in the pores of sPS and sPPMS was calculated by eq. (9). The crystallinity of cast-crystallized sPS was 0.52 after soaking in phenol solution. When this value is used for the calculation, the number of phenol molecules in the pores was calculated to be 0.34.

Daniel et al. reported that the stoichiometric ratio of the styrene monomeric unit and dichloroethane (DCE) in δ-sPS was 3.6 ± 0.3,[46] which corresponds to 1.1 molecules in one pore of the crystal unit. The number of solvent molecules can be predicted by the relationship between the volume of the solvent molecule and the pore size. Milano et al. reported that one DCE molecule of 0.125 nm^3 is incorporated in the cavity volume of 0.151 nm^3 in the sPS/DCE system.[47] We reported that 1.9 ethanol molecules are incorporated in the pore of cast-crystallized δ_e-sPS.[34] Since the volume of ethanol is 0.061 nm^3, 1.9 ethanol molecules can be incorporated in the pore of δ_e-sPS with 0.115 nm^3.[46] As the volume of a phenol molecule is 0.10 nm^3, the pore size of cast crystallized δ_e-sPS is sufficient for one phenol molecule. Therefore the estimate of the incorporation of 0.34 molecules in one pore must be too small. This might be interpreted as due to the difference between δ_e- and δ-sPS. Namely, the δ_e-sPS is prepared by evaporating the solvent molecules from δ-sPS, resulting in the pore size becoming smaller after removing the solvent. The pore size for δ_e-sPS of 0.115 nm^3 is very close to the volume of a phenol molecule. Therefore the pore size of cast-crystallized δ_e-sPS would be too small for a phenol molecule to be incorporated. In fact the crystallinity increased from 38 % to 52 % by soaking in phenol solution. The increase in crystallinity is due to the crystallization to δ-sPS, which is characterized by a complex structure involving sPS and phenol. Therefore, the phenol molecules would be absorbed only in newly crystallized δ-sPS, not in the cast-crystallized δ_e-sPS. When the number of phenol molecules is re-calculated using this hypothesis, the result was 1.2 phenol molecules in one pore. Judging from the pore size of δ-sPS, this is a reasonable value.

In the case of melt-quenched sPS film, only δ-sPS is induced during the measurement. The degree of crystallinity increased from 0 to 7.2 %. Since all phenol molecules are incorporated in newly crystallized δ-sPS, the number of phenol molecules in a pore of δ-sPS was calculated to be 1.2, which is identical with the value estimated for the cast-crystallized sPS. Therefore we can conclude that it was impossible for the phenol molecule to be incorporated in the pores of cast-crystallized δ_e-sPS, and 1.2 phenol molecules were incorporated in the pore of newly crystallized δ-sPS during the measurement.

3.4. Number of phenol molecules in the pore of sPPMS

In contrast to sPS, sPPMS provided a different result. From the calculation using the crystallinity of 6.5 which corresponds to the value after soaking in phenol solution, the cast-crystallized sPPMS incorporated 2.0 phenol molecules in one pore of the crystal, as shown in Table 3. This is almost twice the number of phenol molecules absorbed in sPS. This is because the pore size of sPPMS is larger than that of sPS because of the methyl group attached to phenyl ring in sPPMS. According to Tarallo et al.,[32] the clathrate forms of α- and β-classes contain two and one guest molecules in the pore, respectively. They reported that nanoporous sPPMS cast-crystallized from o-xylene forms a α-class clathrate. So, the m-xylene used in this study should induce the formation of α-class crystals. We also reported that 3.8 ethanol molecules were absorbed in the pores of sPPMS.[34] The volume of the phenol molecule is almost twice as large as that of the ethanol molecule. Since the calculation used the crystallinity obtained after soaking in phenol solution, the phenol molecules were incorporated not only in the pores of cast-crystallized sPPMS but also in those of the newly crystallized material. In the case of melt-quenched sPPMS, 4.4 phenol molecules were incorporated in the pore. This value is too high for the pore in sPPMS. In this melt-quenched material, the crystallinity was as low as 1.3 %. The infrared spectrum in Fig. 2(B) shows that the phenol molecule absorbed also in the non-crystalline region. Those phenol molecules would be hydrogen-bonded, but some of them might be isolated from the others. The contribution of these molecules to the 3540 cm^{-1} band will not be negligible. Therefore the number of phenol molecules in the pore might be overestimated; further study on this topic is required.

3.5. Absorption in a mixed solution of phenol and ethanol

In this section, a mixed solution of phenol and ethanol was used to investigate which solvent is preferentially incorporated in the porous crystals of sPS and sPPMS. Figure 8(A) shows the infrared spectra of (a) phenol/ethanol=5/95 mixed solution, (b) the cast-crystallized sPS after soaking in a phenol/ethanol=5/95 mixed solution and (c) the melt-quenched sPS after soaking in the same mixed solution. The OH stretching mode for the mixed solution of phenol and ethanol exhibited a broad peak. This is because there are interactions between the solvents due to hydrogen bonding. The cast-crystallized sPS showed a peak at 3592 cm^{-1}. This wavenumber is not identical with that of phenol at 3540 cm^{-1}. Rather this peak can be assumed to be from ethanol by referencing to our previous paper.[34] This indicates that only ethanol was incorporated in the pores of cast-crystallized δ_e-sPS. Since the pore size in the cast-crystallized δ_e-sPS is not large enough for phenol molecules, only ethanol molecules would be preferentially incorporated. The phenol is not incorporated in the pore of cast-crystallized δ_e-sPS, but is in the newly crystallized δ-sPS. When the melt-quenched sPS was soaked in the mixed solution, a very small peak around 3540 cm^{-1} was observed, indicating that the phenol molecules had induced the formation of δ_e-sPS. However most of solvent molecules are settled in the non-crystalline region in which there are interactions due to hydrogen bonding.

Figure 8(B) shows the infrared spectra of (a) phenol/ethanol=5/95 mixed solution, (b) the cast-crystallized sPPMS after soaking in a phenol/ethanol=5/95 mixed solution, and (c), the melt-quenched sPPMS after soaking in the same mixed solution. The hydrogen bond-free OH stretching mode was observed for the cast-crystallized sPS. However, it is worth noting that two peaks were also observed at 3592 and 3540 cm^{-1}, which are assignable to the OH stretching modes due to ethanol and phenol, respectively. This indicates that both ethanol and phenol were incorporated in the pores of sPPMS. The fact that the pore size in the cast-crystallized sPPMS is large enough for both ethanol and phenol allowed them both to incorporate.

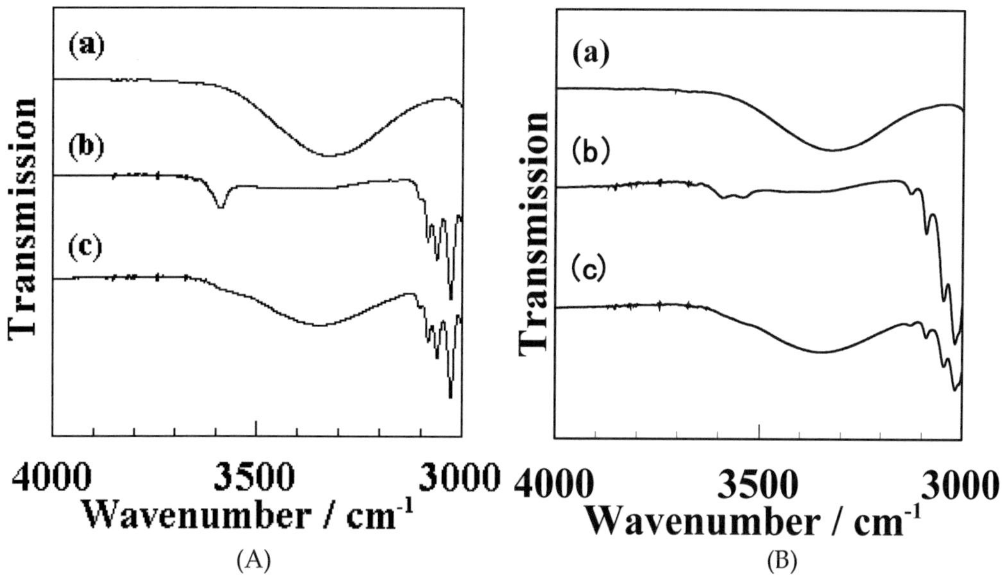

Figure 8. Infrared spectra for the cast-crystallized (a) and melt-quenched (b) sPS (A) and sPPMS (B) after soaking in phenol/ethanol=5/95 for 24 h. The peaks at 3540 cm^{-1} and 3592 cm^{-1} are assignable to the OH stretching mode of phenol and ethanol free from hydrogen bonding, respectively. Nakaoki et al.

In conclusion, sPS preferentially incorporates ethanol, but both ethanol and phenol absorb in sPPMS. This difference resulted from the different pore sizes of sPS and sPPMS.

4. Conclusions

The absorption of phenol into the porous structure of sPS and sPPMS was investigated by infrared spectroscopy. The IR OH stretching mode of phenol for the cast-crystallized sPS and sPPMS films was observed at 3540 cm^{-1}, which corresponds to molecules that are free from hydrogen bonding. This implies that the phenol molecule is isolated from others by incorporation into the nanopores of the crystal. When amorphous films of sPS and sPPMS were soaked in a phenol solution, a hydrogen-bonded peak around 3300 cm^{-1} was observed. This peak was assigned to phenol molecules existing in the non-crystalline phase. In addition, a peak at 3540 cm^{-1} was observed for both sPS and sPPMS, indicating that the formation of a porous crystal was induced by phenol. The diffusion coefficient of phenol into the pores of cast-crystallized sPPMS was larger than that for the cast-crystallized sPS. This is due to the large pore size of sPPMS resulting from the methyl group attached to the phenyl group of the polymer. In contrast, diffusion in non-crystalline phase was much faster than that in the pores of the crystal. This can be interpreted to mean that no hindrance, such as a crystal, to transport exists in the random chain case. The number of phenol molecules in the pore of a crystal was calculated from the crystallinity and the concentration of phenol in the film. One pore in the cast-crystallized sPS and sPPMS contains 1.2 and 2.0 phenol molecules. It is worth noting that the pore in the δ_e-sPS, which was prepared by casting from *m*-xylene was too small for phenol molecules to be incorporated. Therefore the phenol molecules were only incorporated in newly crystallized δ-sPS during the measurement. In contrast, sPPMS can incorporate solvent not only in the nanoporous crystal, but also in the newly induced crystals because its pores are large enough to accommodate a phenol molecule. When these films were soaked in a mixed solution of phenol/ethanol, the cast-crystallized δ_e-sPS only incorporated ethanol molecules, but the cast-crystallized sPPMS incorporated both phenol and ethanol molecules. This can be explained as due to the small pores in sPS but a large enough space for both molecules in sPPMS.

Author details

Kenichi Furukawa and Takahiko Nakaoki*
Department of Materials Chemistry, Ryukoku University, Seta, Otsu, Japan

Acknowledgement

Financial support from a grant from the High-Tech Research Center Program for private universities from the Japan Ministry of Education, Culture, Sports, Science and Technology is gratefully acknowledged.

5. References

[1] Natta G (1955) Une nouvelle classe de polymeres d'α-olefines ayant une régularité de structure exceptionnelle. Journal of Polymer Science 16:143-144.

[2] Natta G, Pino P, Corradini P, Danusso F, Mantica E, Mazzanti G, Moraglio G (1955) Crystalline High Polymers of α-Olefins. Journal of American Chemical Society 77: 1708-1709.

[3] Ishihara N, Seimiya T, Kuramoto M, Uoi M (1986) Crystalline syndiotactic polystyrene. Macromolecules 19:2464-2465.

* Corresponding Author

[4] Immirzi A, de Candia F, Ianneli P, Zambelli A, Vittoria V (1988) Solvent-induced polymorphism in syndiotactic polystyrene Makromol. Chem., Rapid Commun. 9: 761-764

[5] Kobayashi M, Nakaoki T, Ishihara N (1989) Polymorphic Structures and Molecular Vibrations of Syndiotactic Polystyrene. Macromolecules 22: 4377-4382

[6] De Rosa C, Guerra G, Petraccone V, Corradini P (1991) Crystal Structure of the α-Form of Syndiotactic Polystyrene. *Polymer Journal* 23: 1435-1442

[7] De Rosa C (1996) Crystal Structure of the Trigonal Modification (α Form) of Syndiotactic Polystyrene. Macromolecules 29: 8460-8465

[8] Cartier L, Okihara T, Lotz B (1998) The α Superstructure of Syndiotactic Polystyrene: A Frustrated Structure. Macromolecules 31: 3303-3310.

[9] De Rosa C, Rapacciuolo M, Guerra G, Petraccone V, Corradini P (1992) On the crystal structure of the orthorhombic form of syndiotactic polystyrene. Polymer 33: 1423-1428

[10] Chatani Y, Shimane Y, Ijitsu T, Yukinari T (1993) Structural study on syndiotactic polystyrene: 3. Crystal structure of planar form I. Polymer 34: 1625-1629

[11] Tamai Y, Fukui M (2002) Thermally Induced Phase Transition of Crystalline Syndiotactic Polystyrene Studied by Molecular Dynamics Simulation. Macromolecular Rapid Communications 23: 891-895

[12] Rizzo P, Lamberti M, Albunia A R, Ruiz de Ballesteros O, Guerra G, (2002) Crystalline Orientation in Syndiotactic Polystyrene Cast Films. Macromolecules 35: 5854-5860

[13] Rizzo P, Della Guardia S, Guerra G (2004) Perpendicular Chain Axis Orientation in s-PS Films: Achievement by Guest-Induced Clathrate Formation and Maintenance after Transitions toward Helical and Trans-Planar Polymorphic Forms. Macromolecules 37: 8043-8049

[14] Vittoria V, De Candia F, Ianelli P, Immirizi A (1988) Solvent-induced crystallization of glassy syndiotactic polystyrene. Makromol. Chem. Rapid. Commun. 9: 765-769.

[15] Kobayashi M, Nakaoki T, Ishihara N (1990) Molecular conformation in glasses and gels of syndiotactic and isotactic polystyrenes. Macromolecules 23: 78-83

[16] Guerra G, Musto P, Karasz F E, MacKnight W J (1990) Fourier transform infrared spectroscopy of the polymorphic forms of syndiotactic polystyrene (pages 2111–2119)Makromol. Chem., 191, 2111-2119.

[17] Chatani Y, Shimane Y, Ijitsu T, Yukinari T, Shikuma H (1993) Structural study on syndiotactic polystyrene: 2. Crystal structure of molecular compound with toluene. Polymer 34: 1620-1624

[18] Rizzo P, Daniel C, Girolamo Del Mauro A, Guerra G. (2007) New Host Polymeric Framework and Related Polar Guest Cocrystals. Chem. Mater. 19: 3864-3866

[19] Rizzo P, D'Aniello C, Girolamo Del Mauro A, Guerra G (2007)Thermal Transitions of ε Crystalline Phases of Syndiotactic Polystyrene. Macromolecules 40: 9470 -9474.

[20] Petraccone V, Ruiz O, Tarallo O, Rizzo P, Guerra G (2008) Nanoporous Polymer Crystals with Cavities and Channels. Chem. Mater. 20: 3663-3668

[21] Manfredi C, De Rosa C, Guerra G, Rapacciuolo M, Auriemma F, Corradini P (1995) Structural changes induced by thermal treatments on emptied and filled clathrates of syndiotactic polystyrene. Macromol. Chem. Phys. 196: 2795-2808

[22] Handa Y P, Zhang Z, Wong B (1997) Effect of Compressed CO_2 on Phase Transitions and Polymorphism in Syndiotactic Polystyrene. Macromolecules 30: 8499-8509

[23] Reverchon E, Guerra G, Venditto V (1999) Regeneration of nanoporous crystalline syndiotactic polystyrene by supercritical CO2. J. Applied Polym. Sci. 74: 2077-2082

[24] Nakaoki T, Fukuda Y, Nakajima E, Matsuda T, Harada T (2003) Crystallization Condition of Glassy Syndiotactic Polystyrene in Supercritical CO2. Polymer Journal 35: 430-435

[25] De Rosa C, Guerra G, Petraccone V, Pirozzi B (1997) Crystal Structure of the Emptied Clathrate Form (δ_e Form) of Syndiotactic Polystyrene. *Macromolecules* 30: 4147-4152

[26] De Rosa C, Petraccone V, Guerra G, Manfredi C (1996) Polymorphism of syndiotactic poly(p-methylstyrene): oriented samples. Polymer 37: 5247-5253

[27] De Rosa C, Petraccone V, Dal Poggetto F, Guerra G, Pirrozi B, Di Lorenzo M L, Corradini P (1995) Crystal Structure of Form III of Syndiotactic Poly(p-methylstyrene). Macromolecules, 28: 5507-5511

[28] Ruiz de Ballesteros O, Auriemma F, De Rosa C, Floridi G, Petraccone V (1998) Structural features of the mesomorphic form of syndiotactic poly(p-methylstyrene). Polymer 39: 3523-3528

[29] Dell'Isola A, Floridi G, Rizzo P, Ruiz de Ballesteros O, Petraccone V (1997) On the clathrate forms of syndiotactic poly(*p*-methylstyrene). Macromol. Symp. 114: 243-249

[30] Petraccone V, La Camera D, Caporaso L, De Rosa C (2000) Crystal Structure of the Clathrate Form of Syndiotactic Poly(*p*-methylstyrene) Containing o-Dichlorobenzene. Macromolecules 33: 2610-2615

[31] Petraccone V, La Camera D, Pirrozi B, Rizzo P, De Rosa C, (1998) Crystal Structure of the Clathrate Form of Syndiotactic Poly(p-methylstyrene) Containing Tetrahydrofuran. Macromolecules 31: 5830-5836

[32] Tarallo O, Esposito G, Passarelli U, Petraccone V (2007) A Clathrate Form of Syndiotactic Poly(p-methylstyrene) Containing Two Different Types of Cavities. Macromolecules 40:5471-5478

[33] Petraccone V, Esposito G, Tarallo O, Caporaso L (2005) A New Clathrate Class of Syndiotactic Poly(p-methylstyrene) with a Different Chain Conformation. Macromolecules 38: 5668-5674

[34] Furukawa K, Nakaoki T (2011) Comparison of Absorption Kinetics of Ethanol and Butanol into Different Size Nanopores Present in Syndiotactic Polystyrene and Poly(p-Methylstyrene). Soft Materials 9: 141-153

[35] Manfredi C, Del Nobile M A, Mensitieri G, Guerra G, Rapacciuolo M (1997) Vapor sorption in emptied clathrate samples of syndiotactic polystyrene. J. Polym. Sci. Phys. Ed. 35: 133-140

[36] Venditto V, De Girolamo Del Mauro A, Mensitieri G, Milano G, Musto P, Rizzo P, Guerra G (2006) Anisotropic Guest Diffusion in the δ Crystalline Host Phase of Syndiotactic Polystyrene: Transport Kinetics in Films with Three Different Uniplanar Orientations of the Host Phase. Chem. Mater. 18: 2205-2210

[37] Daniel C, Alfano D, Venditto V, Cardea S, Reverchon E, Larobina D, Mensitieri G, Guerra G (2005) Aerogels with a Microporous Crystalline Host Phase. Adv. Mater. 17: 1515-1518

[38] Daniel C, Sannino D, Guerra G (2008) Syndiotactic Polystyrene Aerogels: Adsorption in Amorphous Pores and Absorption in Crystalline Nanocavities. Chem. Mater. 20: 577-582

[39] Daniel C, Giudice S, Guerra G (2009) Syndiotatic Polystyrene Aerogels with β, γ, and ε Crystalline Phases. Chem. Mater. 21: 1028-1034

[40] Mohri S, Rani D A, Yamamoto Y, Tsujita Y, Yoshimizu H (2004) Structure and properties of the mesophase of syndiotactic polystyrene—III. Selective sorption of the mesophase of syndiotactic polystyrene. J. Polym. Sci., Part B: Polym. Phys. 42: 238-245

[41] Mahesh K P O, Sivakumar M, Yamamoto Y, Tsujita Y, Yoshimizu H, Okamoto S (2004) Structure and properties of the mesophase of syndiotactic polystyrene. VIII. Solvent sorption behavior of syndiotactic polystyrene/*p*-chlorotoluene mesophase membranes. J. Polym. Sci.: Part B: Polym. Phys. 42: 3439-3446

[42] Sivakumar M, Mahesh K P O, Yamamoto Y, Yoshimizu H, Tsujita Y (2005) Structure and properties of the δ-form and mesophase of syndiotactic polystyrene membranes prepared from different organic solvents. J. Polym. Sci.: Part B: Polym. Phys. 43: 1873-1880

[43] Mahesh K P O, Tsujita Y, Yoshimizu H, Okamoto S, Mohan D J (2005) Study on δ-form complex in syndiotactic polystyrene–organic molecules systems. IV. Formation of complexes with a mixture of solvents and structural changes during the sorption of solvents by syndiotactic polystyrene mesophase membranes. J. Polym. Sci.: Part B: Polym. Phys. 43: 2380-2387

[44] Nakaoki T, Goto N, Saito K (2009) Selective Sorption and Desorption of Organic Solvent for δ-Syndiotactic Polystyrene. Polymer J. 41: 214-218

[45] Nakaoki T, Kobayashi M (2003) Local Conformation of Glassy Polystyrenes with Different Stereoregularity. J. Mol. Struct. 655: 343-349

[46] Daniel C, Guerra G, Musto P (2002) Clathrate Phase in Syndiotactic Polystyrene Gels. Macromolecules 35: 2243-2251

[47] Milano G, Venditto V, Guerra G, Cavallo L, Ciambelli P, Sannino D (2001) Shape and Volume of Cavities in Thermoplastic Molecular Sieves Based on Syndiotactic Polystyrene. Chem. Mater. 13: 1506-1511

Chapter 6

Nonconventional Method of Polymer Patterning

Oleksiy Lyutakov, Jiri Tuma, Jakub Siegel, Ivan Huttel and Václav Švorčík

Additional information is available at the end of the chapter

http://dx.doi.org/10.5772/46143

1. Introduction

The unical properties of polymers, their cheap coast and possibility of easy chemical or physical modification, make these materials ideal building blocks for nano- or micro-patterning. Techniques for polymers fabricating on nano- and micro-length scales span a wide range, from improved conventional lithographic methods to more recent materials and chemical advances that rely on self-organization of block copolymer. In addition to traditional methods, there are a number of techniques used exclusively in polymer materials processing. The most famous of them include molding, writing and printing, laser scanning, self-organization and surface instabilities utilization.

Nano-imprint lithography (NIL), first proposed in work [1], and represents a high resolution and high throughput polymer lithography technique. The general principle of NIL consists in mechanical deformation of the polymer layer with a stamp presenting a surface topography. Different modification of NIL includes ultraviolet or temperature assisted polymer curing, application of solvent vapor for material softening and so on. In a research laboratory, nano-imprint lithography works very well and the structure with nanometer scale can be prepared by this way. However, for industrial application, some problem must be solved. Main disadvantages of NIL can be attributed to flatness of stamp and substrate, local roughness of prepared patterns, stamp design issue, stamp contamination, and material shrinkage. Generally, nano-imprint lithography is still considered as a next generation lithography technique.

An alternative method introduced in work [2] has been proven to overcome most of drawbacks of traditional lithography. This method is named soft lithography and relies on using elastomeric polymers as soft molds to obtain patterned surfaces on other polymeric materials. The soft elastomeric stamp can be used either as a vehicle for molecular surface patterning (in the case of so-called micro-contact printing (μCP)) or to create three dimensional reliefs, as in micro-molding in capillaries, micro-transfer molding or solvent-assisted molding.

Schematic representation of different variants the soft lithography process is given in the work [3]. Soft lithography technique was successfully applied in the fabrication of polymer patterns with dimensions down to the sub-100nm scale.

A direct-write assembly is a fabrication method that employs a controlled translation stage, which moves a pattern-generating device, for example, ink deposition nozzle, to create materials with controlled architecture and composition [4]. Arbitrary 3D structures can be prepared by this method. For the polymeric ink solidification occurs either upon deposition into a coagulating reservoir or UV exposure. This method can be characterized to be bio-inspired because they copy a nature process - direct ink writing in the form of spider webs.

There are several processes which can play a role in polymer modification under laser scanning. Major processes occurring during polymer modification by excimer laser can be subdivided to: material transport under ablation threshold, ablation and polymer surface chemistry modification [5-7]. When polymer surfaces are exposed to a polarized pulsed laser to more than several hundred pulses with a sub-threshold flounce for ablation, ripple structures appear if the photons are absorbed with a high absorption coefficient in the surface layer only several hundred angstroms thick [5]. Possible explanation of observed phenomena consists in the materials transport under light interference pattern. The incident laser beam can interact with the reflected or scattered by surface inhomogeneities light and form the light interference pattern. Then the polymer flow occurs due to effects of temperature or electrical gradients. This method can be applied to the most of conventional polymers. The resulting periodicity is limited by the applied laser wavelength and by the complex response of the material to the laser illumination [7].

When polymer films are exposed to pulsed laser beams at a proper wavelength with beam energy above the ablation threshold energy, photoablation takes place [8]. In a low-fluence range an incubation period exists, i.e. after absorbing the laser energy the polymer starts to swell, and subsequently ablated materials are ejected. In a high-fluence range a shock wave front is formed and ablated material is ejected instantaneously with high speed. Usually the ablation of polymers includes depolymerisation of materials and evaporation of low-molecular weight products [9,10].

Another example of polymer modification by pulsed laser exposure is the surface chemistry modification [12.13]. For this purpose pulse fluencies of the UV laser should be higher than those used for the periodic ripple formation [10]. By exposing polymer surfaces to pulsed UV lasers, chemical properties of polymer surfaces can be improved to more desirable ones.

In our experiments we used the periodical laser scanning for polymer surface modification.

2. Polymer modification by laser scanning

At the first stage the optical properties of polymers were changed in desired way by suitable dotation. We used a porphyrine, or disperse red (DR1) as dopants with large absorption coefficients.

Thin polymer films doped by organic dye were prepared by spin-coating technique. For the polymer dotation two ways were used. In the case of bulk dotation the solution of polymer was mixed with organic dye, dissolved in dichlorethane. In the case of DR1 the organic dye was added as dopant or chemically connected with polymer. Fig. 1 gives the chemical structure of used materials.

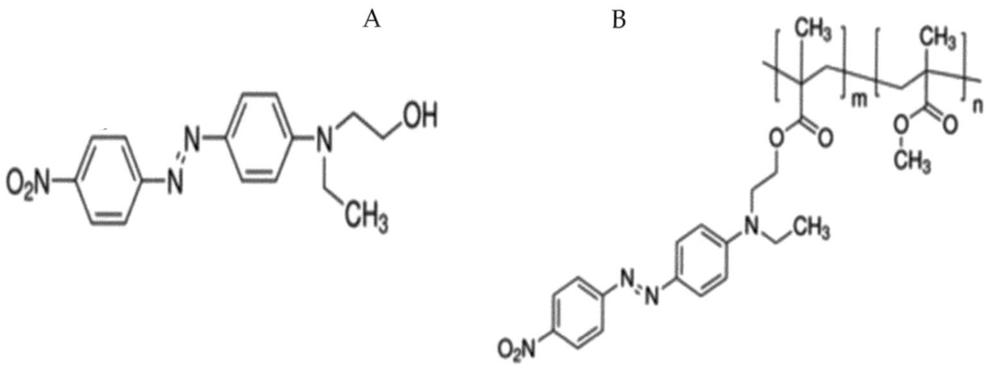

Figure 1. Chemical structure of Disperse Red 1 (A) and DR1 covalently linked to PMMA.

For surface dotatiton the vacuum deposition technique was applied. In this case the layer of pristine polymer was prepared by spin-coating and served as a substrate for next vacuum deposition of porphyrine. Vacuum deposition of porphyrine was performed under 10^{-6} Torr pressure and at 1 A electric current. For preparation of more complex structures the combination of both methods was used.

The doped polymer films were modified by laser scanning and simultaneous mechanical sample movement. This technique of polymer modification was recently proposed by our team [14]. Chosen polymer area is scanned by continual laser beam line by line. Laser beam was focused into spot with approximate diameter 0.5 μm. Laser operating at 405 nm wavelength and 0.1 W laser power was applied. The laser light of applied wavelength is expected to be absorbed by porphyrine molecules. Sample mechanical movement is added to the laser scanning. The velocity of the sample movement was 2 μm/sec [14]. The process is schematically depicted in Fig. 2.

For preparation of more complex structures both techniques of dotation were used. Firstly, polymer films were doped by bulk dotation and modified by laser scanning. Then, in a second step, the porphyrine was deposited on the surface of modified polymer, the sample was turned by 90° regarding to previous orientation and scanned again.

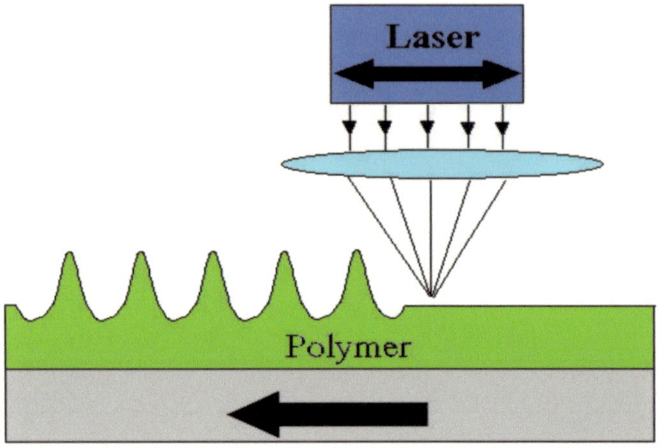

Figure 2. Schematic of the present experimental arrangement. Polymer is deposited by spin coating onto glass wafer, polymer surface layer is doped with porphyrin by deep coating. Then the surface is scanned by focused laser beam, together with simultaneous mechanical movement of polymer film. Direction of the mechanical movement is given by arrow.

Fig. 3 gives the typical structure prepared by laser scanning and simultaneous sample movement. Thin polymer film doped by porphyrine in bulk was exposed to focused beam of laser light with wavelength appropriate to maximum of porphyrine molecules absorption. Part A of the figure corresponds to application of laser scannning and gives the optical image and surface distortion at the boundary of exposed and measured by AFM. It is evident, that after the laser modification the polymer tends to form two protruding surface structures on the boundary of scanned area. It should be noted, that appearance of the structure occurs only in the direction of laser scanning (left and right sides of image). Formation of the structure in perpendicular direction (top and bottom sides of image) was never observed. It could be assumed, that polymer tends to flow in the direction of more pronounced temperature gradient. When the simultaneous mechanical movement of sample is added the periodical structure is formed (see Fig. 4B). Prepared structure exhibits the system of well ordered surface maximum and minimum peaks. In other words, the surface profile represents periodical array of lines along which the polymer mass was pulled above initial flat surface.

Surface patterning of pure polymer films was performed by covering the surface of polymer by suitable dye (porphyrine or phthalocyanine) and subsequent laser scanning. From the practical point of view it seems to be interesting to choose the polymer photoresist for patterning. The structure, prepared by laser scanning in our case or by another method can be strengthened and fixed by following UV light illumination. One of examples is the application of soft lithography techniques for patterning of commonly used photoresist – Su-8 [15-17]. It must be noted, that in our case bulk dotation of Su-8 is not applicable, during the laser illumination two concurrent processes will occur – polymer flow and crosslinking.

At the fist stage the pure Su-8 film was deposited onto Si substrate. In the next step, Su-8 layers covered by porphyrine were exposed to laser scanning with simultaneous sample movement. Fig. 4A shows the structure obtained after the first stage of patterning. It is evident, that prepared structure shows some tendency to be „periodical", but the quality of this periodicity is „bad" (Fig. 4A). However, the next laser illumination results in material redistribution and sufficient improvement of the structure quality. The Fig. 6B gives the surface profile of prepared structure after laser scanning and subsequent simple, homogenous laser illumination. The differences between Figs. 4A and B are evident, but the nature of observed phenomenon, the driving forces for material flow and structure improvements, are not clear now. In the next step prepared structures were fixed by UV light illumination and developed in dichlorethane. As can be expected, cross-linked Su-8 is not soluble and only top layers of porphyrine are dissolved. Typical surface profile of prepared structure is given in the Fig. 4C.

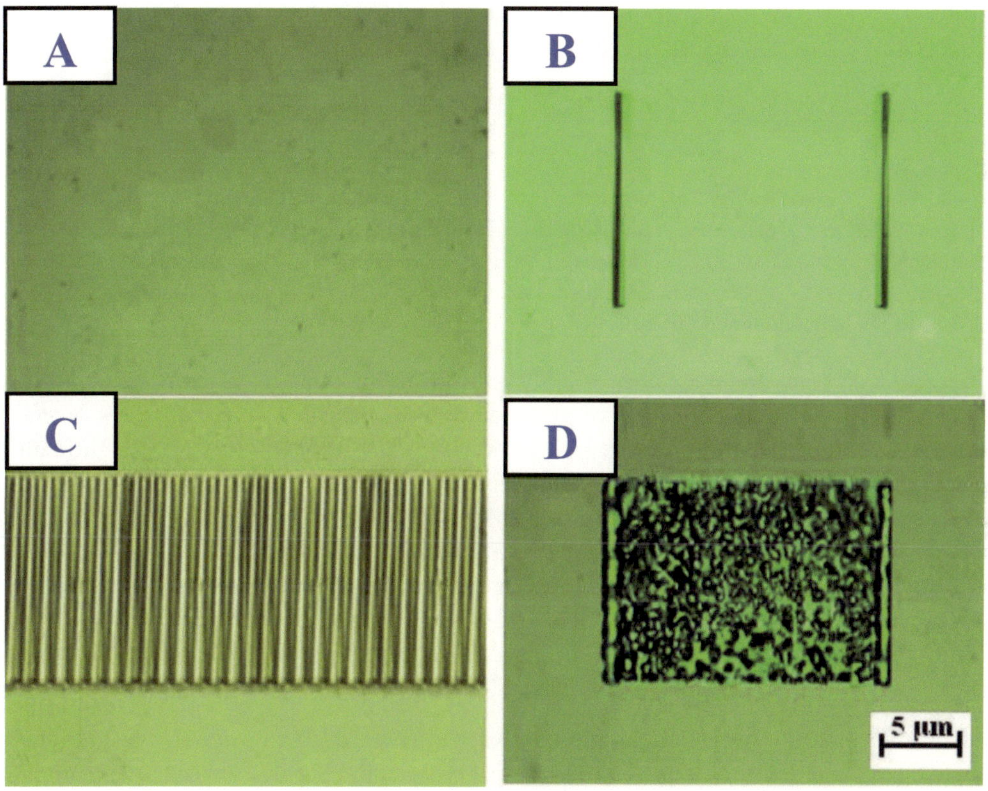

Figure 3. Various stages of the surface patterning as a function of the laser intensity and the velocity of the mechanical movement. Laser scanning has been applied accross 20×20 μm area at the centre of the image. Part A - correspond to laser intensity of 0.03 mW, B – 0.12 mW, C – 0.12 mW and simultaneous mechanical movement, and D – 0.25 mW.

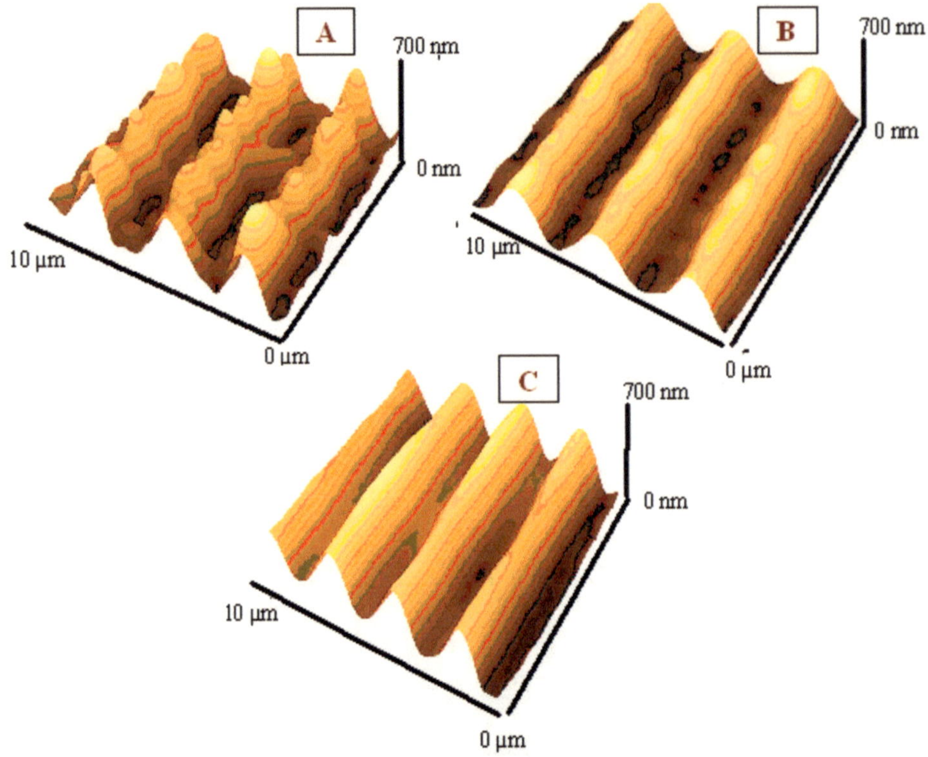

Figure 4. AFM images of surface of Su-8 with doped with porphyrine after modification by laser scanning and simultaneous mechanical movement: A- surface after application of laser scanning and sample movement, B- surface after subsequent laser illumination and C- surface of modified film exposed to UV light and treated in dichlorethane.

Proposed technique opens-up a possibility to utilize wider range of materials as dopant, regardless of their solubility. Class of phthalocyanines, affined with porphyrines, seems to be very interesting from the sensor application [18] point of view, however it is difficult to dissolve phthalocyanine in common solvents. Application of high vacuum deposition can help to avoid bad solubility of phthalocyanine. Different dyes, deposited on the top of polymer can serve as light absorber for patterning process and as „active" material in next utilization of prepared structures e.g. in sensor application. In these case properties of polymer matrix remain invariable; which may be important is some specific applications. Additionally, evaporation of dyes onto formerly patterned polymer surface opens-up new way for more complex 2D polymer surface patterning.

Another technique of polymer dotation consists in chemical bounding of dye to polymer chain. Chemical structure of used molecules is given on the Fig. 1. Two sets of experiments were performed in this case. At first case the dye was added as dopant and in second case covalently bounded. The results of surface measurement by AFM are given on the Fig. 5. It is evident, that covalently bounding results in sufficiently highest amplitude of prepared structure. Additionally, this technique of polymer dotation opens a way for process reversibility.

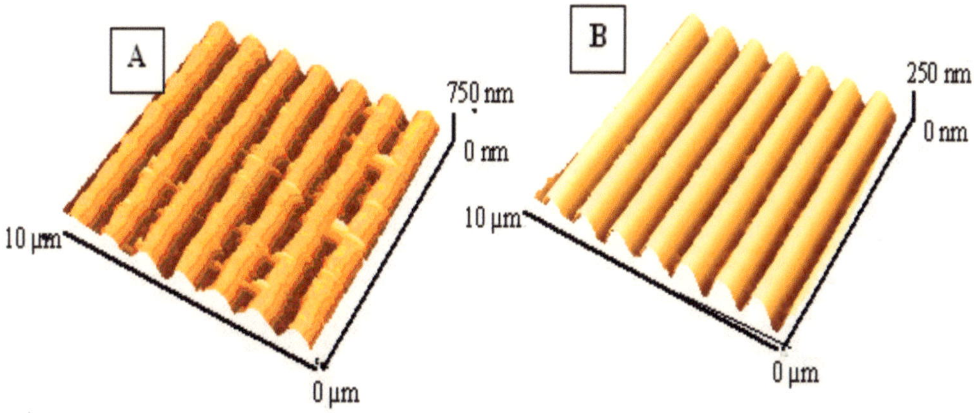

Figure 5. Surface structures of PMMAcoDR1 (A) and PMMA doped by DR1 (B). Surfaces were modified under laser intensity of 0.03 mW and 2 µm/sec speed of simultaneous mechanical movement.

2.1. Mechanism of polymer flow

Proposed mechanism of the structure formativ consists in polymer mass re-distribution governed by surface tension gradient (so-called Marangoni effect [19]). The classical mechanism of the Marangoni effect expressed in the formation of surface structures at the interface of two phases is the nonuniformity of surface tension due to temperature gradient [20]. In our experiments absorption of light during the scanning generates spatial variability of surface tension, usually responsible for Marangoni effect. To our opinion in the first step the polymer matrix is molted. At the next step periodical temperature gradient and the surface tensions gradient are created. As can be predicted by Marangoni phenomenon the polymer tends to flow from the region with smaller surface tension to the region with higher surface tension. On the boundaries of scanned region the polymer flow must be stopped because the materials outside of this region are not in the molten state. When continual mechanical movement of sample is added to the laser scanning, processes of polymer melting, flow and stopping achieves equilibrium and 2D periodical structure is created. The driving force for the structure formation is the magnitude of thermal gradient and related inhomogeneity in the surface tension. The initiation and propagation of this process at "stationary conditions" is well described in several works [31, 32] from the theoretical point of view. Nevertheless, in the present case we deal with nonequilibrium conditions because of continuous sample movement.

2.2. Surface chemistry

For verification of surface or bulk chemistry changes occurring during laser irradiation the UV-VIS, IR and XPS analysis were performed. Fig. 6 gives the UV-VIS optical spectra obtained before and after laser patterning. Peak at the 405 nm corresponds to so-called

Sorett band. From the figure is evident, that porphyrine molecules retain their configuration during laser modification.

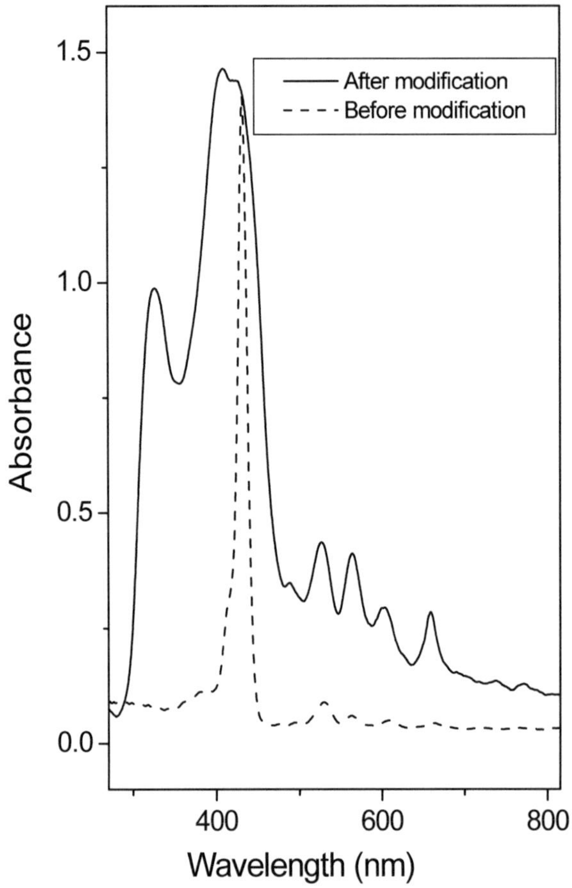

Figure 6. Absorption spectra of PMMA film doped by porphyrine before and after laser modification.

Surface chemistry of modified sample was measured by XPS analysis. Formation of surface patterns can arise not only due to the local heating of the polymer surface, but also due to local heat release in chemical and physico-chemical processes in the laser irradiated polymer. XPS was performed on PMMA samples doped on surface and in bulk. Relative element concentrations before and after surface modification are given in Table 1. In the case of bulk doped PMMA the slight decrease in C concentration and increase in N and O concentrations were found. This result indicates absence of sufficient changes in polymer surface chemistry. Observed changes in element concentrations can be attributed to redistribution of polymer and dopant molecules in fused state. In the case of surface doped PMMA the considerably smaller concentration of O on the pristine surface was found, in comparison with that doted in bulk. This result could be expected, because the top layer of porphyrine screens the polymer molecules. Similarly, as well as in the case of bulk doted samples only minor alterations in surface chemistry after the laser modification were found. Small increase in C concentration and small decrease in N concentration can be explained in the terms of polymer flow and mixing of porphyrine/polymer layer.

	Element concentration (in at. %)		
	C(1s)	O(1s)	N(1s)
Bulk doped PMMA			
before patterning	64.3	24.8	10.9
after patterning	60.8	26.0	13.2
Surface doped PMMA			
before patterning	83.8	5.6	10.6
after patterning	85.7	6.7	7.6

Table 1. Surface concentration of elements (C(1s), O(1s) and N(1s)) in bulk and surface doped PMMA before and after the laser modification determined from XPS analysis.

Since the chemical changes of the polymer surface initiated by the laser irradiation are not substantive one can conclude that the surface patterning is mostly due to classical temperature effects.

Standard procedure for theoretical analysis of surface pattern in thin polymer films under temperature gradient is linear stability analysis [22, 23] according which the formation of surface structures is governed by the surface tension and material redistribution ratio due to heat transfer, where the temperature gradient once again destabilizes the situation, while the viscous drag damps disturbances. There are several cases, which can lead to periodical structure appearance under temperature gradient: (i) nonuniform density distribution over the layer of viscous liquid heated from one side so-called gravitational instability [35], (ii) so-called concentration instability appears in multi-component system due to difference in component concentration [22] and (iii) surface tension instability corresponds to differences in surface tension of homogeneous thin film due to temperature gradient, presented in these films [23].

In case of gravitational instability, stability of thin films limit is determined by the value of the Rayleigh dimensionless criterion and doesn't depend on the forces of surface tension. Rayleigh dimensionless criterion is strongly related to thickness of treated film. For verification of this mechanism the series of samples with different layer thickness were prepared and examined in the same way. Dependences of amplitude of prepared periodical structure on the sample thickness are given in Fig 3. In Fig. 3 the dependence for PMMA with two different molecular weights is shown. Samples were prepared with the same porphyrine concentrations and modified at the same laser light intensity. From Fig. 3 it is evident, that the amplitude of the patterns increases linearly with increasing sample thickness up to the thickness of about 2100 nm in both investigated cases. Above this threshold thickness the increasing rate slows down considerably. Appearance of threshold thickness and linear character of amplitude increase leads us to the conclusion that the polymer flow is not governed by gravitational instability. Rising part of curves can be explained by the amount of absorbed light energy increasing with the film thickness. The upper part of the curves can be attributed to limited light penetration - absorption in polymer when only thin surface layer of the polymer is affected.

Surface instability may also be attributed to nonuniform concentrations in multi-component system. In our experiments, where we used ether the pristine films or films with homogeneous concentration of dopant, this mechanism can be excluded.

Another reason for outbreak of surface instability may be the presence of surface tension gradients due to nonuniform temperature field of a system on the upper free surface. This mechanism seems to be more reasonable in the present case. In this case the structure formation must depend on the surface tension and viscous damp. With the aim to investigate the viscous damp influence, two PMMA with different molecular weights were examined since it is well known, that viscous damp of polymer is strongly related to their molecular weight. Linear polymer with the larger molecular weight will have greater value of viscous damp, than its low-molecular homolog. In our work we investigated PMMA with 1500K and 15K molecular weights. Sufficiently sharper structure with higher amplitude was formed in the case of PMMA with lower molecular weight. One can say, that utilizing of polymers with different molecular weight can result in differences in surface tension too. It should, however, be noted that surface tension of thin films is much more dependent on the chemistry of macromolecules rather than on molecular weight [23].

3. Application

3.1. Optical components

Periodical or ordered structures can serve as different optical components. In our works we examined the polymer lattice prepared onto polymer surface as light diffraction elements of waveguide coupler.

At the first stage the diffraction of light through the lattice was estimated. Because the spot of applied laser possesses the sizes of an order of square millimeters, system of lattices deposited abreast was prepared. Optical image of prepared structures is given in Fig. 7. Considering difficulties in precise positioning of each lattice, some disagreement in their mutual orientation is observed. All lattices had identical periodicity, however were displaced relative to each other on some small step. For the next experiments the areas of 1×1 cm2 were patterned on the sample surface. In the next stage the diffraction intensity of light passing through prepared structures were observed and photographed. Obtained image is shown in Fig. 8. From the Fig. 8 it is evident, that prepared structure exhibits satisfactory diffractive properties. However, some illegibility and vagueness in diffraction intensity maxima is observed. This discrepancy in diffraction can be attributed to disagreement in mutual ordering of both lattices.

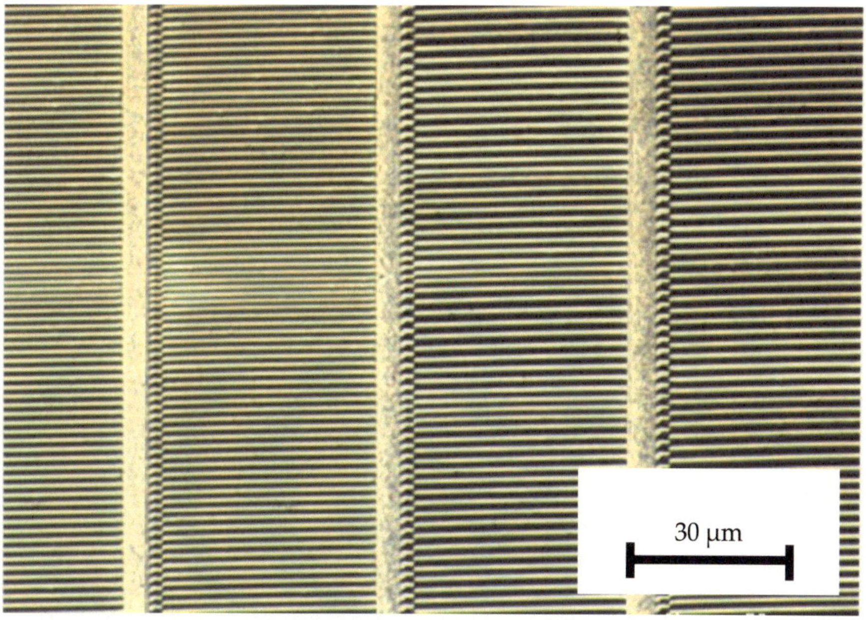

Figure 7. Optical image of system of lattices, prepared abreast on PMMA doped by 3% porphyrine.

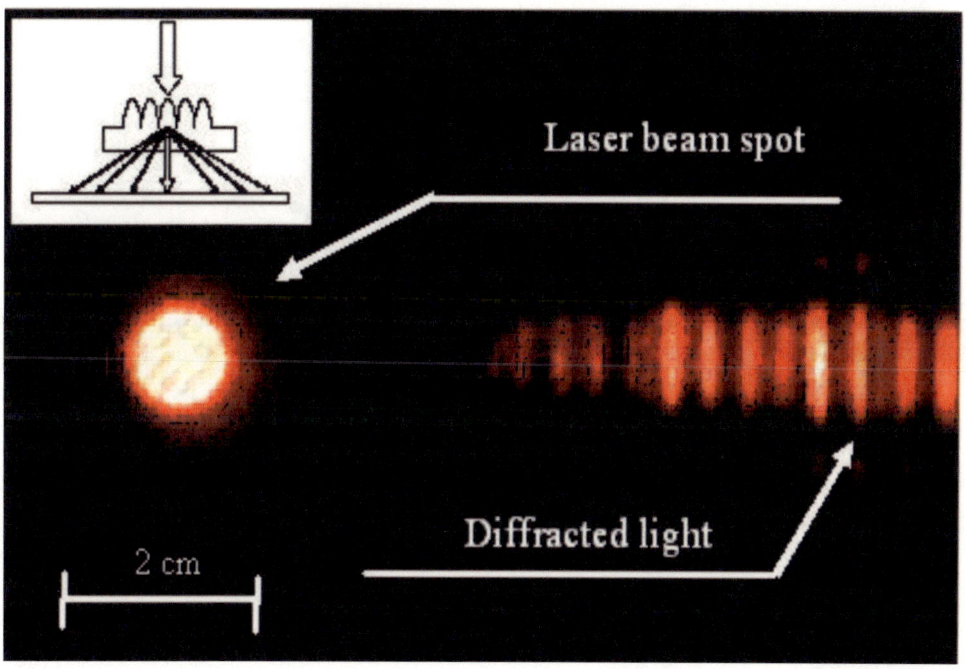

Figure 8. Diffraction pattern of laser light (633 nm) passed through a system of lattices deposited abreast on PMMA (1500k) doped by 3% porphyrine. Presented process is schematically depicted in the inset.

One of the possible applications of prepared structure is coupling of light into waveguide. Example of utilization of prepared structure as waveguide coupler is given in Fig. 9. As well as in previous experiment the sample with system of lattices was used. In this experiment the sample was deposited onto rotation stage and the optimal angle of laser beam incident was founded. This angle corresponds to the most effective light coupling into planar waveguide. From Fig. 9 it is apparent, that a part of laser energy was tapered by lattices and travelled along a waveguide. The prepared profile of a surface pattern had sinusoidal character. From literature it is well known, that for light coupling the rectangular structure profile is more effective [24]. However, for laboratory conditions the light coupling exibits sufficient efficiency. Additionally, presence of good diffractive optical properties of prepared structures opens-up a way for their utilization in sensing and plasmonics.

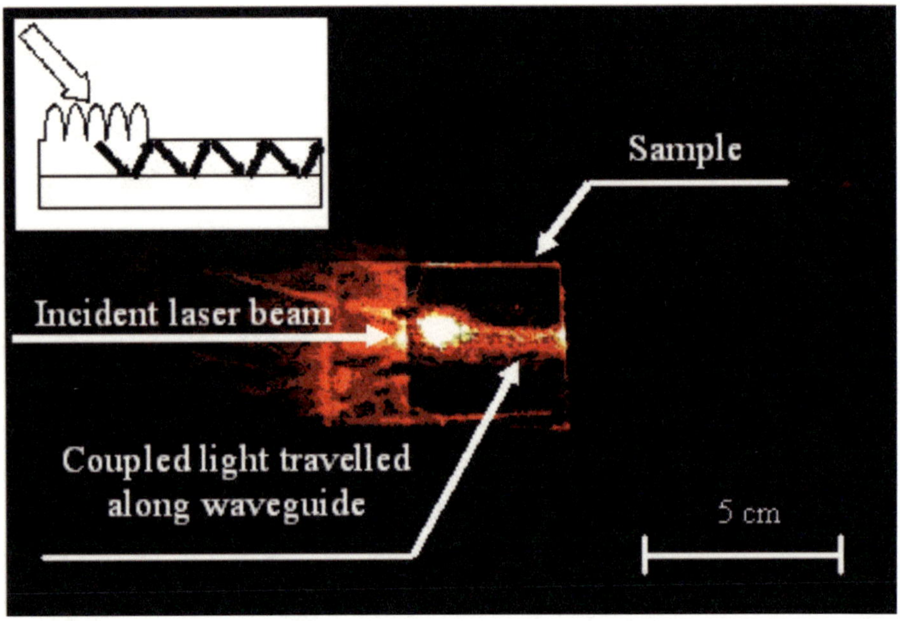

Figure 9. Coupling of light in planar waveguid through system of lattices deposited abreast on PMMA (1500k) doped by 3% porphyrine. Presented process is schematical depicted in the inset.

3.2. Surface wettability

Wetting phenomena on structured solid surfaces are of both fundamental and technological interest. The enhancement of hydrophobicity by roughness is described by either Wenzel or Cassie model, which includes water or air trapped in the structure below the drop. Engineered surfaces exhibiting superhydrophobicity or superhydrophility have been reported [24]. Periodical ordered surface patterning can be used for preparation of surface with anisotropic wetting behavior. This anisotropic wetting behaviour has been observed on one-dimensional (1D) patterned surfaces achieved either through chemical patterning or surface roughness. Surfaces with controlled anisotropic wetting, confining liquid flow to a single direction, have potential applications in microfluidic devices, evaporation-driven formation of patterns, and easy-clean coatings. A relatively strong anisotropic wettability was measured on imprinted hierarchical structures wchich combines micro and nanosized ordering of surface. Weak anisotropic wetting characteristics were observed on periodic polymer structures.

Some results of our work in this field is summarized in the table 2 Results is presented in the form of difference in contact angle of water and glycerol, when observation is performed from two perpendicular point of view.

As was reported in the literature, the drop contact anle can be a function of drop volume. The phenomenon of dependence of wettability on the drop evaporation time is given, for example in the work [25]. From this reason in the next step the wettability of surface for „small" volume of deposited drop was studied. Glycerol was chooses as probe liquid due to hear great sensitivity and lowest evaporation time. Because of experimental restriction we can't direct measure the contact angle and demonstrate the view from the top from optical microscope. Fig. 10 gives the behavior of glycerol droop deposited onto substrate patterned substrate. The anisotropy in drop behavior is apparent. From figure can be concluded, that substrate are sufficiently more wettable in the direction paralel to prepared lattice pattern..

Sample	Amplitude (nm)	Periodicity (μm)	Difference in contact angle	
-	-	-	Glycerol	Water
1	100	1.35	0.8	0.2
2	300	1.35	2.5	0.6
3	300	3.5	7.5	6.5

Table 2. Anisotropic wetting of modified PMMA surface.

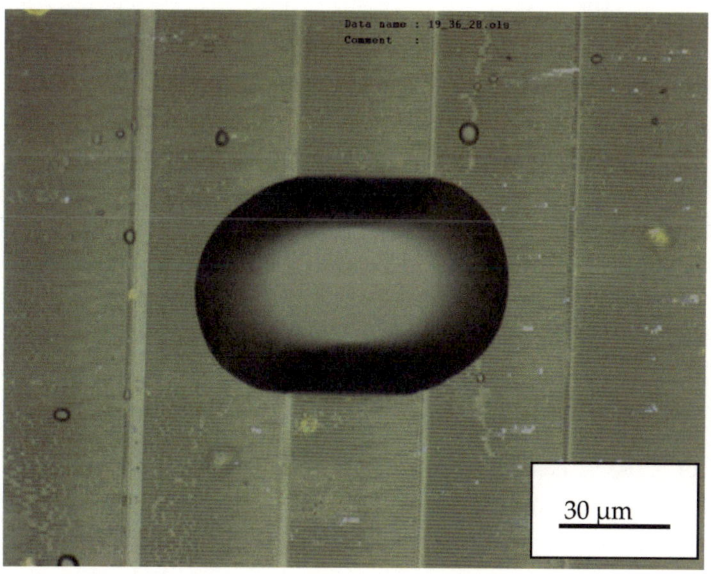

Figure 10. Glycerol drop deposited onto modified PMMA surface.

3.3. Surface coating

Ordered array of hybrid polymer-metal structures can found their application in areas such as plasmonics [26], new photonics metamaterials [27], sensing [28] and catalysts [29]. Large electromagnetic field enhancements near anisotropic nanoparticles make metal nanowires attractive candidates as substrates for surface-enhanced Raman scattering [30]. Observed colossal increase in the effective scattering cross section of a molecule deposited on nanostructure metal surface (by 4-6 orders of magnitude) can lead to single-molecule detection. Additionally, various types of near field optical lithography utilizing enhancement of optical field make use of different types of ordered metal structures. Light interaction with specially designed metallic structures results in effects unattainable with naturaly occurring materials, e.g. negative permeability [30], negative refractive index [28] and nonlinear effects in magnetic meta-materials [29]. These observations led to a new area of photonics called optical meta-materials that have been exponentially growing over the last few years.

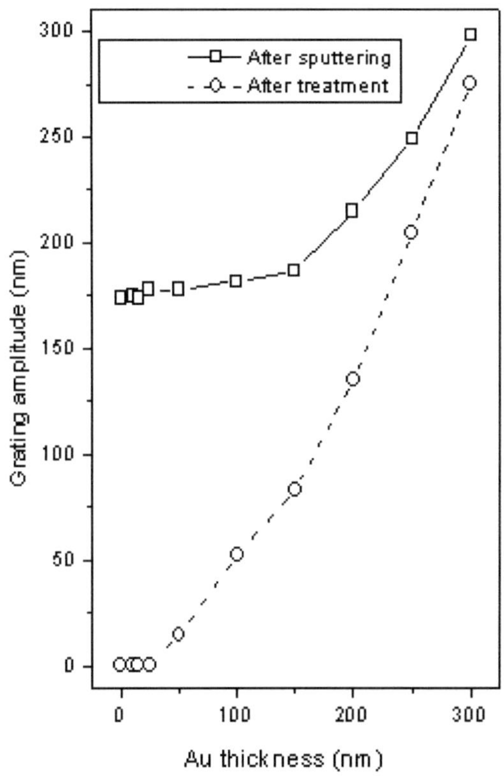

Figure 11. Change of grating amplitude during the Au deposition and thermal annealing.

Examination of surface coating of structure prepared by two different methods was performed. The metals [Ag and Ga] were deposited by sputtering technique. Additionally, it is well known, that annealing of thin metal layer can results in material rearrangement and formation of nano- and micro-clusters. Dependences of amplitude of prepared structure on the deposited thickness of metal and subsequent thermal treatment are given in Fig. 11 and

Fig.12. From figures is evident, that in both case amplitude of structures tends to slightlly increase during metal deposition. Possible explanation consists in primary deposition of metal on the top of sinusoidal pattern due to shadow effect. Subsequent annealing, however leads to fundamentally different results in the case of Au and Ga coating. Results of thermal treatment at 200° during 10 hours for coated polymer surface are given in Figs. 13 and 14.

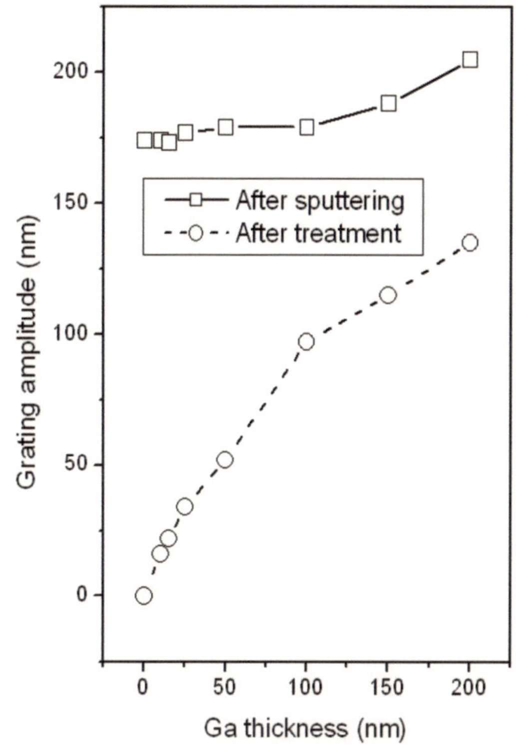

Figure 12. Change of grating amplitude during the Ga deposition and subsequent thermal annealing.

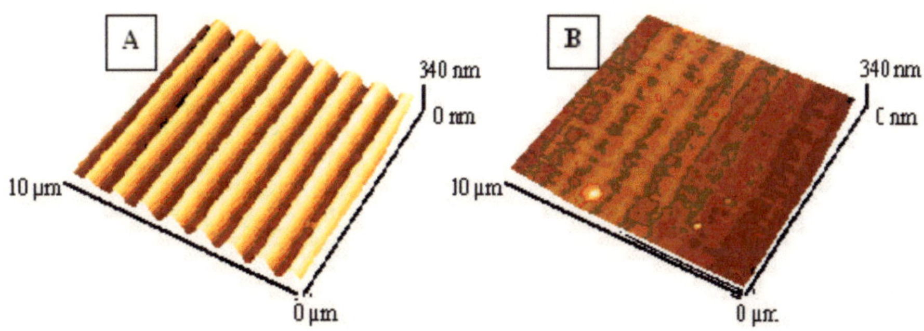

Figure 13. Thermal treatment of patterned PMMA surface covered by 50 nm Au layer. Part A – polymer surface before treatment. Part B – after treatment

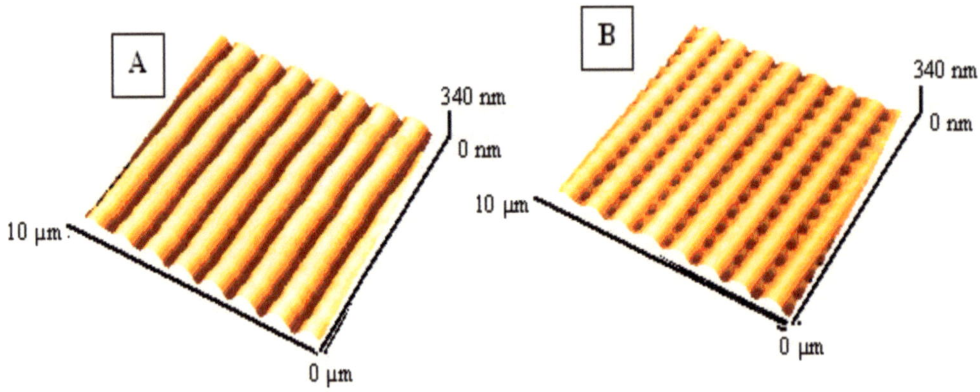

Figure 14. Thermal treatment of patterned PMMA surface covered by 50 nm Ga layer. Part A – polymer surface before treatment. Part B – after treatment

Additionally, changes in optical properties of doped polymer were observed. Fig. 15 shows the transmission spectra of PMMA doped by porphyrine before and after thermal annealing. Possible explanation of observed phenomenon consists in dopant molecule migration and selforganization into crystal structure.

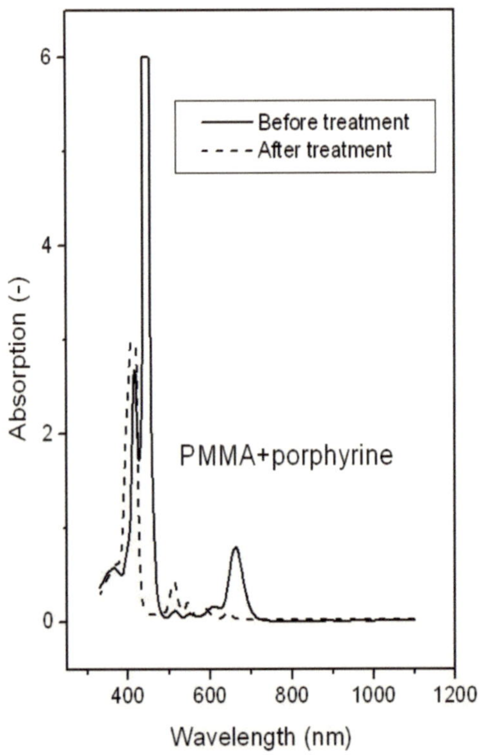

Figure 15. Optical absorption of PMMA film doped by 3% porphyrine before and after thermal treatment at 200o C.

4. Thin film instability enhanced by external electric field

One relatively new, promising technique for polymer patterning is a process based on hydrodynamic instabilities of liquid polymer surface induced by external electric field [31, 32]. Recently thin film instability was reported which could be used for creation of well-controlled patterns on thin polymer films [33,34]. This technique allows creation, with a high accuracy and reproducibility, of well defined polymeric patterns with dimensions ranging from a few tens of nm to a few mm. A lot of experimental work was done by Russel et al. and by Chou et al. [31,32]. By technique proposed by Chou and called LISC (lithography induced self-construction) remarkable uniformity can be obtained over a large surface, but inevitable mechanical contact between polymer and mask introduces an additional degree of roughness to the waveguides which adds a significant scattering loss contribution to the overall waveguide loss. Another mechanism of polymer flow under external electric field leads to appearance of ordered non-continuous structure. In the work [32] this mechanism was called LISA - lithography induced self-assembling.

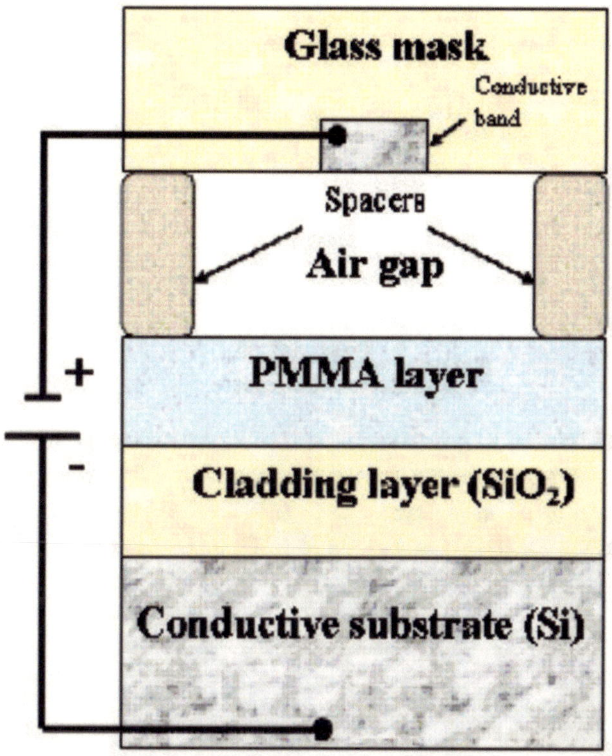

Figure 16. Scheme of polymer surface modification. Electric field effect the heated PMMA layer in the space between conductive silicon substrate with a SiO_2 layer and a metallic (chrome) stripe 50 μm wide on a glass lithographic mask in distance 25 μm from the PMMA layer.

The present experiments were performed on polymethylmethacrylate (PMMA) and novolak resin (Su-8). Experimental set-up is schematically given on Fig. 16. Polymer films were sandwiched between conductive substrate and glass mask with conductive strip (50 μm wide and few mm long). The part of the sample was exposed to electric field created between strip electrode and Si backing. The 25 μm long distance from the upper electrode to the polymer film was choose and the field intensities of electric field do not exceed a breakdown limit of about $5 \cdot 10^6$ V/m. The whole assembly under the electric field was heated in an oven to temperature 160°C for Su-8 and 275°C for PMMA films. After exposure, the sample was allowed to cool down spontaneously to room temperature under the continued effect of the electric field. In the case of Su-8 films were subsequently irradiated with UV light.

Fig. 17 represents the dynamism of channel growth in the case of PMMA. The polymer layers were exposed to the field under increased temperature for various periods of time. The homogeneity of the prepared structures depended on the period of the electric field effect. From figure is obvious that the prepared structure were discontinuous and inhomogeneous until 60 minutes of field effect. In figure there is also visible that the channel growth starts at "random" spots along the full length of the future channel. These discontinuities get connected after a longer effect of the electric field (under increased temperature). Continuous and homogeneous structure appears after 60 minutes.

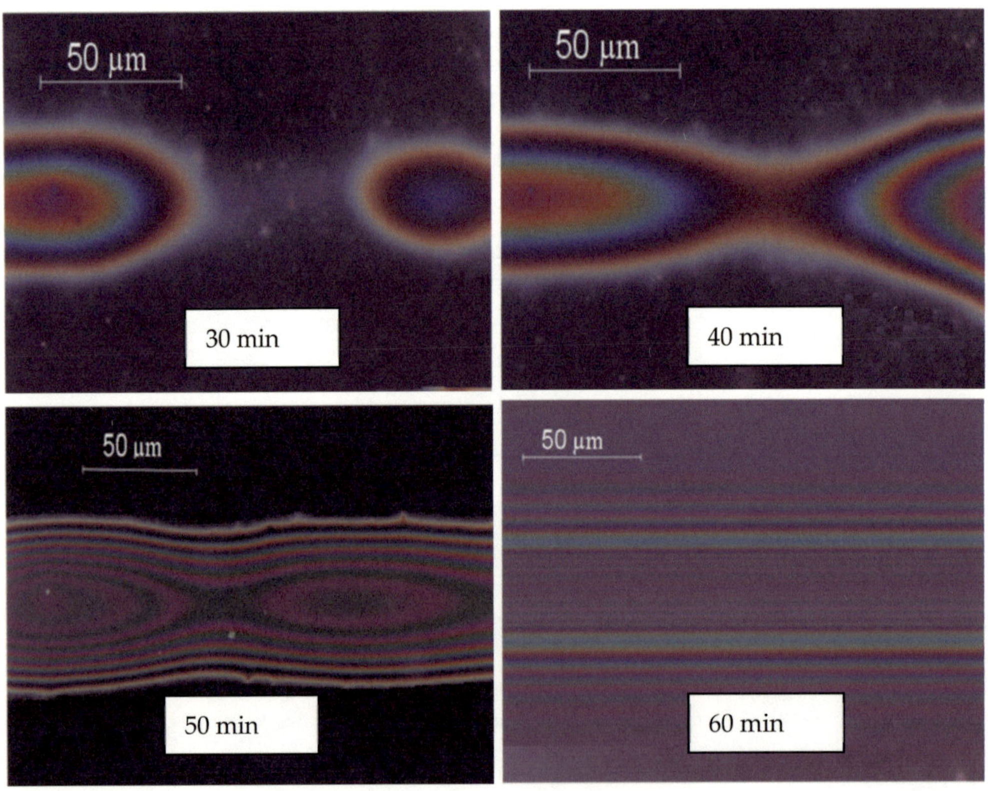

Figure 17. Photo of PMMA structures prepared under the electric field ($2.5 \cdot 10^6$ V/m, at 275±5°C) for various time (min), from optical microscope.

A typical profile of the surface of the structure prepared by heating the entire surface and under local effect of the electric field, as measured perpendicular to the structure, is presented in Fig. 18. We can see that a redistribution of material occurs during the interaction of the electric field with dielectric material and a part of the polymer layer is pulled out in the direction given by the electric field vector. Fig. 18 shows that "pulling" the polymer layer in the area exposed to the electric field makes the layer around to the prepared structure thinner. Fig. 18 also presents the dependence of height of the prepared structures on applied electric field intensity. The thickness of the prepared structures increased with the intensity of the electric field, as we expected. Higher intensity of the electric field results in an increased electrostatic force and subsequent elongation of the structure along the direction electric field. The prepared channels were homogeneous and continuous.

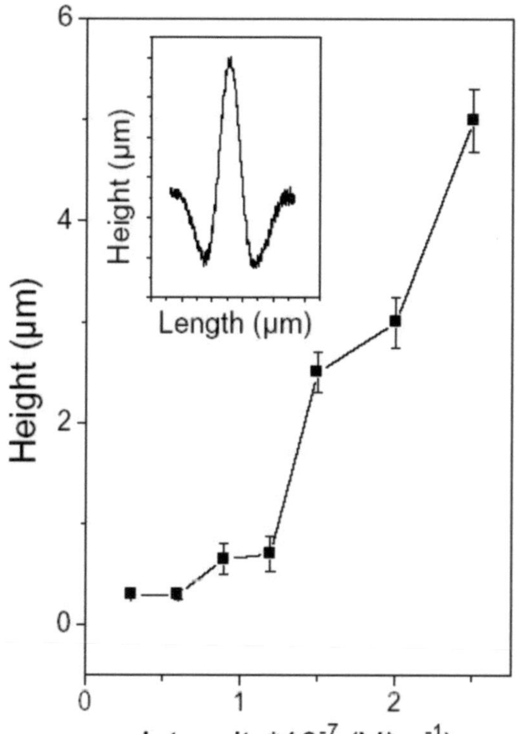

Figure 18. Dependence of height of structures prepared under temperature 275±5°C on an applied electric field. The insert shows the typical profile of a channel prepared under 1.2·106 V/m.

Fundamentally different results were obtained in the case of Su-8 growing. Fig. 19 represents two series of experiments performed at two electric field intensities of 5.0 and 2.5x10^6 V/m and with exposure times ranging from 3-15 min. For short exposures longitudinal homogeneous channel structures are formed. For longer exposures the homogenous structures disappear and system of dots is created. It may therefore be concluded that LISC mechanism prevails in initial stages of the pattern formation which is replaced by LISA mechanism at longer exposures. The LISC to LISA transition comes earlier for higher field intensity.

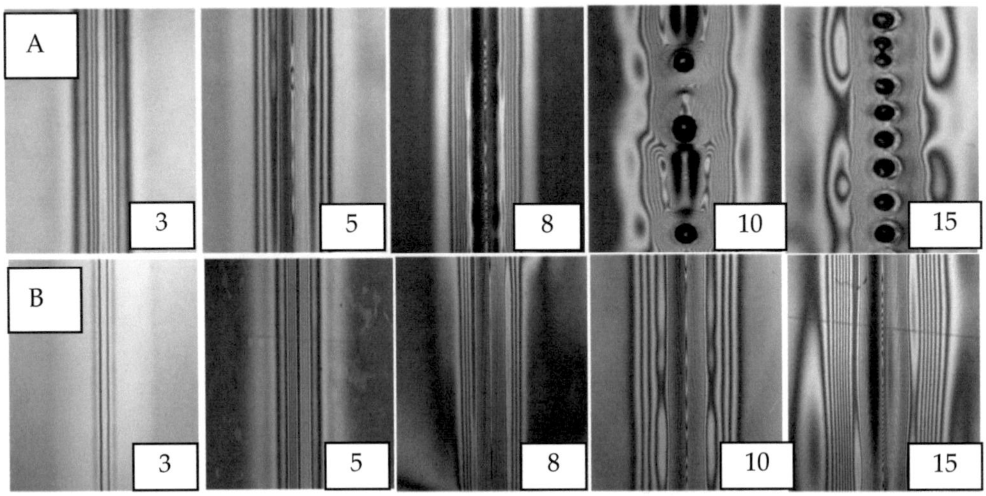

Figure 19. Confocal microscope images illustrating the evolution of the structures in dependence on the exposure time in s. The numbers are the exposure times in minutes. The structures were created under two electric field intensities of $5x10^6$ (A) and $2.5x10^6$ V/m (B) and with the initial distances between the mask and the polymer surface of 40 (A) and 5 μm (B), respectively.

5. Application

The above described technique seems to be prospective for fabrication of polymer based optical devices. The method is simple, free of disadvantages of conventional techniques. It gives highly reproducible results and is suitable for production of high-quality polymer waveguides. The above described LISC technique can be applied for preparation of optical channel waveguide. Waveguiding structure was prepared under following conditions: electric field intensity of $2.5x0^6$ V/m, substrate temperature of 160°C, distance between mask and polymer 50 μm, exposure time 5 min. After pattern formation sample was irradiated by UV light with the aim to improve mechanical properties and stability of the polymer structures. Upon exposure to UV radiation, the oligomer systems form highly cross-linked networks, which exhibit low intrinsic absorption in the wavelength range extending from 400–1600 nm. The longitudinal parameters of the created linear structure were controlled by profilometry and the height and width of the structure were found to vary within ± 5 % along the whole structure length. Within these limits the produced structures can be considered as homogeneous and continuous. The refractive index profile in a cut of the typical rib waveguide structure is shown in Fig. 20A. It is evident, that the technique enables simple fabrication of rib-type waveguide avoiding main disadvantages of traditional photolithography. Wave-guiding properties of the structure were examined by feeding it with 632.8 nm laser light. Distribution of light intensity radiated from the end of channel waveguide was measured and is given in the Fig. 20B as function of space coordinates.

Previous results were used for computer simulation of optical properties of the fabricated structures. By substitution to the RSoft software of measured polymer refractive index and the structure surface profile the inside and output optical field distribution were calculated using beam propagation method. The calculated output optical field is given in Fig 20C. In addition, calculation proved that the effective refractive index for fundamental mode is approximately 1.6276 at 632.8 nm. The prepared multimode structure supported 30 modes at wavelength 632.8 nm.

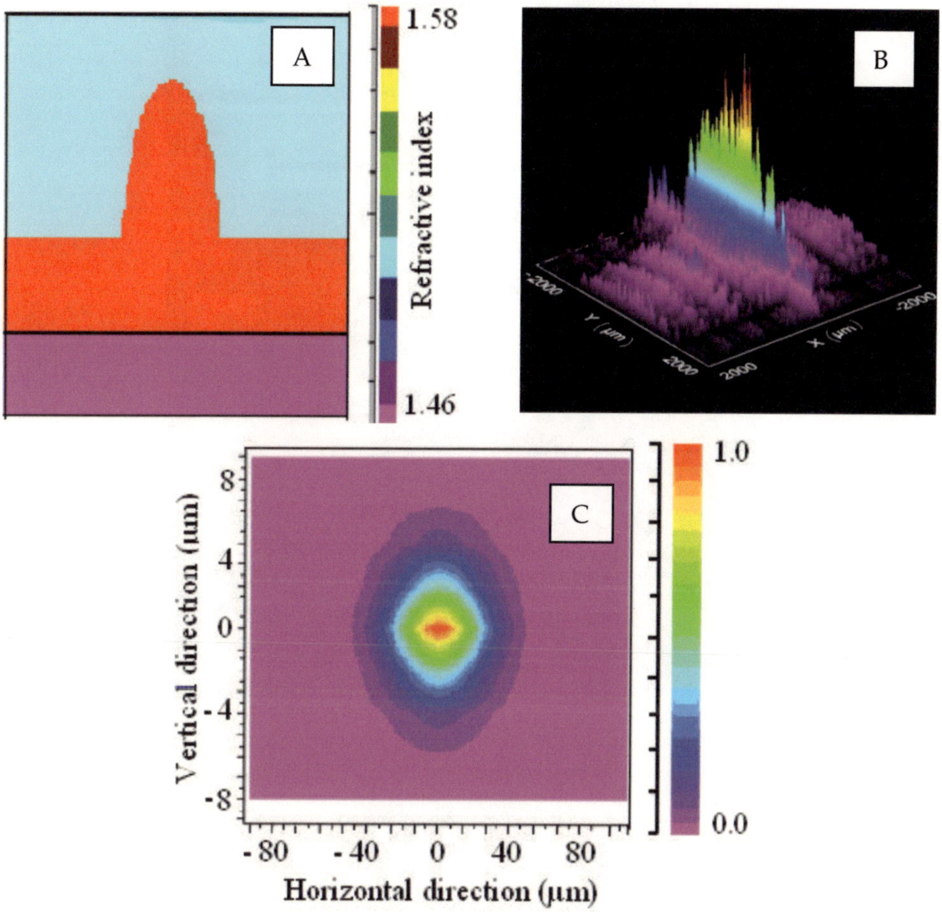

Figure 20. (A) Cross-section refractive index profile for typical rib waveguide, (B) output optical field (632 nm).distribution radiate from the end of rib channel waveguide, (C) optical far field distribution for rib channel waveguide calculated in RSOFT software.

Example of prepared Y-splitter of waveguide power is given on Fig. 21. Part A represents the microscopy image of applied top mask. Part B show the microscope image of prepared structure. Part C gives the light throughput prepared structure. It is well visible, that light are guided by prepared structure and divided into two beams according to structure geometry.

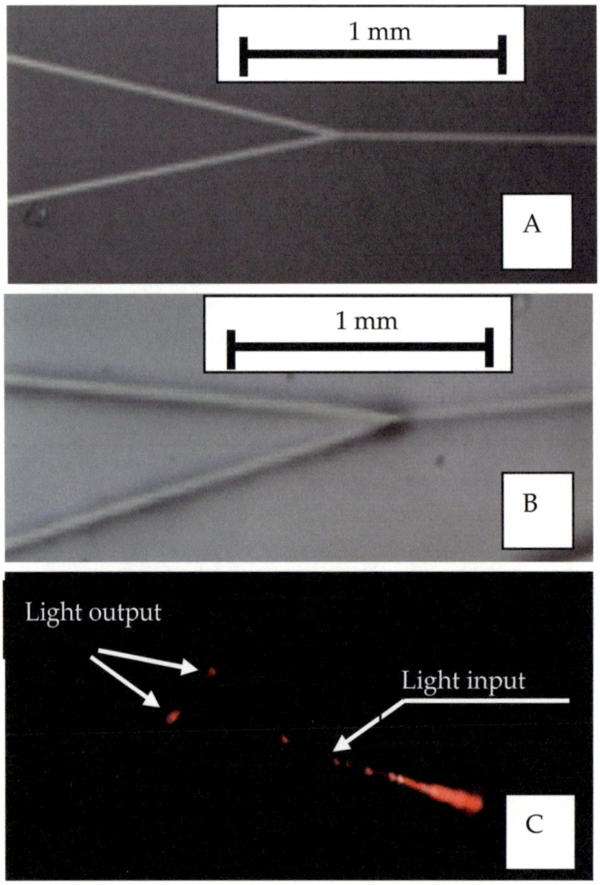

Figure 21. Junction of waveguide power. Part A represents the microscope image of applied top mask Part B show the microscope image of prepared structure. Part C give the light througput prepared structure.

6. Conclusions

In summary, this chapter gives a comprehensive insight into the problematic of examined new method for polymer surface patterning. For surface modification we used PMMA with different molecular weight and polymer photoresist. Mechanism of structure creation was analyzed and found to be dependent on surface tension gradient introduced by laser light scanning. Applications of prepared structures in photonics as diffraction elements or waveguide coupler are also given.

There is a considerable technological significance in the experiments we performed. It was shown that a nontactile nonablative method without wet chemistry is able to create surface structures of interesting properties. The polymer patterning step can take place in ambient air and can be strongly modulated by change of speed of polymer sample movement, molecular weight of polymer and dopant concentration.

Another described method of polymer surface modification consists in application of thin film instability enhanced by external electric field. Linear structures were created by the effect of locally limited electric field in thin polymer films heated to fluid temperature. The form and the shape of the structures were examined as a function of the exposure time, electric field intensity and the initial distance between mask and the film. It was shown that both continuous or dot structures can be created in dependence on the process parameters. For short processing times continuous structures are produced and with increasing times the continuous structures disintegrate into a system of dots. With increasing intensity of the electric field the height of the structures increases. The technique was successfully applied for preparation rib, channel waveguides. Optical properties of prepared light-guiding structures were measured and calculated using beam propagation method.

The above described technique seems to be prospective for fabrication of polymer based optical devices. The method is simple, free of disadvantages of conventional techniques. It gives highly reproducible results and is suitable for production of high-quality polymer waveguides. However, it should be noted, that the response to external electric field is not the same for different polymers. Apparently the quality and the shape of the structures created by the effect of the external electric field depend on the polymer type, its molecular weight, temperature of flow, molecular structure ect. Therefore the conditions of pattern preparation must be optimalized for each particular case.

In addition, surface coating of patterned polymer by thin metal layer were also described. Thermal treatment of prepared structure results in pattern disappearance or conservation of surface pattern, depending on the thickness and properties of metal coating.

Author details

Oleksiy Lyutakov, Jiri Tuma, Jakub Siegel, Ivan Huttel and Václav Švorčík
Department of Solid State Engineering, Institute of Chemical Technology, Prague, Czech Republic

Acknowledgement

This work was supported by the GA CR under the projects 108/11/P840, 108/11/P337, and 108/12/1168 and MPO CR under the project FR-TI3/797.

7. References

[1] Chou SY, Krauss PR, Renstrom PJ (1995) Appl. Phys. Lett., 67: 3114-3117
[2] Kumar A, Whitesides GM (1993) Appl. Phys. Lett., 63: 2002–2004.
[3] Quake SR, Scherer A (2000) Science, 290: 1536-1538.
[4] Lewis JA, Gratson GM (2004), Mater. Today, 7: 32-47.
[5] Gratson GM, Xu MJ, Lewis JA (2004) Nature, 428: 386-388.
[6] Yablonovitch E, (1987) Phys. Rev. Lett., 58: 2059-2061.
[7] Jin, HJ, Kaplan DL, (2003) Nature, 424: 1057-1060.
[8] Yokoyama S, Nakahama T, Miki H, Mashiko S (2003) Thin Solid Films, 15: 438-439,
[9] Li CF, Dong XZ, Jin F, Jin W, Chen WQ, Zhao ZS, Duan XM (2007) Appl. Phys. A, 89: 145–148.
[10] Klein S, Barsella A, Leblond H, Bulou H, Fort A, Andraud C, Lemercier G, Mulatier JC, Dorkenoo K (2005) Appl. Phys. Lett., 86: 211118-211120.
[11] Sherwood T, Young AC, Takayesu J, Jen AKY, Dalton LR, Chen AT, (2005) IEEE Phot. Tech. Lett., 17: 2107–2109.
[12] Guo R, Xiao SZ, Zhai XM, Li JW, Xia AD, Huang WH (2006) Opt. Express, 14: 810–816.
[13] Serbin J, Ovsianikov A, Chichkov B (2004) Opt. Express, 12: 5221–5228.
[14] Lyutakov O, Huttel I, Siegel J, Svorcik V (2009) Appl. Phys. Lett. 95: 17-19.
[15] Ho K, Tsou Y, (2001) Sensors Actuat. B 77: 253-256.
[16] Melloni A, Martinelli M (2002), J. Lightwave Techn. 20: 296-299.
[17] Lyutakov O, Tuma J, Prajzler V, Huttel I, Hnatowicz V, Svorcik V (2010) Thin Solid Films 519: 452-456
[18] Kafesaki M, Tsiapa I, Katsarakis N, Koschny T, Soukoulis CM, Economou EN (2007) Phys. Rev. B 75: 235114-235117.
[19] Malkin AY (2008) Colloid J. 70: 673-676..
[20] Bestehorn M, Pototsky A, Thiele U (2003) Europ. Phys. J. 33: 457-461.
[21] Reichenbach J, Linde H (1981) J. Colloid Interface Sci. 84, 433-432.
[22] Perez-Garsia C, Carneiro G (1991) Phys. Fluids A 3: 292-297.
[23] Ashley KM, Raghavan D, Douglas JF, Karim A (2005) Langmuir 21: 9518-9521.
[24] Pollock C, Lipson M (2003) Integrated Photonics, Kluwer Academic Publishers, Boston.
[25] Erbil HY, Meric RA (1997) J. Phys. Chem., 101: 6867-6873.
[26] Murphy C, Gole A, Hunyadi S, Orendorff C (2006) Inorg. Chem., 45:7544-7547.
[27] Nie SM, Emery SR, Science 1997, 275:1102-1108.
[28] Zhou J, Zhang L, Tuttle G, Koschny T, Soukoulis C (2006) Phys Rev B, 73:041101-041105.
[29] Zhang S, Fan WJ, Panoiu NC, Malloy KJ, Osgood RM, Brueck SRJ (2005) Phys. Rev. Lett., 95:137404-137407.
[30] Linden S, Enkrich C, Dolling G, Klein M, Zhou J, Koschny T, Soukoulis C, Burger S, Schmidt F, Wegener M Europ. Phys. J. (2006) 12:1097-1101.
[31] Schaffer E., Thurn-Albrecht T, Russel T, Steiner U (2000) Nature 403: 874-877.
[32] Schaffer E, Thurn-Albrecht T, Russel T, Steiner U (2001) Eur. Phys. Lett. 53: 518-521.
[33] Verma R, Sharma A, Kargupta K, Bhaumik J (2005) Langmuir 21: 3710-3712.
[34] Shankar V, Sharma A (2004) J. Colloid. Interf. Sci. 274: 294-298.

Chapter 7

The Properties and Application of Carbon Nanostructures

Petr Slepička, Tomáš Hubáček, Zdeňka Kolská, Simona Trostová, Nikola Slepičková Kasálková, Lucie Bačáková and Václav Švorčík

Additional information is available at the end of the chapter

http://dx.doi.org/10.5772/51062

1. Introduction

Nanocomposite carbon-based substrates are a large group of materials promising for medicine and various biotechnologies, particularly for coating biomaterials designed for hard tissue implantation, constructing biosensors and biostimulators or micropatterned surfaces for creation of cell microarrays for advanced genomics and proteomics. These substrates comprise nanocomposite hydrocarbon plasma polymer films, amorphous carbon, pyrolytic graphite, nanocrystalline diamond films, fullerene layers and carbon nanotube and nanoparticles-based substrates. Polymer/carbon composites have attracted increasing interest owing to their unique properties and numerous potential applications in the automotive, aerospace, construction and electronic industries.

Amorphous carbon, also referred to as diamond-like carbon (DLC), possesses a number of favourable properties, such as high hardness, a low friction coefficient, chemical inertness and high corrosion resistance, which is due to its particular structure, i.e. cohabitation of the sp^2 and sp^3 phases [1]. These properties make DLC attractive for coating bone and dental implants coating bone and dental implants in order to improve the resistance of these devices against wear, corrosion, debris formation and release of metallic ions, which can act as cytotoxic, immunogenic or even carcinogenic materials [2,3]. However, unmodified amorphous carbon usually acts as bioinert, i.e. not promoting its colonization with cells, which property prevents hemocoagulation, thrombosis and inflammatory reaction on the surfaces [4]. DLC coated materials have been utilized for construction of articular surfaces of joint prostheses [2] or blood-contacting devices (intravascular stents, mechanical heart valves, pumps).

From all nano-sized carbon allotropes, diamond has been often considered as the most advantageous material for advanced biomedical an biosensoric applications, which is mainly due to the absence of its cytotoxicity, immunogenicity and other adverse reactions [5,6]. Other remarkable properties of nanodiamond, enabling its application in biotechnologies and medicine (particularly in hard tissue surgery), are high hardness, a low friction coefficient, and also high chemical, thermal and wear resistance. In our earlier studies and in studies by other authors, nanodiamond has proven itself as an excellent substrate for the adhesion, growth, metabolic activity and phenotypic maturation of several cell types in vitro, including osteogenic cells [5,7]. An interesting issue is doping of NCD films with boron. This doping renders the NCD films electroconductive [8]. Boron-doped NCD films have been applied in electronics and sensorics, e.g. for he construction of sensors for DNA hybridization [9], bacteria [10] or glucose [11].

Graphite is one of the most common allotropes of carbon, and the most stable form of carbon under standard conditions. However, despite its electrical conductivity, which is usually associated with the stimulatory effects on cell colonization and functioning, unmodified graphite is rather bioinert, i.e., less adhesive for cells [12]. It is due to a relatively low ability of graphite to adsorb cell adhesion-mediating proteins from the serum supplement of the culture medium [13] and also bone morphogenetic proteins (BMP), i.e. factors promoting the osteogenic cell differentiation [14]. Fullerenes are spheroidal molecules and are made exclusively of carbon atoms (e.g. C_{60}, C_{70}). Their unique hollow cage-like shape and structural analogy with clathrin-coated vesicles in cells support the idea of the potential use of fullerenes as drug or gene delivery agents [15]. Fullerenes display a diverse range of biological activity, which arises from their reactivity, due to the presence of double bonds and bending of sp^2 hybridized carbon atoms, which produces angle strain. Fullerenes can act either as acceptors or donors of electrons. When irradiated with ultraviolet or visible light, fullerenes can convert molecular oxygen into highly reactive singlet oxygen. Thus, they have the potential to inflict photodynamic damage on biological systems, including damage to cellular membranes, inhibition of various enzymes or DNA cleavage. This harmful effect can be exploited for photodynamic therapy of tumors [16], viruses including HIV-1 [17], broad spectrum of bacteria and fungi [18]. On the other hand, C_{60} is considered to be the world's most efficient radical scavenger. This is due to the relatively large number of conjugated double bonds in the fullerene molecule, which can be attacked by radical species. Thus, fullerenes would be suitable for applications in quenching oxygen radicals, and thus preventing inflammatory and allergic reactions [19] and damage of various tissues and organs, including blood vessels [20] and brain [21]. Finally, carbon nanotubes are formed by a single cylindrically-shaped graphene sheet (single-wall carbon nanotubes, referred usually to as SWNT or SWCNT) or several graphene sheets arranged concentrically (multi-wall carbon nanotubes, referred to as MWNT or MWCNT). Carbon nanotubes have excellent mechanical properties, mainly due to sp^2 bonds. The tensile strength of single-walled nanotubes is about one hundred times higher than that of the steel, while their specific weight is about six times lower [22]. Thus, carbon nanotubes could be utilized in hard tissue surgery, e.g., to reinforce artificial bone implants, particularly scaffolds for bone tissue engineering made of relatively soft synthetic or natural polymers. In our earlier studies, nanotubes were combined with termoplasts of polytetrafluoroethylene, polyvinyldifluoride and polypropylene, which significantly enhanced its attractiveness of nanotube-based substrates for colonization with bone-derived cells [23].

Recently, nanotechnology has gained much attention in research to develop new carbon-based materials with unique properties. Nanotechnology can be broadly defined as the creation, processing, characterization and use of materials, devices, and systems with dimensions in the range 0.1–100 nm, exhibiting novel or significantly enhanced physical, chemical, and biological properties, functions, phenomena, and processes due to their nanoscale size [24]. Ultrathin carbon films can be used for analytical applications, e.g. carbon micro-arrays for transmission electron microscopy [25], high resolution microscopy [26], microelectromechanical systems [27] or electrodes for corrosion sensor applications at high temperatures [28]. "Carbon composites" has attracted increasing interest owing to their unique properties and numerous potential applications in the automotive, aerospace, construction and electronic industries [29]. Diamond-like carbon based films on polymer substrates can strongly influence gas barrier performance [30]. Polymer/carbon nanoparticle systems can be used as polar vapour sensors [31]. Thin films on a polymer-fullerene base are used for a hybrid solar cells construction [32]. The intensive investigations of carbon nanolayers and carbon/polymer nanocomposites [33] stimulated remarkably by the discovery of carbon nanotubes [34], fullerenes [35] and graphene layers [36] resulted in the conclusion that the character of the carbon atom connections in the carbon layer has crucial importance for the structure and the properties of carbon nanoparticles and thin layers. Sputtered or evaporated [37] carbon structures can create nanostructures of different electrical or morphological properties. The opportunities for systematic investigation of nanolayers structures [38-42] are very promising for new application both in electronics or nanoengineering and biomedicine.

2. Carbon nanolayers

Thin carbon layers are considered as a prospective material for a wide range of biomedical application [39,43-45], e.g. tissue regeneration [46], controlled drug delivery [47], surface coating for bone-related implants [48], increase of resistance to microbial adherence, blood interfacing implants applications [49] or neuronal growth.

2.1. Carbon layer flash evaporation

Little attention has been devoted to the study of carbon layers prepared by the simplest deposition technique – flash evaporation. By flash evaporation carbon layers of different thickness can be produced for routine SEM and TEM electron microscopy. In general, there is a need for these layers to be fine grain, even coating, with uniform and controllable layer thickness. Flash deposition is distinguished from other techniques (e.g. vacuum evaporation, ion beam) by short deposition time and low total power input. The thickness of flash prepared carbon layer should be controlled, but at present none of the conventional methods in general use allows precise and reproducible deposition and layer thickness control. Flash deposition can be accomplished either by rapid evaporation of a carbon filament or by pulsed laser vaporization of a carbon target [50]. The former technique is based on rapid evaporation of carbon filament caused by an electric discharge. The majority of the evaporation material is believed to be in the form of molten globules. The carbon layers can be deposited by flash evaporation onto polymer substrates (e.g. PET and PTFE). Their properties are of interest for many potential applications mentioned above and, in our case, for their usage in the study of interaction of living cells with carbonaceous materials and carbon based structures with potential applications in medicine. Physical, chemical and electronic properties of the deposited carbon layer were studied as a function of the distance between the substrate and the carbon source and the layer thickness.

2.1.1. Layer thickness

The dependence of the layer thickness, measured by the scratch technique, on the distance between the substrate and the carbon filaments is shown in Fig. 1 for arrangements with one and two filaments. It is obvious that the layer thickness decreases with deposition distance monotonically, as expected. It should be noted that the deposition process is affected by several phenomena. The filament does not represent a point-like carbon source [51] and during the evaporation process it breaks without fully evaporating along its complete length. The flash evaporation proceeds in rather low vacuum and evaporated carbon particles under a large number of inelastic scattering events with molecules of residual gas on the way from the filament to the substrate, by which their energy is reduced and flight direction is changed randomly.

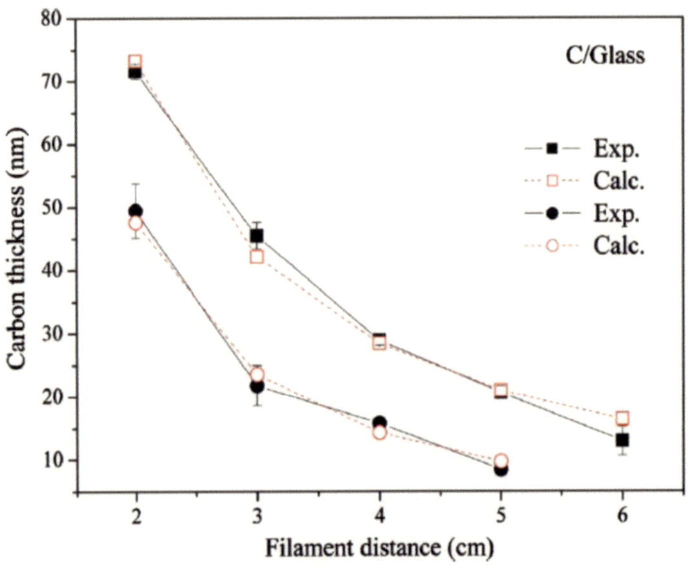

Figure 1. Dependence of the thickness of the flash-evaporated carbon layer on the distance of glass substrate from 1 (circle points) and 2 (square p.) filaments, determined by AFM technique. The values calculated according to formulae presented in [37].

2.1.2. Chemical composition and structure

The typical XPS spectrum of carbon layer deposited onto PTFE is introduced in Fig. 2. One can see that besides of carbon the oxygen is also observed. The presence of oxygen is explained by oxygen absorption from residual atmosphere during deposition process. C1s and O1s peaks correspond to about 94.1 and 5.8 at. % of carbon and oxygen concentration, respectively. The presence of carbonyl, carboxyl and hydroxyl structures in the carbon layer were proved. Hydrogen depth profile was determined by ERDA. It was found that the concentration of carbon and hydrogen decreases with increasing depth while the concentration of fluorine increases. No hydrogen is detected beyond 75 nm. This observation may indicate that the thickness of deposited carbon layer is about 75 nm (in accordance with measurements performed on glass samples) and at larger depths the composition approaches pristine PTFE [37].

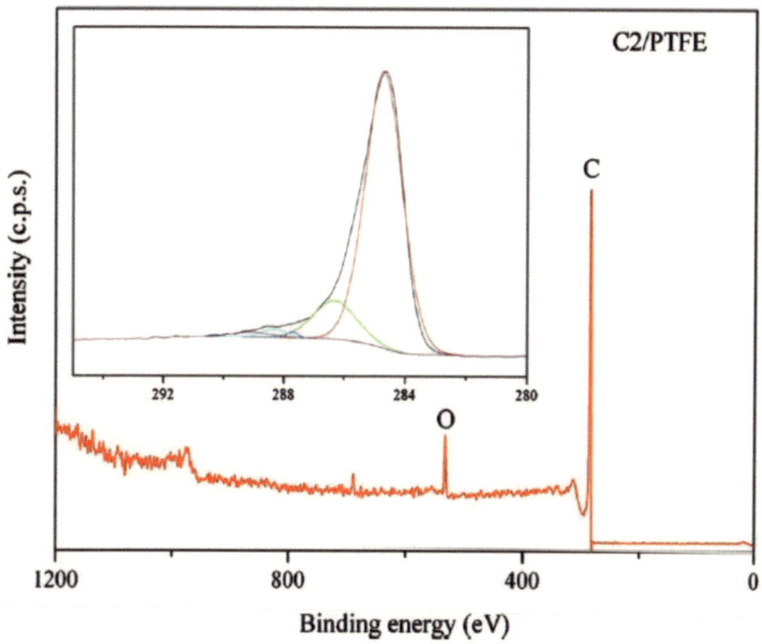

Figure 2. XPS spectrum of the carbon layer deposited from 2 cm distance on PTFE; decomposition of the O1s band is shown too [37].

2.1.3. Surface properties

Surface wettability depends on surface chemical structure [52] and is commonly characterized by contact angle. Contact angle on polymers was studied as a function of the distance of PET and PTFE substrates from the filament. It was observed on PTFE substrate that the contact angle increases with an increasing deposition distance (thinner layer thickness), while on PET the contact angle does not change within experimental errors for deposition distances from 2 to 7 cm. According to XPS analyses chemical composition of the deposited carbon layer is the same for both polymers with low concentration of oxidized, polar structures. It is supposed that the wettability might also be affected by surface morphology and roughness of both polymers. While the roughness of PET after carbon coating remains unchanged within experimental error, a dramatic change of the roughness is observed on PTFE. With increasing thickness of the carbon layer the PTFE roughness decreases from 13.3 nm for pristine PTFE to 2.9 nm for thickest layer (deposition from 2 cm). Significant differences in the surface morphology are found between both polymers before and after carbon deposition (see Fig. 3). For pristine PET the surface is composed of tiny, rounded formations, homogenously distributed over the sample surface. Carbon deposition does not result in any significant change in the surface morphology and roughness. Surface of the pristine PTFE is markedly wrinkled and its roughness is higher comparing to PET. Carbon deposition results in a dramatic morphology change and roughness declines indicating a preferential carbon accumulation into holes.

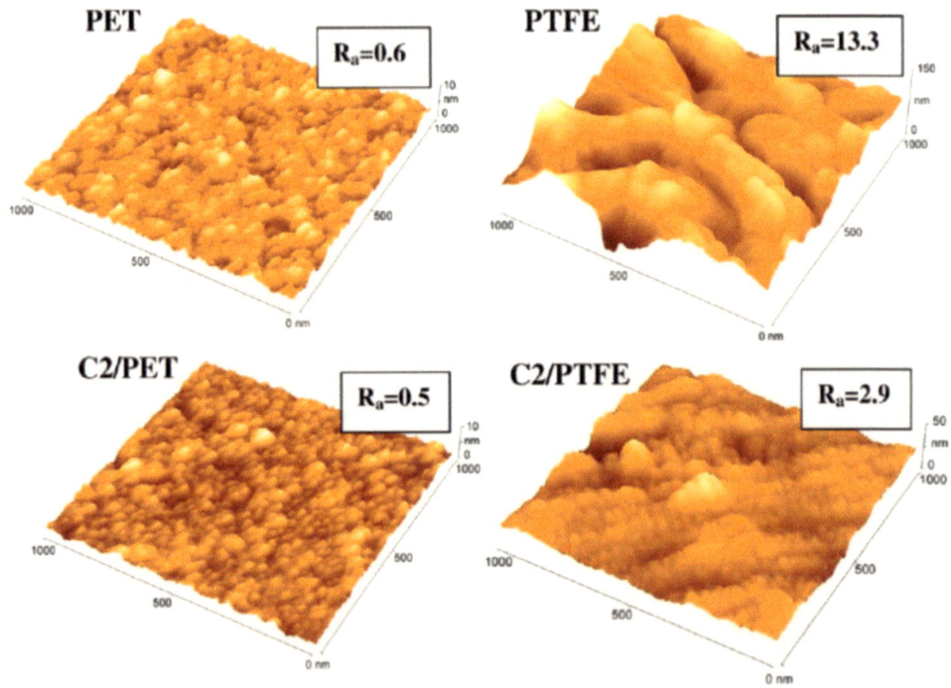

Figure 3. AFM images of pristine PET and PTFE and those coated with carbon from the deposition distance of 2 cm (C2/polymer). R_a is surface roughness in nm [37].

2.1.4. *Electrical properties*

Carbon deposition results in a rapid resistance decrease (comparing to pristine polymers) indicating formation of continuous carbon layer on the polymer. The decrease of electrical sheet resistance (R_s) is more pronounced on PET, probably due to lower surface roughness. According to our previous results [37] the conditions of the layer deposition were chosen to obtain about 70 nm thick carbon layers. For the measurement of the sheet resistance R_s as a function of the temperature, the C/PTFE samples were placed in the cryostat cooled by LN_2.

Temperature dependence of the sheet resistance, measured in the temperature range from 80 to 350 K, is shown in Fig. 4a. One can see that over a broad temperature range the sheet resistance decreases rapidly with increasing temperature. The decrease is typical for semiconductors; "semiconductance" of amorphous carbon (a-C) was reported earlier, e.g. in [53]. It is believed that the mechanism of the charge transport in carbon layers proceeds according to variable range hopping (VRH) mechanism, suggested by Mott [54], which depends on the density of the states present near the Fermi level. The same dependence as in Fig. 4a is shown in Fig. 4b in ln $R_s(T)$ vs. $T^{-1/4}$ representation. One can see that the VRH model describes well the charge transport in the temperature range from 80 to 350 K. It is supposed that the electron states are localised on carbon clusters sp^2, dispersed randomly within the carbon layer. After application of an external electrical field electron hopping between the clusters takes place. The electron movement proceeds via phonon-assisted tunneling and with decreasing temperature the electrons tend to hop to larger distances on sites which are energetically closer than the nearest neighbour.

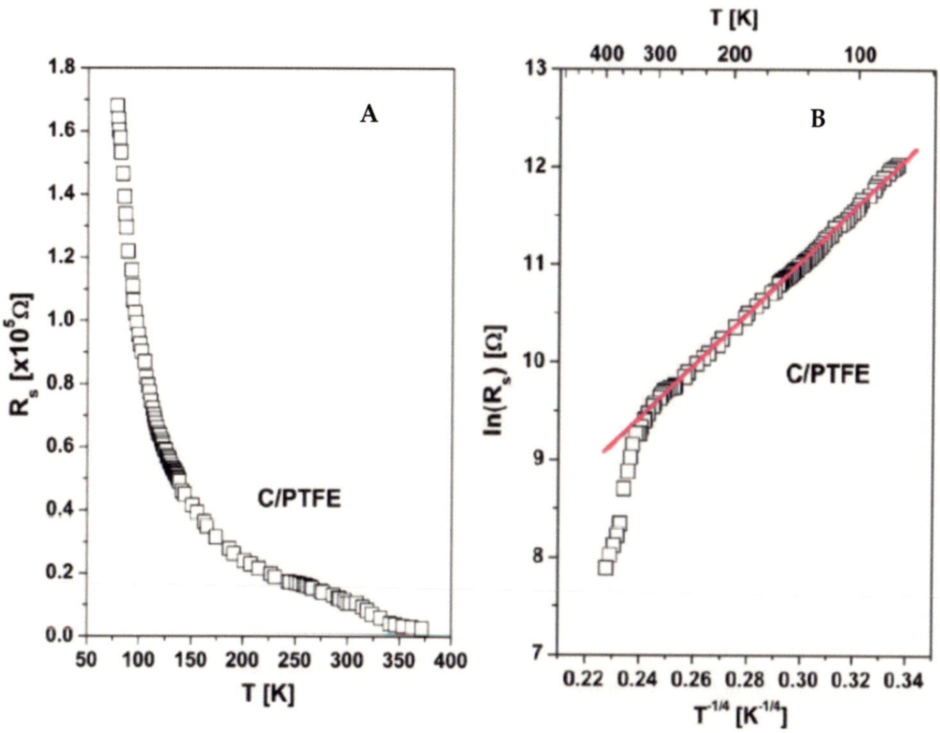

Figure 4. Temperature dependence of the sheet resistance (R_s) for carbon layer, 70 nm thick, flash evaporated onto PTFE (a) and temperature dependence of the sheet resistance R_s in form ln R_s vs. $T^{-1/4}$ (b) for the same sample [41].

2.1.5. Zeta potential

Electrokinetic analysis results of flash deposited carbon layers on PTFE are presented in Fig. 5 [41]. It is clear, that after carbon deposition decreases zeta potential obtained by

Helmholtz-Smoluchowski equation due to creation of carbon layer. Carbon layers embody the similar behavior as gold layer [55]. The thicker carbon layer the lower zeta potential value. Results for sample distance of 4 and 7 cm are the same due to similar value of surface roughness [41]. After deposition from distance 2 cm zeta potential dramatically decreases due to significant decrease of thickness of carbon layer, surface roughness and sheet resistance. Difference between zeta potential obtained by both of used methods and equations (streaming current, Helmholtz-Smoluchowski eq. and streaming potential, Fairbrother-Mastins eq. [41]) is significant for pristine PTFE due to great surface roughness. For other samples this difference increases with decreasing distance of filaments. It can be explained by increasing surface conductivity, which plays an important role in zeta potential calculation and comparison. Zeta potential obtained by Fairbrother-Mastins equation increases due to increasing polarity of surface, which is explained by creation of polar groups on surface. It can be concluded from Fig. 5 that carbon layer deposited from distance 2 cm is the most conductive.

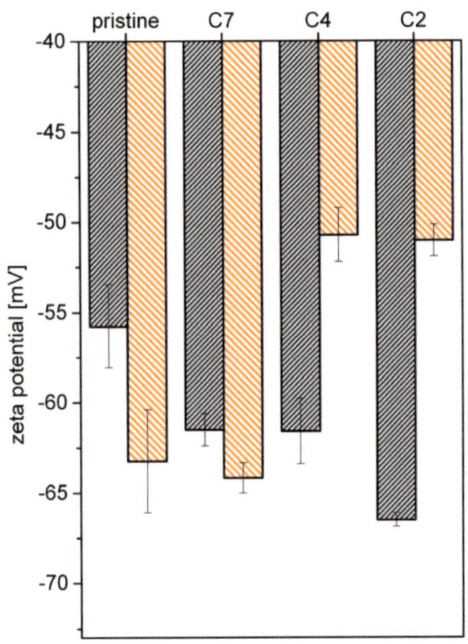

Figure 5. Zeta potential of pristine PTFE (pristine) and PTFE with carbon flash evaporated layer from distances 2, 4 and 7 cm (C2, C4 and C7). Black columns represent data obtained by streaming current method (Helmholtz-Smoluchowski equation), the orange ones by streaming potential (Fairbrother-Mastins equation) [41].

2.1.6. Cell adhesion and proliferation

Cytocompatibility of samples was determined from *in vitro* experiments on adhesion and proliferation of LEP cells (human diploid fibroblastoids) performed on pristine and carbon coated PTFE. For comparison the same experiments were performed on „tissue polystyrene" (PS) too. The results are presented in Fig. 6. It is seen that the carbon coating increases adhesion and proliferation (Fig. 6) of LEP cells significantly in comparison with pristine PTFE [41].

Cell proliferation 3 days after the seeding is comparable with tissue PS. For both cell adhesion and proliferation maximum positive effect is seen on the samples carbon coated from the distance 4 cm. In this case the carbon layer is about 32 nm thick and it exhibits the higher roughness, good electrical conductivity and contact angle of about 80°. Low surface roughness and wettability seem to have negative effect on the cell adhesion and proliferation [41].

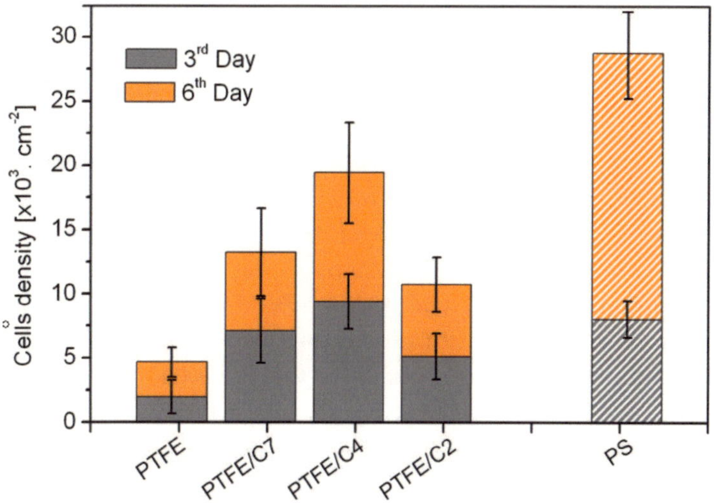

Figure 6. The proliferation of LEP cells proliferation (3rd and 6th day) on pristine PTFE, PTFE with carbon layers prepared by carbon flash evaporation from the distances 2, 4 and 7 cm (C2, C4 and C7) and on the „tissue" polystyrene PS [41].

2.2. Carbon layer sputtering

Different types of carbon layers can be preferentially prepared by various types of carbon deposition. DLC films can be deposited using DC plasma chemical vapor deposition, radio frequency magnetron sputtering or ion beam-based methods. DLC (polycrystalline diamond) needs high temperatures to be deposited [56]. Amorphous carbon can be prepared at low temperatures by different techniques, but its physical, chemical and mechanical properties depend on the deposition conditions, mainly on the temperature and hydrogen content [57]. Hydrogenated amorphous carbon (a-C:H) is usually prepared by plasma-assisted CVD of hydrocarbons (i.e. methane or ethylene). Amorphous carbon (a-C) is prepared by PVD techniques such as sputtering, arc discharge or pulsed laser deposition. Amorphous hydrogenated carbon is unstable under thermal treatment since it tends to eliminate hydrogen and transform in to a more stable graphitic structure. As an alternative method for thin layer preparation the sputtering method was chosen. Carbon layers on polyethyleneterephtalate (PET) backing were prepared by sputtering from graphite target. The deposited layers were characterized by different techniques (UV-VIS, Raman spectroscopy, RBS, AFM) and the biocompatibility of the layers was studied by cultivation of 3T3 mouse fibroblasts.

2.2.1. *Structure of sputtered carbon layers*

UV-VIS spectrometry is used frequently to follow the changes in chemical structure of polymers. Absorbance increase indicates an increase of the concentration of structures with certain length (number) of conjugated double bonds. Longer structures absorb on longer wave lengths [58]. It was determined that the amount of π bonds (sp^2 hybridization) and the length of conjugated double bonds are increasing functions of the sputtering time. The thickness of the deposited carbon layer increases with increasing sputtering time as could be expected. The structure of the carbon layers can be characterized by Raman spectroscopy [59]. Raman peak at 1360 cm^{-1} is attributed to disordered mode of graphite and that at 1500–1550 cm^{-1} corresponds to an amorphous-like structure with sp^3 + sp^2 bonding [59]. Fig. 7 shows the Raman spectra from pristine PET and PET with carbon layer deposited for 30–90 min. The carbon deposition for the times up to 30 min does not result in any observable changes in the spectra. The deposition for longer times leads to appearance of a signal in 1100–1700 cm^{-1} region, the intensity of the signal being an increasing function of the deposition time. All spectra exhibit a broad peak at 1530 cm^{-1} indicating that the deposited layers are composed mostly of amorphous carbon with sp^3 and sp^2 bonds. Small peak at 1360 cm^{-1}, which is also present in all spectra, is due to the presence of disordered graphite. The results of Raman and UV-VIS spectroscopy show that the thickness of the deposited layers increases with increasing sputtering time.

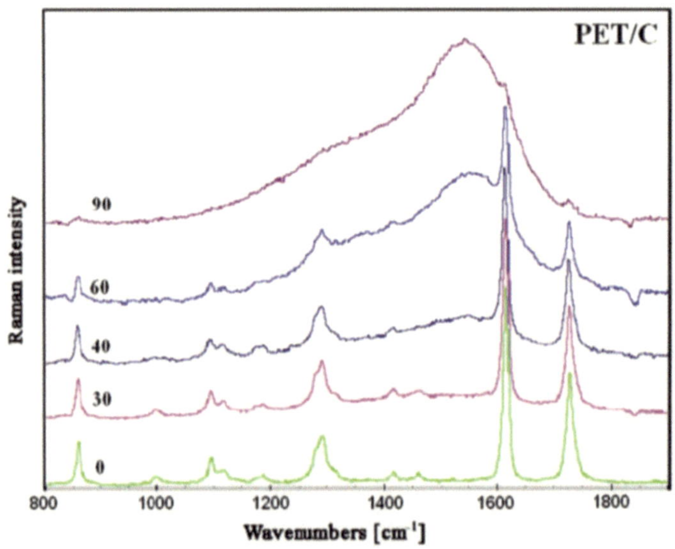

Figure 7. Raman spectra from pristine PET and PET samples with carbon layers deposited for different sputtering times (min) as indicated in the figure [45].

ERDA (Elastic Recoil Detection Analysis) revealed information on hydrogen concentration and its depth profile in the deposited layers. RBS spectra provided information on carbon and oxygen concentration and on the layer total thickness. It was observed that for the sputtering times above 45 min the composition of the deposited layers does not depend, within RBS and ERDA experimental errors, on the deposition time. The measured concentrations varied from 7 to 9 at. % for oxygen and from 16 to 26 at. % for hydrogen. These concentrations were significantly lower than those in pristine PET. The layer thickness increased roughly linearly with increasing deposition time.

The layer morphology changes as a function of the sputtering time and the layer thickness. It was found that after 15 s of deposition carbon creates rounded, regular grains the size of which is larger compared to those observed for longer sputtering times [45]. For the sputtering times above 30 s the carbon grains become smaller but some irregularities arise. The surface roughness is an increasing function of the deposition time.

2.2.2. *Cell growth*

The carbon layers-PET structures were used as a substrates for cultivation of 3T3 mouse fibroblasts (Fig. 8) [38]. The number of adhering cells increases with increasing deposition time for deposition times up to 30 min (Fig. 9). For longer deposition times on the contrary the cell number decreases with increasing deposition time (Fig. 9). Possible explanation of the decline may be found in unfavourable surface morphology and roughness of the layers deposited for longer times. Possibly the adhering cells prefer smooth surface without sharp irregularities. An effect of the layer continuity or discontinuity can not be excluded, too. It should also be noted that in the present case a non-polar material (carbon) is deposited onto polar substrate. The surface modification of PET by carbon has the positive influence on cells adhesion and sample sputtered 30 min has the greatest amount of adhered cells.

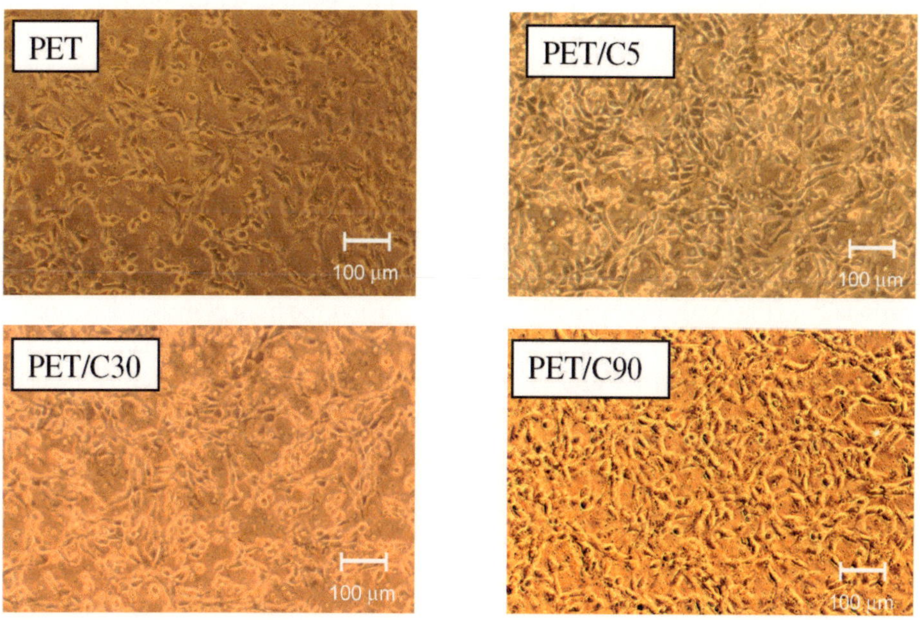

Figure 8. Cells distribution on PET with different time of carbon deposition. The equable distribution of the 3T3 fibroblasts on the PET after carbon deposition is shown [38].

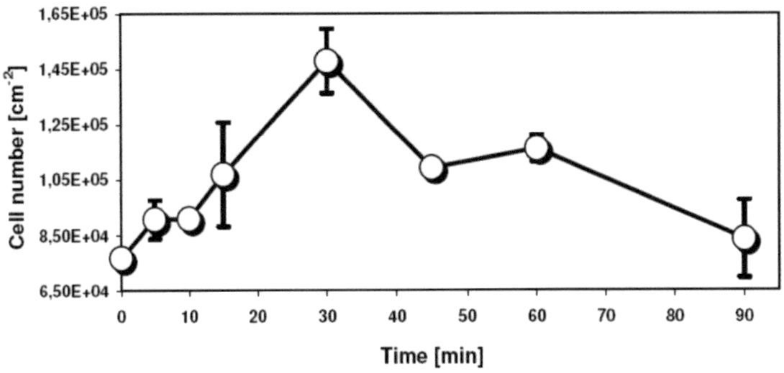

Figure 9. Dependence of number of cells on the time of carbon deposition by magnetron sputtering on the PET [38].

2.3. Fullerenes

Above other interesting properties, fullerenes emit photoluminescence which could be utilized in advanced imaging technologies [60]. In their pristine unmodified state, fullerenes are highly hydrophobic and water-insoluble. On the other hand, they are relatively highly reactive, which enables them to be structurally modified. Fullerenes can form complexes with other atoms and molecules, e.g. metals, nucleic acids, proteins, synthetic polymers as well as other carbon nanoparticles, e.g. nanotubes. In addition, fullerenes can be functionalized with various chemical groups, e.g. hydroxyl, aldehydic, carbonyl, carboxyl, ester or amine group, as well as amino acids and peptides. This usually renders them soluble in water and intensifies their interaction with biological systems [61]. Despite all these exciting findings, relatively little is still known about the influence of fullerenes, particularly when arranged into layers and used for biomaterial coating, on cell-substrate adhesion, subsequent growth, differentiation and viability of cells, especially bone-forming cells.

2.3.1. Physical and chemical properties of fullerene C_{60} layers

Fullerenes C_{60} were deposited onto microscopic glass coverslips by evaporation of C_{60} in the Univex-300 vacuum system (Leybold, Germany). The thickness of the layers increased proportionally to the temperature in the Knudsen cell and the time of deposition. Four types of layers of different thickness and morphology were prepared: thin continuous, thick continuous, thin micropatterned and thick micropatterned. The micropatterned layers were created by deposition of fullerenes through a metallic mask with rectangular holes [62].

Raman analysis was performed on micropatterned samples immediately after deposition and then after sterilization with ethanol. Immediately after deposition, the Raman spectra showed that the fullerene films were prepared with high quality [62]. After sterilization with ethanol, the thin micropatterned fullerene layers were almost intact, and a considerable amount of fullerenes was found not only on sites underlying the openings of the grid, but also below its metallic part. However, in thick micropatterned layers, an analysis of the vibration mode showed that the C_{60} molecules reacted with oxygen or polymerized [62].

The proportion of C_{60} molecules involved in these chemical changes and reached about 50%. Moreover, the amount of fullerenes below the metallic bars of the grid was very low, though still detectable. The color intensity increased with layer thickness, while the transparency of the layers in a conventional light microscope decreased [62]. Despite of this, the cells on both continuous layers were well observable, even those native and non-stained (Fig. 9A–D). On thick micropatterned layers, the bulge-like prominences were relatively dark, and the contrast between the bulges and grooves was relatively high (Fig. 9E). In addition, it was not possible to focus the cells on bulges and in grooves simultaneously, whereas the fluorescence signal from both groups of cells was observable. Thus, the presence and morphology of cells on bulges and in grooves was evaluated using fluorescence microscopy (Fig. 9F).

Reflection goniometry showed that all fullerene C_{60} layers were relatively highly hydrophobic. The continuous and micropatterned layers had similar water drop contact angles ranging from 95.3 ± 3.1° to 100.6 ± 6.8°.

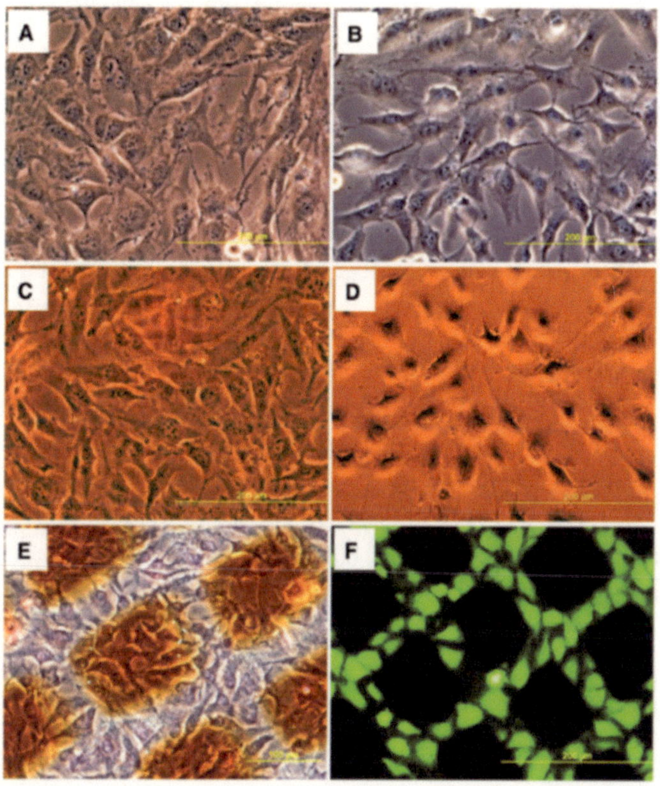

Figure 10. Morphology of human osteoblast-like MG 63 cells in cultures on a polystyrene dish (A) a microscopic glass coverslip (B), thin continuous (C), thick continuous (D), thin micropatterned (E) and thick micropatterned (F) fullerene C60 layers. A–D: native cultures; E: a culture stained with hematoxylin and eosin; F: a culture stained with LIVE/DEAD viability/cytotoxicity kit (Invitrogen). A–D: day 5 after seeding, E–F: day 7 after seeding. Olympus IX 50 microscope, DP 70 digital camera, obj. 20x, bar=200 μm except E, where bar=100 μm [62].

2.3.2. Adhesion and proliferation of cells

On day 1 after seeding, the cells on both continuous thin and thick fullerene layers adhered at similar numbers (3420±420 cells cm^{-2} and 2880±440 cells cm^{-2}, respectively), which was comparable to the values found on standard cell culture substrates, represented by the tissue culture polystyrene dish (3080±290 cells cm^{-2}) and the microscopic glass coverslip (2560±310 cells cm^{-2}) [62]. On both micropatterned thin and especially thick fullerene layers, the average cell population densities tended to be lower (by 11 to 43 %) than both polystyrene and glass, but these differences were not statistically significant. The cells colonized practically exclusively the grooves (Fig. 9F), thus they used less space for their proliferation. Although the grooves occupied only 41±1 % of the material surface, they contained from 80±4 % to 98±1 % of the total cells on the material surface [62]. The cell population density in the grooves was about 5 to 57 times higher than on the bulges, and these differences increased with time of cultivation. On the other hand, on the thin micropatterned films, the cells colonized homogeneously the entire surface of the sample (Fig. 9E) and the percentage as well as the population density of cells in the grooves and on the bulges were similar.

2.3.3. Presence and spatial arrangement of β1-integrins, talin, β-actin and osteopontin

As revealed by immunofluorescence, MG 63 cells on both continuous and micropatterned fullerene layers were intensively stained for β1 integrins and talin, i.e. molecules participating in cell-substrate adhesion, β-actin, an important component of the cytoplasmic cytoskeleton, as well as osteopontin, a marker of osteogenic cell differentiation. This staining intensity was similar as in cells on the control polystyrene culture dish and microscopic glass coverslips (Fig. 10). All these molecules (particularly extracellular matrix protein osteopontin) were found in fine granular distribution throughout the cells, often preferentially located in the perinuclear region [62]. In addition, both β1 integrins and talin also formed dot- or streak-like focal adhesion plaques, visible mainly on the cell periphery.

Beta-1 integrin-containing focal adhesion plaques were particularly well developed and were often located on fine long protrusions formed by cells, which was accompanied by the formation of a fine mesh-like β-actin cytoskeleton. No apparent differences in the staining intensity and distribution of all molecules mentioned here were found between cells growing on thin and thick micropatterned fullerene layers or in cells in grooves and on bulges [62].

2.4. Carbon nanoparticles

Carbon nanoparticles, nanotubes and nanodiamonds, are considered as promising building blocks for the construction of novel nanomaterials [63,64] for emerging industrial technologies, such as molecular electronics, advanced optics or storage of hydrogen as a potential source of energy. In addition, they are considered as promising materials for biomedical applications, such as photodynamic therapy against tumors and infectious agents, quenching oxygen radicals, biosensor technology, simulation of cellular components, such as membrane pores or ion channels, as well as controlled drug or gene delivery,

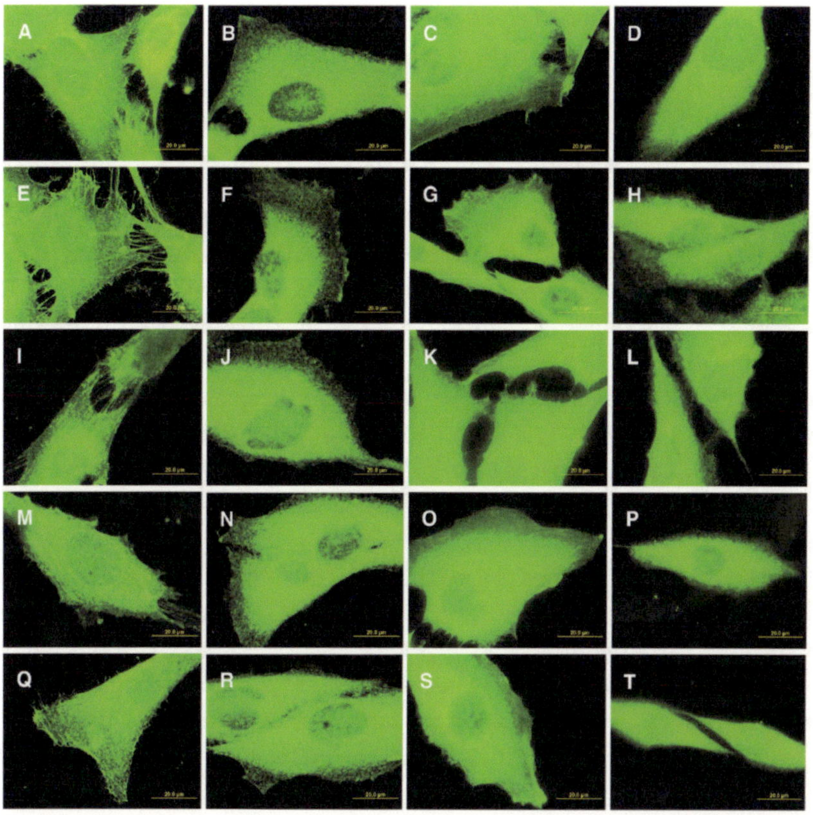

Figure 11. Immunofluorescence staining of β1 integrins (A, E, I, M, Q), talin (B, F, J, N, R), β-actin (C, G, K, O, S) and osteopontin (D, H, L, P, T) in human osteoblast-like MG 63 cells on day3 after seeding on microscopic glass coverslips (A–D), thin continuous (E–H), thick continuous (I–L) thin micropatterned (M–P) and thick micropatterned (Q–T) fullerene C60 layer. Olympus epifluorescence microscope IX 51, digital camera DP 70, obj. 100×, bar=20 μm [62].

particularly targeting the mineralized bone tissue [65]. Despite these exciting perspectives, relatively little is known about the influence of carbon nanoparticles present on the biomaterial surface on the adhesion and growth of cells.

Therefore the three types of materials modified with carbon particles were prepared: (i) carbon fibre-reinforced carbon composites (CFRC), materials promising for hard tissue surgery, coated with a fullerene C_{60} layer, (ii) terpolymer of polytetrafluoroethylene, polyvinyldifluoride and polypropylene mixed with 4 wt. % of single or multi-walled carbon nanotubes and (iii) nanostructured or hierarchically micro- and nanostructured diamond layers deposited on silicon substrates [23].

The materials were seeded with human osteoblast-like MG 63 cells (density from 8500 cells cm^{-2} to 25 000 cells cm^{-2}) [23]. On the fullerene layers, the cells (day 2 after seeding) adhered in number from 2.3 to 3.5 times lower than those on control non-coated CFRC or polystyrene dishes. However, their spreading area was larger by 68 % to 145 % than that on the control surfaces. On diamond layers, the number of initially adhered cells was higher on the nanostructured layers, whereas the subsequent proliferation was accelerated on the layers with a hierarchical micro-and nanostructure [23].

2.4.1. Carbon fibre-reinforced carbon composite (CFRC)

Carbon fibre-reinforced carbon composite (CFRC) was coated with fullerene layer. The fullerene coating did not significantly change this surface microroughness but created a nanostructured pattern on the pre-existing microarchitecture of the CFRC surfaces. The contact angle was unmeasurable due to a complete absorption of the water drop into the fullerene layer, which suggested a certain non-compactness or porosity of this layer. The contact angle of the non-coated CFRC was 99.5±1.0°. The release of fullerenes into the culture media and their cytotoxic action seemed to be less probable in our experiments. The cell attachment and spreading on the uncoated regions of them fullerene-modified CFRC, and also on the bottom of the polystyrene dishes containing fullerene-coated samples, were similar as in the control polystyrene dishes without fullerene samples. At the same time, the fullerene layer was resistant to mild wear, represented by swabbing with cotton, rinsing with liquids (water, phosphate-buffered saline, culture media) and exposure to cells and proteolytic enzymes used for cell harvesting. After these procedures and/or one-year-storage at room temperature in dark place, the Raman spectra did not change significantly [23]. On day 2 after seeding, the cell population density (Fig. 11A) on the CRFC surfaces was lower than that on the control uncoated material and TCPS, which could be due to the relatively high hydrophobicity of the non-functionalized fullerenes. In addition, the fullerene-coated CFRC surfaces were stronger and less prone to release carbon particles, which is an important limitation of the potential biomedical use of CFRC [66]. Moreover, the spreading area of cells on the fullerene-coated samples amounted to 3,182 ± 670 mm^2, while on both control surfaces it was only 1888 ± 400 and 1300 ± 102 mm^2 (Fig. 11B). This could be explained by the low cell population density on the fullerene layer, which provided the cells with more space for them to spread [23].

2.4.2. Carbon nanotube–polymer composites

Similarly as on the fullerene layer, the cells on PTFE/PVDF/PP mixed with single-wall carbon nanohorns (SWNH) or multiwall nanotubes (MWNT-A) were well spread, polygonal, and contained distinct beta-actin filament bundles, whereas most cells on the pure terpolymer were less spread or even round and clustered into aggregates [23]. The enzyme-linked immunosorbent assay (ELISA) revealed that the cells on the material with SWNH contained a higher concentration of vinculin and talin, i.e. components of focal adhesion plaques. While on day 1 after seeding the initial cell population density was similar on the terpolymers with or without MWNT-A, on day 7, the cells on the MWNT-A-modified terpolymer reached a density 4.5 times higher than the density on the unmodified samples (Fig. 11C). The improved adhesion and growth of MG 63 cells on the nanotube-modified terpolymer could be attributed to changes in its surface roughness rather than to its surface wettability, which remained unchanged and relatively low [23].

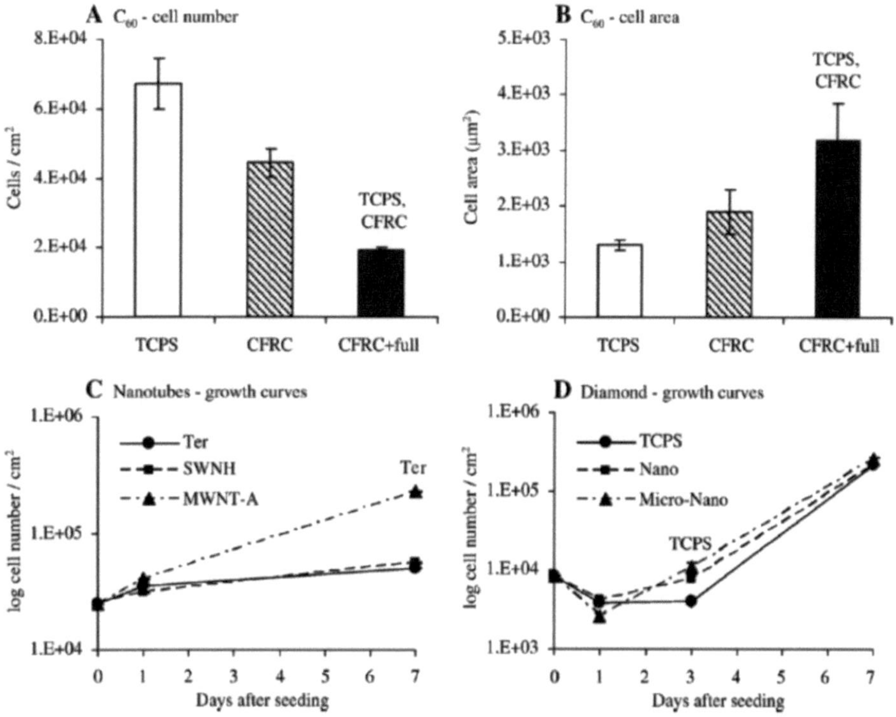

Figure 12. Population density (A) and adhesion area (B) of osteoblast-like MG 63 cells on day 2 after seeding on tissue culture polystyrene dish (TCPS), carbon fibrereinforced carbon composites (CFRC) and CFRC coated with a fullerene layer (CFRC+full). C: Growth curves of MG 63 cells on a terpolymer of polytetrafluoroethylene, polyvinyldifluoride and polypropylene (Ter), terpolymer mixed with 4 wt. % of single-wall carbon nanohorns (SWNH) or 4 wt.% of high crystalline electric arc multi-wall nanotubes (MWNT-A). D: Growth curves of MG 63 cells on TCPS, a nanostructured diamond layer (Nano) and a layer with hierarchically organized micro-and nanostructure (Micro-Nano). Mean±S.E.M. from 4–12 measurements, ANOVA, Student–Newman–Keuls method. Statistical significance: TCPS, CFRC, Ter: $p \leq 0.05$ compared to the values on tissue culture polystyrene, pure CFRC and pure terpolymer [23].

2.4.3. Diamond layers

On nanostructured diamond layers, the number of initially adhered cells on day 1 after seeding was similar to that found on the control TCPS (Fig. 11D), whereas on the layers with a combined micro and nanoarchitecture, this number was significantly lower. In addition, the cells on the latter samples were distributed non-homogeneously [23]. However, from day 1 to 3 after seeding, the cells on the hierarchically micro- and nanostructured layers, which are considered to resemble the architecture of natural tissues, showed the quickest proliferation. Their doubling time was 23.4 h, whereas on the nanostructured layers, it was 56.7 h, and on TCPS, the cells still remained in the lag phase and have not yet started their proliferation [23]. As a result, on day 3 after seeding, the cells on the diamond layers with a combined micro-and nanostructure reached the highest cell population density compared to nanostructured diamond and TCPS, respectively [23].

2.5. Chemical vapour deposition

Another type of carbon layer useful for the preparation of biocompatible surfaces includes chemical and physical vapor deposition. The preparation of the carbon layers on polytetrafluoroethylene (PTFE) by photoinduced CVD from acetylene and their physical properties and chemical structure have been studied. These properties related to the adhesion and proliferation of human umbilical endothelial cells (HUVEC) seeded thereon were characterized [39].

2.5.1. Surface morphology and layer thickness

The surface morphology was changed during the photo-deposition by carbon. The images suggest that in a first step the surface is covered by carbon and then the holes present in the material are filled. However, one should keep in mind that due to the production process the surface of pristine PTFE foils are typically not very homogeneous regarding surface roughness. The main difference is between the pristine PTFE and PTFE/C. The surface roughness is decreasing after more than 20 min deposition [39].

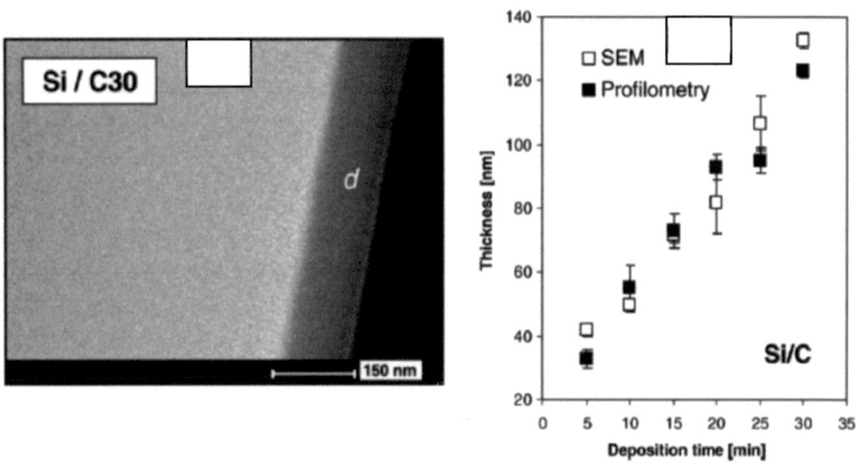

Figure 13. SEM scan of the carbon layer (d=thickness, gray shadow part) deposited 30 min on a silicon substrate (bright white part, left) (A) and the thickness dependence of the carbon layers on the deposition time. (B) The thickness was measured on Si substrate by SEM and profilometry [39].

The thickness of the deposited layers was measured by SEM microscopy and profilometry [39]. For both methods silicon platelets were used as substrates. For SEM microscopy, the coated Si platelets were broken and such a cross-section can be obtained. Fig. 14A shows the SEM cross-section of a broken Si sample coated with carbon for 30 min. The silicon substrate is represented by the bright white part on the left of the image, while the gray shadow part represents photo-deposited carbon layer. The dependence of the carbon layer thickness on deposition time measured by SEM and by profilometry is presented in Fig. 14B. Data of both methods show a nearly linear increase of the thickness with the deposition time.

2.5.2. *Surface wettability and chemical stability*

Surface wettability can be characterized by water contact angle, mainly influencing adhesion of cells on the modified polymer [45]. Pristine PTFE is a strongly hydrophobic material with a very high contact angle. With increasing carbon deposition time, the water contact angle strongly decreases to values considerable below 90°, which is a typical value for hydrocarbons. As discussed later, we attribute this decrease of the contact angle below 90° to polar groups in the carbon layers. There may be also effects of the surface roughness.

2.5.3. *Surface chemistry*

The results from Raman spectroscopy (whole layer thickness) and XPS (only surface layer) indicated, that the photoinduced deposition from acetylene results in layers consisting of C-H, C-C, C=C, C=O and O-H it means also C-O-H) bonds. The oxygen containing groups are probably either formed due to reactions layer with residual gases in reaction chamber or due to oxidation of unsaturated radicals by the exposure of samples to air after the deposition. The occurrence of polar oxygen containing groups and remaining radicals are suggested to be the main reason for the low water contact angle [45].

2.5.4. *Cell-surface interaction*

HUVEC were seeded on the pristine PTFE and PTFE coated with different carbon layers [45]. The adhesion was studied after 1 day, the proliferation after 3 and 7 days. The amount of cells was determined by counting from images and MTS test [39]. The positive influence of the photo-deposited carbon layer for cytocompatibility is more significant after 3 and even more after 7 days of proliferation. The highest cell densities were detected after 7 days on the sample with a deposition time of 20 min. The phase-contrast micrographs of HUVEC on various samples 7 days after seeding revealed only a relative small number of cells adhered pristine PTFE. The cells have a small diameter and a round shape and seem to try to avoid the contact with the surface. On the other hand on carbon coated PTFE, the number of cells was much higher. The highest amount of cells is significant on the sample coated for 20 min by carbon. Here the cells were spreaded onto surface and had a polygonal shape. The large difference in cell adhesion and proliferation between pristine PTFE and carbon coated PTFE allows to confine the cells to certain areas at the surface. This is demonstrated in Fig. 15, where the sample was covered during carbon deposition by a contact mask with 1.5 mm diameter holes. The cells practically only adhere and proliferate at the carbon coated spot.

2.6. One substrate, three deposition methods

2.6.1. *Thickness, contact angle and resistance of deposited C-layers*

For the comparison of deposition methods the samples with approximately the "same" thickness (interval 73-85 nm) were chosen. From the values of contact angle measured with water drop presented we can resume that the most hydrophobic is pristine PET. Due to coverage of the substrate by the carbon layers we observed lower values of contact angle. This

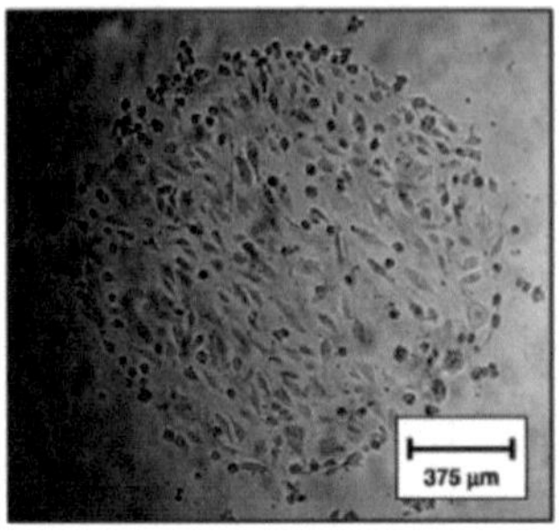

Figure 14. Phase-contrast micrographs of HUVEC (7 days after cell seeding) on a sample coated selectively with carbon for 20 min by photo-induced CVD through a contact mask (spot diameter 1.5 mm) [39].

fact shows an increase of hydrophilic character of the carbon layers. The concentration of carbon, oxygen and nitrogen in PET and carbon layer deposited by evaporation, sputtering and CVD deposition was determined. The samples prepared by sputtering contained the most amount of oxygen in comparison with others deposited carbon samples. This means that sputtered layer is containing more polar oxygen groups, but still it is showing higher contact angle. It is known that polarity or wettability of the surface [67] is determined also by other parameters for example surface roughness and morphology which will be discussed later.

2.6.2. Surface morphology of deposited C-layers

It was observed that carbon layer evaporation did not cause significant change of the surface morphology of pristine PET. The values of roughness R_a were comparable for the pristine PET and evaporated PET. The higher values of R_a were typical for the samples prepared by sputtering and CVD method. The surface morphology was changed dramatically during the photo-deposition by carbon. It is seen that the morphology of the evaporated and sputtered layers are similar instead of layer after CVD from acetylene. The sharp areas on the sample prepared by CVD method were not found on the others.

2.6.3. Chemical structure of deposited carbon layers

The Raman spectra of pristine PET and carbon layers on PET prepared by evaporation, sputtering and CVD methods are shown in Fig. 16. The spectra have differences especially between 3000-2750 cm^{-1} where some characteristic peaks are present in the spectrum of CVD: 2931 cm^{-1} is typical for C-H vibration from $-CH_2-$ groups, a weak band at ca. 3040 cm^{-1} is assigned to C-H stretching vibration on unsaturated carbons (sp^2). Raman peak for CVD spectrum at 1360 cm^{-1} is attributed to disordered mode of graphite and that at 1500-1550 cm^{-1} corresponds to an amorphous-like structure with sp^3 and sp^2 bonding [61]. CVD-carbon layer has a broad peak at 1530 cm^{-1} indicating that the deposited layers are composed mostly of amorphous carbon with sp^3 and sp^2 bonds [68]. Peak at 1360 cm^{-1}, which is also present in evaporated and sputtered C-layers, is due to the presence of disordered graphite [68]. Band position at 1617 cm^{-1} is typical for plane stretching C=C vibration (sp^2), called G ("graphite")-band in the case of carbon materials. From spectrum is obvious that layer deposited by evaporation and sputtering doesn't contain valence C-H vibration, which is normally appearing around 2950 cm^{-1}.

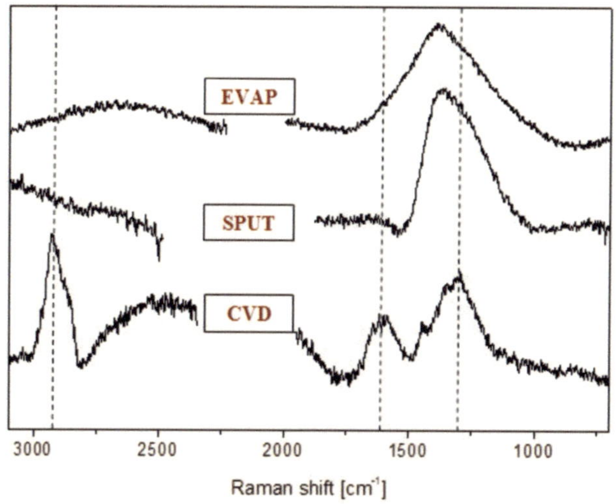

Figure 15. Raman spectra of pristine PET and carbon layers on PET prepared by evaporation, sputtering and CVD method. The layers have comparable thickness (approx. 80 nm).

Longer "structures" absorb on longer wave lengths. It was observed that in comparison with pristine PET the sputtered and CVD deposited C-layers are containing more conjugated double bonds. Dramatically an increase of amount of double bonds is present at the evaporated layers. This result is confirmed by measurement of sheet resistance. The sputtered and CVD layers show small decrease of R_s in comparison with pristine PET.

Significant decrease of 10 orders was measured for the evaporated C-layers. It was published that evaporated C-layers deposited by flash-evaporation were showing high electrical conductivity [41]. The concentrations of carbon, oxygen and nitrogen in the surface layer with the detection thickness around 1 μm were measured by RBS and ERDA After deposition of C-layers the increase of the carbon content was observed. The highest oxygen concentration being observed for sputtered C layers on PET.

2.6.4. *Cells adhesion and proliferation*

We have observed by the Raman spectroscopy, that there are no significant changes of the carbon layer after sterilization [39]. The sterilization in autoclave of the samples was performed 1 day before the experiment and then they were kept under sterile conditions. Adhesion and proliferation of 3T3 fibroblasts on the pristine PET and evaporated, sputtered and CVD coated foils with approximate C-layer thickness around 80 nm. The amount of adhered cells after 1 day of cultivation and proliferating cells after 3 and 5 days of cultivation is shown in Fig. 17. It is seen that the cells adhere similarly on all the surfaces without any significant difference. The proliferation after 3 days is showing certain differences in the amount of the cells especially increase of the amount of the cells on the evaporated and sputtered layers.

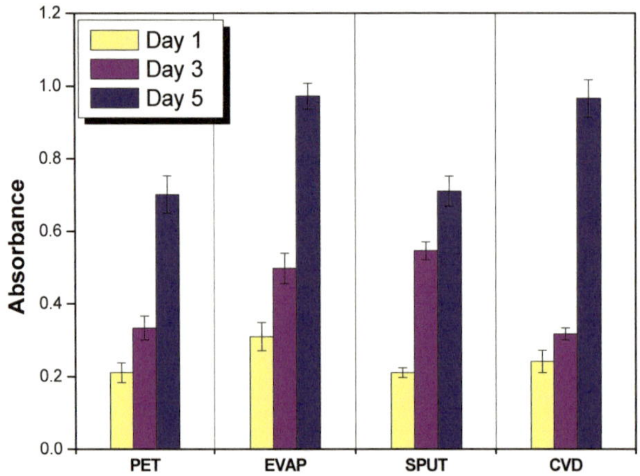

Figure 16. The amount of the 3T3 cells measured from absorbance of MTT test after adhesion (1st day) and proliferation (3rd and 5th day) on pristine PET and carbon layers on PET prepared by evaporation, sputtering and CVD method. The layers have comparable thickness (approx. 80 nm). The amount was calculated in respect to the tissue PS.

As was observed, the values of the water contact angle on the surface of evaporated and CVD deposited coatings are similar and slightly higher for sputtered layers. The amount of adhered cells after 24 hours is equal for all coatings, so there is no significant influence of the water contact angle (wetting properties) on the adhesion of the cells. The evaporated layers were showing high electrical conductivity in comparison with other techniques which is due to the creation of the conjugated systems of double bonds. An increase of electrical conductivity can be taken as the main factor positively influencing the proliferation of the 3T3 fibroblast cells on the evaporated surfaces compared to pristine PET.

3. Conclusions

The present results can be summarized as follows:

- carbon nanolayers for enhancing of surface biocompatibility can be prepared by sputtering, evaporation, CVD method or by nanoparticles deposition on polymer. The scheme in Fig. 19 represents the idea of our research,
- carbon layers deposited onto polymer substrate strongly influences the surface morphology, wettability and chemical structure of polymer's surface,
- PVD methods (sputtering, evaporation) and CVD deposition of carbon layers leads to the contact angle decrease and the surfaces wettability increase,
- carbon layer deposition leads to the significant decrease of surface sheet resistance,
- the sputtered carbon layers consist of amorphous hydrogenated carbon (a-C:H) containing an oxygen admixture,

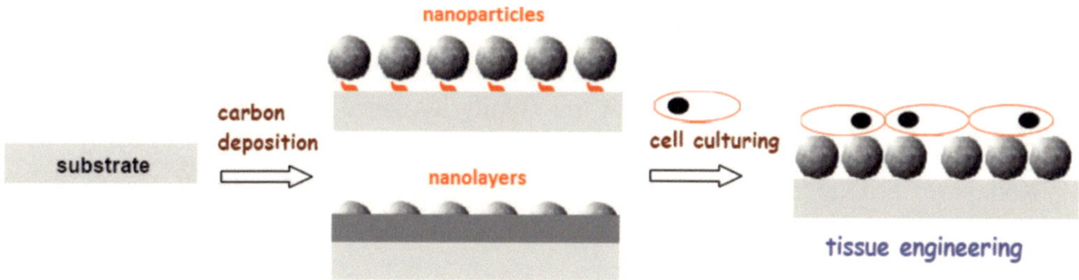

Figure 17. Vizualization of the individual steps of our research.

- the presence of sp^3 and sp^2 bonding declare the presence of double bonding system between the C atoms in case of CVD technique,
- surface morphology and carbon layer thickness is an important parameter influencing adhesion and short-term proliferation of cells,
- the chemical composition and surface wettability seems to be important parameter especially for long-term proliferation of the 3T3 fibroblasts,
- fullerenes C_{60} deposited as continuous films or layers micropatterned with grooves and bulges, Ti and C_{60}/Ti films used in this study gave good support to the adhesion, spreading, growth and viability of human osteoblast-like MG 63 cells,
- the large difference in cell adhesion and proliferation between pristine PTFE and carbon coated PTFE allows to confine the cells to certain areas at the surface, which opens a wide field of possible biomedical applications,
- carbon nanoparticle-containing materials supported adhesion and growth of bone-derived cells. The carbon nanoparticle layers, especially hard diamond coatings, could be used for surface modification of bone implants

Author details

Petr Slepička, Tomáš Hubáček, Simona Trostová,
Nikola Slepičková Kasálková and Václav Švorčík
Department of Solid State Engineering, Institute of Chemical Technology, Prague, Czech Republic

Zdeňka Kolská
Faculty of Science, J.E. Purkyně University, Ústí nad Labem, Czech Republic

Lucie Bačáková
Institute of Physiology, Academy of Sciences of the Czech Republic, Prague, Czech Republic

Acknowledgement

This work was supported by the GACR under projects P108/10/1106 and P108/12/1168.

4. References

[1] Chai F, Mathis N, Blanchemain N, Meunier C, Hildebrand HF (2008) Osteoblast Interaction with DLC-Coated Si Substrates. Acta biomater. 4: 1369-1381.

[2] Santavirta S (2003) Compatibility of the Totally Replaced Hip - Reduction of Wear by Amorphous Diamond Coating. Acta orthop. scand. suppl. 74: 310-316.

[3] Kobayashi S, Ohgoe Y, Ozeki K, Hirakuri K, Aoki H (2007) Dissolution Effect and Cytotoxicity of Diamond-Like Carbon Coatings on Orthodontic Archwires. J. mater. sci. mater. med. 18: 2263-2268.

[4] Roy RK, Lee KR (2007) Biomedical Applications of Diamond-Like Carbon Coatings: A Review. J. biomed. mater. res. B 83: 72-84.

[5] Schrand AM, Huang H, Carlson C, Schlager JJ, Osawa E, Hussain SM, Dai L (2007) Are Diamond Nanoparticles Cytotoxic? J. phys. chem. 111: 2-7.

[6] Bačáková L, Grausová L, Vandrovcová M, Vacík J, et al. (2008) In: Nanoparticles: New Research. Lombardi SL (Ed.). Hauppauge, New York: Nova Science Publishers, Inc. pp. 39-107.

[7] Amaral M, Dias AG, Gomes PS, Lopes MA, Silva RF (2008) Nanocrystalline Diamond: In Vitro Biocompatibility Assessment by MG63 and Human Bone Marrow Cells Cultures. J. biomater. res. A 87: 91-99.

[8] Gajewski W, Achatz P, Williams OA, Haenen K, Bustarret E, Stutzmann M, Garrido JA (2009) Electronic and Optical Properties of Boron-Doped Nanocrystalline Diamond Films. J. phys. rev. B 79: 045206.

[9] Nebel CE, Shin D, Rezek B, Tokuda N, Uetsuka H, Watanabe H (2007) Diamond and Biology. J. r. soc. interface 4: 439-461.

[10] Majid E, Male KB, Luong JHT (2008) Boron Doped Diamond Biosensor for Detection of Escherichia Coli. J. agric. food chem. 56: 7691-7695.

[11] Zhao J, Wu L, Zhi J (2009) Non-Enzymatic Glucose Detection Uusing As-Prepared Boron-Doped Diamond Thin-Film Electrodes. analyst 134: 794-799.

[12] Watari F, Takashi N, Yokoyama A, Uo M, Akasaka T, Sato Y, Abe S, Totsuka Y, Tohji K (2009) Material Nanosizing Effect on Living Organisms: Non-Specific, Biointeractive, Physical Size Effects. J. r. soc. interface 6: S371-S388.

[13] Aoki N, Akasaka T, Watari F, Yokoyama A (2007) Carbon Nanotubes as Scaffolds for Cell Culture and Effect on Cellular Functions. Dent. mater. j. 26: 178-185.

[14] Li X, Gao H, Uo M, Sato Y, Akasaka T, Abe S, Watari F (2009) Maturation of Osteoblast-Like SaoS2 Induced by Carbon Nanotubes. Biomed. mater. 4: 015005.

[15] Gonzalez KA, Wilson LJ, Wu W, Nancollas GH (2002) Synthesis and in Vitro Characterization of a Tissue-Selective Fullerene: Vectoring C-60(OH)(16)AMBP to Mineralized Bone. Bioorg. med. chem. 10: 1991-1997.

[16] Liu J, Tabata Y (2010) Photodynamic Therapy of Fullerene Modified with Pullulan on Hepatoma Cells. J. drug target. 18: 602-610.

[17] Marchesan S, Da Ros T, Spalluto G, Balzarini J, Prato M (2005) Anti-HIV Properties of Cationic Fullerene Derivatives. Bioorg. med. chem. lett. 15: 3615-3618.

[18] Huang LY, Terakawa M, Zhiyentayev T, Huang YY, Sawayama Y, Jahnke A, Tegos,GP, Wharton T, Hamblin MR (2010) Innovative Cationic Fullerenes as Broad-Spectrum Light-Activated Antimicrobials. Nanomedicine 6: 442-452.

[19] Dellinger A, Zhou Z, Lenk R, MacFarland D, Kepley CL (2009) Fullerene Nanomaterials Inhibit Phorbol Myristate Acetate-Induced Inflammation. Exp. dermatol. 18: 1079-1081.

[20] Maeda R, Noiri E, Isobe H, Homma T, Tanaka T, Nagishi K, Doi K, Fujita T, Nakamura E (2008) A Water-Soluble Fullerene Vesicle Alleviates Angiotensin II-Induced Oxidative Stress in Human Umbilical Venous Endothelial Cells. Hypertens. res. 31: 141-151.

[21] Tykhomyrov AA, Nedzvetsky VS, Klochkov VK, Andrievsky GV (2008) Nanostructures of Hydrated C-60 Fullerene (C(60)HyFn) Protect Rat Brain Against Alcohol Impact and Attenuate Behavioral Impairments of Alcoholized Animals. Toxicology 246: 158-165.

[22] Iijima S (2002) Carbon Nanotubes: Past, Present, and Future. Physica B 323: 1-5.

[23] Bačáková L, Grausová L, Vacík J, Fraczek A, Blazewicz S, Kromka A, Vaněček M, Švorčík V (2007) Improved Adhesion and Growth of Human Osteoblast-Like MG 63 Cells on Biomaterials Modified with Carbon Nanoparticles. Diamond relat. mater. 16: 2133-2140.

[24] Thostenson ET, Li C, Chou TW (2005) Nanocomposites in Context. Compos. sci. technol. 65: 491-516.

[25] Chester DW, Klemic JF, Stern E, Sigworth FJ, Klemic KG (2007) Holey Carbon Micro-Arrays for Transmission Electron Microscopy: A Microcontact Printing Approach. Ultramicroscopy 107: 685-691.

[26] Yubuta K, Hongo T, Atou T, Nakamura KG, Kikuchi M (2007) High-Resolution Electron Microscopy of Microstructure of MnF2 Subjected to Shock Compression at 4.4 GPa. Solid state commun. 143: 127-130.

[27] Sullivan JP, Friedman TA, Hjor K (2001) Diamond and Amorphous Carbon MEMS. MRS bull. 26: 309-311.

[28] Chiang KT, Yang L, Wei R, Coulter K (2008) Development of Diamond-Like Carbon-Coated Electrodes for Corrosion Sensor Applications at High Temperatures. Thin solid films 517: 1120-1124.

[29] Yang YL, Gupta MC, Dudley KL, Lawrence RW (2005) A Comparative Study of EMI Shielding Properties of Carbon Nanofiber and Multi-Walled Carbon Nanotube Filled Polymer Composites. J. nanosci. nanotechnol. 5: 927-931.

[30] Abbas GA, Roy SS, Papakonstantinou P, McLaughlin JA (2005) Structural Investigation and Gas Barrier Performance of Diamond-Like Carbon Based Films on Polymer Substrates. Carbon 43: 303-309.

[31] Bouvree A, Feller JF, Castro M, Grohens Y, Rinaudo M (2009) Conductive Polymer nano-bioComposites (CPC): Chitosan-Carbon Nanoparticle a Good Candidate to Design Polar Vapour Sensors. Sens. actuator B 138: 138-147.

[32] Masuda K, Ogawa M, Ohkita H, Benten H, Ito S (2009) Hybrid Solar Cells of Layer-by-Layer Thin Films with a Polymer/Fullerene Bulk Heterojunction. Sol. ener. mater. sol. cells 93: 762-767.

[33] Moniruzzaman M, Winey KI (2006) Polymer Nanocomposites Containing Carbon Nanotubes. Macromolecules 39: 5194-5205.

[34] Coleman JN, Khan U, Blau WJ, Gun'ko YK (2006) Small but Strong: A Review of the Mechanical Properties of Carbon Nanotube-Polymer Composites. Carbon 44: 1624-1652.

[35] Wang CC, Guo ZX, Fu S, Wu W, Zhu DB (2004) Polymers containing fullerene or carbon nanotube structures. Prog. polym. sci. 29: 1079-1141.

[36] Kuilla T, Bhadra S, Yao D, Kim NH, Bosed S, Lee JH (2010) Recent Advances in Graphene Based Polymer Composites. Prog. polym. sci. 35: 1350-1375.

[37] Švorčík V, Hubáček T, Slepička P, Siegel J, Kolská Z, Bláhová O, Macková A, Hnatowicz V (2009) Characterization of Carbon Nanolayers Flash Evaporated on PET and PTFE. Carbon 47: 1770-1778.

[38] Kubová O, Bačáková L, Švorčík V (2005) Biocompatibility of Carbon Layer on Polymer. Mater. sci. forum 482: 247-250.

[39] Kubová O, Švorčík V, Heitz J, Moritz S, Romanin C, Macková A (2007) Characterization and Cytocompatibility of Carbon Layers Prepared by Photo-Induced Chemical Vapor Deposition. Thin solid films 515: 6765-6772.

[40] Švorčík V, Kasálková N, Slepička P, Záruba K, Bačáková L, Pařízek M, Lisá V, Ruml T, Macková A (2009) Cytocompatibility of Ar(+) Plasma Treated and Au Nanoparticle-Grafted PE. Nucl. instrum. meth. B 267: 1904-1910.

[41] Hubáček T, Lyutakov O, Rybka V, Švorčík V (2010) Electrical Properties of Flash-Evaporated Carbon Nanolayers on PTFE. J. mater. sci. 45: 279-281.

[42] Stein A, Wang Z, Fierke MA (2009) Functionalization of Porous Carbon Materials with Designed Pore Architecture. Adv. mater. 21: 265-293.

[43] Narayan RJ (2005) Nanostructured Diamond-Like Carbon Thin Films for Medical Applications. Mater. sci. eng. 25: 405-416.

[44] Wang DY, Chang YY, Chang CL, Huang YW (2005) Deposition of Diamond-Like Carbon Films Containing Metal Elements on Biomedical Ti Alloys. Surf. coat. technol. 200: 2175-2180.

[45] Švorčík V, Kubová O, Slepička P, Dvořánková B, Macková A, Hnatowicz V (2006) Structural, Chemical and Biological Properties of Carbon Layers Sputtered on PET. J. mater. sci. mater. med.17: 229-234.

[46] Zhang L, Webster TJ (2009) Nanotechnology and Nanomaterials: Promises for Improved Tissue Regeneration. Nano today 4: 66-80.

[47] Kim S, Shibata E, Sergiienko R, Nakamura T (2008) Purification and Separation of Carbon Nanocapsules as a Magnetic Carrier for Drug Delivery Systems. Carbon 46: 1523-1529.

[48] Schroeder A, Francz G, Bruinink A, Hauert R, Mayer J, Wintermantel E (2000) Titanium Containing Amorphous Hydrogenated Carbon Films (a-C: H/Ti): Surface Analysis and

Evaluation of Cellular Reactions Using Bone Marrow Cell Cultures in Vitro. Biomaterials 21: 449-456.

[49] Ma WJ, Ruys AJ, Mason RS, Martin PJ, Bendavid A, Liu Z, Ionescu M, Zreiqat H (2007) DLC Coatings: Effects of Physical and Chemical Properties on Biological Response. Biomaterials 28: 1620-1628.

[50] Greer JA, Tabat MD (1995) Large-Area Pulsed Laser Deposition: Technique and Application. J. vac. sci. technol. A 13: 1175-1181.

[51] McLaughlin JA, Maguire PD (2008) Advances on the Use of Carbon Materials at the Biological and Surface Interface for Applications in Medical Implants. Diam. relat. mater. 17: 873-877.

[52] Kotál V, Švorčík V, Slepička P, Bláhová O, Šutta P, Hnatowicz V (2007) Gold Coating of PET Modified by Argon Plasma. Plasma proc. polym. 4: 69-76.

[53] Pradhan D, Sharon M (2007) Opto-Electrical Properties of Amorphous Carbon Thin Film Deposited from Natural Precursor Camphor. Appl. surf. sci. 253: 7004-7010.

[54] Mott N (1990) Metal Insulator Transitions. London: Taylor & Francis. p.1-286.

[55] Švorčík V, Kolská Z, Luxbacher T, Mistrík J (2010) Properties of Au Nanolayer Sputtered on Polyethyleneterephthalate. Mater. lett. 64: 611-613.

[56] Avigal Y, Kalish R (2001) Growth of Aligned Carbon Nanotubes by Biasing During Growth. Appl. phys. lett. 78: 2291-2293.

[57] Cappelli E, Orlando S, Mattei G, Zoffoli S, Ascarelli P (2002) SEM and Raman Investigation of RF Plasma Assisted Pulsed Laser Deposited Carbon Films. Appl. surf. sci. 197-198: 452-457.

[58] Švorčík V, Ročková K, Dvořánková B, Hnatowicz V, Ochsner R, Ryssel H (2002) Cell Adhesion on Modified Polyethylene. J. Mater. Sci. 37: 1183-1188.

[59] Yang P, Kwok SCH, Fu RKY, Huang N, Leng Y, Chu PK (2004) Structure and properties of Annealed Amorphous Hydrogenated Carbon (a-C : H) Films for Biomedical Applications. Surf. coat. technol. 177: 747-751.

[60] Levi N, Hantgan RR, Lively MO, Carroll DL, Prasad GL (2006) C60-Fullerenes: detection of intracellular photoluminescence and lack of cytotoxic effects. J. nanobiotechnol. 4: 14.

[61] Nakamura E, Isobe H (2003) Functionalized Fullerenes in Water. The First 10 Years of Their Chemistry, Biology, and Nanoscience. Acc. chem. res. 36: 807-815.

[62] Grausová L, Vacík J, Švorčík V, Slepička P, Bílková P, Vandrovcová M, Lisá V, Bačáková L (2009) Fullerene C(60) Films of Continuous and Micropatterned Morphology as Substrates for Adhesion and Growth of Bone Cells. Diamond rel. mater. 18: 578-586.

[63] Vandrovcová M, Vacík J, Švorčík V, Slepička P, Kasálková N, Vorlíček V, Lavrentiev V, Voseček V, Grausová L, Lisá V, Bačáková L (2008) Fullerene C-60 and Hybrid C-60/Ti Films as Substrates for Adhesion and Growth of Bone Cells. Phys. stat. sol. A 205: 2252-2261.

[64] Endo M, Hayashi T, Kim YA, Terrones M, Dresselhaus MS (2004) Applications of Carbon Nanotubes in the Twenty-First Century. Philos. transact. r. soc. A math. phys. eng. sci. 362: 2223-2238.

[65] Kohli P, Martin CR (2005) Smart Nanotubes for Biotechnology. Curr. pharm. biotechnol. 6: 35-47.

[66] Bačáková L, Starý V, Kofroňová O, Lisá V (2001) Polishing and Coating Carbon Fiber-Reinforced Carbon Composites with a Carbon-Titanium Layer Enhances Adhesion and Growth of Osteoblast-Like MG63 Cells and Vascular Smooth Muscle Cells in Vitro. J. biomed. mater. res. 54: 567-578.

[67] Švorčík V, Kolářová K, Slepička P, Macková A, Novotná M, Hnatowicz V (2006) Modification of Surface Properties of High and Low Density Polyethylene by Ar Plasma Discharge. Polym. degr. stab. 91: 1219-1225.

[68] Li DJ, Cui FZ, Gu HQ (1999) Studies of Diamond-Like Carbon Films Coated on PMMA by Ion Beam Assisted Deposition. Appl. surf. sci. 137: 30-37.

Chapter 8

Electrokinetic Potential and Other Surface Properties of Polymer Foils and Their Modifications

Zdeňka Kolská, Zuzana Makajová,
Kateřina Kolářová, Nikola Kasálková Slepičková,
Simona Trostová, Alena Řezníčková, Jakub Siegel and Václav Švorčík

Additional information is available at the end of the chapter

http://dx.doi.org/10.5772/46144

1. Introduction

In most of heterogeneous systems consisting of solid material in liquid medium the phase interface exhibits an electrical charge. This charge can play an important role in stability of these systems and in their behaviour in liquid surrounding. The knowledge of this behaviour is necessary for next usage of these materials. The electrical charge at solid-liquid interface is a key parameter for surface science and engineering. When the surface is charged, then it can attract ions of opposite charge (counter-ions) from the liquid. In this way the electrical double layer is created, consisting of two oppositely charged layers. Between the surface and volume liquid phase is a potential gradient and the potential of volume liquid phase is equal to zero. When the liquid flows along the solid surface, the electrical double layer divides. Inner layer (the solid phase) does not move, and adsorbed or bonded counter ions, while the outer layer flows along. The potential created between solid surface and this mobile interface is known as an electrokinetic potential or zeta potential (ζ-potential) [1]. The electrical charge at the interface can be caused by several mechanisms: (i) ionization or dissociation of surface groups of surface layers, e.g. either dissociation of a proton from a carboxylic group, which leaves the surface with a negative charge; (ii) preferential solution of some ions of crystal lattice in contact with liquid; (iii) preferential adsorption of some ions from the solution on initially uncharged surface, e.g. adsorption of either hydroxide or hydronium ions created by the enhanced autolysis of water at the surface; (iv) isomorphic substitution of ions of a higher valence by ions of lower valence in the case of clay minerals (e.g. Si^{4+} for Al^{3+}); (v) accumulation of electrons in the case of metal-solution interface, etc. [1-3].

Dissociation of the surface groups is a common charging mechanism in the case of latex particles, which are frequently used in biomedical applications [1]. The adsorption of hydroxide ions created by enhanced autolysis of water at hydrophobic surfaces may be the reason for the hitherto unexplained charge of hydrophobic micro-fluidic substrates [2-4]. The behavior at hydrophobic and hydrophilic polymer surfaces is quite different. While on hydrophobic surface the surfactants displace water from the solid surface, it cannot displace water from the hydrophilic surface. In that case adsorption occurs mainly by electrostatic interactions between the surfactant "head" group and the surface groups [5]. Pristine polymers exhibit the first (polymers without any functional groups on surface) or the third (when surface presents any functional groups able to dissociate in liquid) type of mechanisms depending on their surface chemistry. When polymers are deposited by metal nanostructures, the accumulation of electrons begins to play a role.

Zeta potential is a characteristic parameter for description of solid surface chemistry. It is an important physico-chemical property of cellulose [6], glass, metals, textile fabrics [7,8], wafers, ceramics [9], etc., giving information about chemistry, polarity, swelling, porosity and other surface characteristics. Zeta potential is used for characterization of natural and synthetic fibres and in investigation of their hydrophilicity, which is important for their interactions with dyes and surfactants [5,6,8]. It is also important for research of membranes, filters [10,11], pristine polymers [12-15] or polymers modified by plasma [12,15-18], by laser or by chemical coatings, e.g. by chitosan [16], biphenyldithiol [17], polyethyleneglycol [18] or by gold coatings [14,17]. It can also be used in research of textiles, where ζ-potential plays an important role in the electrical characterization in wet processing [7], hairs [19], biomaterials [20] and glass [21,22] and as a parameter of colloid stability (paints, printing inks, drilling muds etc.). Zeta potential of human enamel is of physiological importance since it affects interactions between enamel surfaces and the surrounding aqueous medium of saliva [23]. It is also of importance for sewage treatment, especially for industrial and domestic wastewater treatment [24]. Zeta potential can further be used for examination of micro-environment effects on bioactivity of compounds at the solid–liquid interface. Zeta potential, combined with pH of liquid phase, determines electrostatic interactions between the polymer surface and the immobilized bioactive compound. These interactions affect the kinetics between the bioactive compound and its target (metabolite, antigen/antibody, enzyme substrate, etc.) in the liquid [25]. Also properties of different thin layers deposited on solid substrates (glass, polymers, silica, etc.) can be characterized by ζ-potential, e.g. thin carbon layers deposited on silica [26,27] or Ag dopped hydroxyapatite layers on silica [28]. Also gold nanoparticles stabilized in solution by citrate were characterized by ζ-potential [29]. Research effort has been devoted to the suppression of ζ-potential by different techniques including coatings with organosilanes, and deposition of polymer films [30]. There are fundamental differences between silica and polymer surfaces. Silica surfaces usually have high charge density (pronounced hydrophilicity) and polymers with lower charge density, are mostly hydrophobic and their surfaces can be affected by fabrication procedure [13,31].

Our team provides the study of different polymer foils and especially their variable modification for consequent usage in optics, electronics, tissue engineering, etc. For description and characterization of surface properties and their changes we apply a wide variety of analyses, such as X-ray photoelectron spectroscopy (XPS), atomic force microscopy (AFM), goniometry, infrared spectroscopy (IR), sheet resistance, *in vitro* living cell growth and proliferation study and many other methods. During last years we have also applied the electrokinetic analysis as a first, easy and fast identification and characterization of changes of surface properties before and after modifications. In this chapter we present the collection of results for zeta potential determination of pristine polymer foils and modified ones by different techniques and their comparison with other analyses. In comparison with these methods, the electrokinetic analysis is the fast, the easy and it can give us the first information if the surface was successfully modified.

Zeta potential was determined for many polymer foils at constant pH value and also in the pH range for an isoelectric point study (zeta potential equals to zero). Polymer foils were than exposed to plasma, laser, gold sputtering and different chemical grafting. Changes of zeta potential values help us quickly recognize the chemical changes on modified surfaces due to different behaviour of surfaces in liquid surrounding. Polymer foils were modified in variable ways: by (i) plasma treatment, (ii) laser exposure, (iii) gold sputtering and consequent annealing, (iv) polyethylene glycol grafting, (v) dithiols grafting, (vi) cysteamine grafting, (vii) subsequent Au nanoparticles grafting, etc. All zeta potential results indicating different changes were subsequently confronted with other surfaces analyses.

2. Zeta potential determination

The zeta potential of planar samples represents the surface behaviour which occurs in the presence of an aqueous solution. It gives information about the nature and dissociation of reactive (functional) groups, about polarity, hydrophilicity or hydrophobicity of the solid surface, indirectly the chemical nature of studied sample and about ion or water sorption too [7,32]. This information is important for other polymer foils usage. Zeta potential can be determined using electrokinetic measurements, inclusive electrophoresis, sedimentation potential, electroosmosis, streaming current or streaming potential [7,24,30-35]. Measurements using multiple techniques are invaluable, since the errors with different techniques are often of opposite sign [30]. The more detailed research and comparison of different techniques was presented previously [36]. Most suitable for the planar samples is the streaming current or streaming potential approach [8,24,30,32,36,37]. Streaming current or streaming potential is generated by pressure-driven flow through a conduit [30]. It is accomplished by applying a pressure, and the potential is induced as a result [24,35]. The zeta potential derived from the streaming potential or the streaming current is considered as ζ-potential at the hydrodynamic phase boundary against the bulk liquid.

Streaming potential, streaming current and Helmholtz-Smoluchowski equation approach is correct and valid when electrolyte solution is forced through a narrow slit formed by two similar measured surfaces. This ensures that the thickness of the electrochemical double layer of the studied surface is smaller than the slit width and an overlapping of the double layers can be excluded [8,35].

While the determination of ζ-potential on colloidal samples is quite frequent there are not much studies accomplished on flat samples. The data on ζ-potential of polymer foils are important for physics and chemistry of surfaces, for electronics, material science, for packaging of food, even for biology and medicine [38]. Survey of ζ-potentials for some polymers are given in [31]. There are summarized data obtained by many authors measured and by several techniques for PDMS, PC, PET, PMMA, PE, PS, PVC and PTFE along with the data for colloidal samples and for polymer foils.

Zeta potential depends on many solution and substrate parameters, such as pH, counter-ions concentration, valence and size, temperature, substrate material and its surface properties, etc. [30,31]. Strong effect on the ζ-potential has the valence and concentrations of applied electrolytes [30]. Therefore some of authors tested various electrolytes and different electrolyte concentrations [39]. The effects of counter-ion valence and size do not play role in the case of potassium and sodium ions, differences being small for silica and also polymer substrates and are only important in cases where larger ions dominate the counter-ion concentration [30,31]. The bivalent cations exhibit specific adsorption which can be ignored for monovalent cations, such as Li^+, Na^+ and K^+ [30]. It can be concluded that monovalent cations are more suitable and there are no significant differences between e.g. KCl and NaCl. Therefore we used KCl and KNO_3 for our studies.

The study of pH effect has been accomplished by some investigators. Some of them find a linear variation with pH, while others reported a plateau at some pH values [30]. Therefore we studied zeta potential of polymers at constant pH, only for isoelectric point determination we provided pH dependence. Also the effect of temperature should not be neglected, e.g. the ζ-potential of silica increases approximately 1.75% per 1°C [30]. It is therefore important to preserve the constant temperature during zeta potential determination.

Zeta potential of polymer foils is also strongly affected by their surface roughness and other surface properties [13,32]. The roughness leads to geometry-induced changes, which can affect an electroosmotic flow and ζ-potential. Surface roughness affects the position of the charged surface and bulk solution, which is not homogenous. It can lead to higher surface conductivity [30,35].

2.1. Principle

Our ζ-potential determination of planar samples is based on the measurement of the streaming potential dU and/or the streaming current dI as a function of continuously increasing pressure dp of electrolyte circulated through the measuring cell containing the solid sample. The relationships between ζ-potential and the streaming potential dU or the streaming current dI are linear, with the slope dU/dp, resp. dI/dp [9,24,30,32,34,36,37]. Streaming potential dU or alternatively the streaming current dI are detected by electrodes placed at both sides of the sample. The electrolyte conductivity, temperature and pH values are measured simultaneously [34,37].

The measured values of Δp and ΔU or ΔI serve to calculate ζ-potential by the following Helmholtz-Smoluchowski (HS) and Fairbrother-Mastins (FM) equations [8,34,37] (eqs. 1, resp. 2):

$$\zeta = \frac{dI}{dp} \cdot \frac{\eta}{\varepsilon \cdot \varepsilon_0} \cdot \frac{L}{A} \tag{1}$$

$$\zeta_{\text{apparent}} = \frac{dU}{dp} \cdot \frac{\eta}{\varepsilon \cdot \varepsilon_0} \cdot \kappa_B \tag{2}$$

where dU/dp, resp. dI/dp are slopes of streaming potential, resp. streaming current versus pressure p, η is an electrolyte viscosity, ε is a dielectric constant of electrolyte, ε_0 is a vacuum permittivity, L is a length of the streaming channel, A is a cross-section of the streaming channel (the rectangular gap between the planar samples), κ_B is an electrolyte conductivity.

Comparison of ζ (eq. 1) and $\zeta_{apparent}$ (eq. 2) potentials reveals additional information about excessive conductivity that is present in the measuring cell representing solid surface properties. Properties that influence the conductivity inside the streaming channel involve sample porosity, surface swelling, surface roughness and morphology and bulk material conductivity. The ratio L/A in eq. 1 is given by equation: $L/A=\kappa_B \cdot R$, where R is a cell resistance inclusive the electrolyte and sample properties [34]. The width of measuring space between samples should be much higher than the thickness of shear plane (the space at surface when the double layer is created due to arrangement of ions) [30]. More detailed description of the ζ-potential determination, applied equations and their limitations, was given in [30].

Because of strong dependence of the measured values on experimental conditions discussed above, it is strongly recommended that all data on ζ-potential should be supplemented with detailed information on particular measuring conditions (pH, ionic strength, counter-ion type and concentration, temperature, etc).

We study zeta potential of several densities of polymer foils. Therefore zeta potential depends on counter-ion type and valence, but these effects do not play a role in the case of potassium and sodium counter-ion and observed differences are small for all polymer substrates [30], we finally used only potassium ions. The standard electrolyte was 0.001 mol/dm^3 solution of KCl. The low electrolyte concentration ensures high sensitivity of the method [30,31,34]. To investigate the influence of electrolyte type and concentrations on ζ-potential determination, two monovalent (symmetric) electrolytes KCl and KNO_3 of concentration 0.001 mol/dm^3 and KCl also of 0.005 mol/dm^3 we used [40].

2.2. Materials

The following polymers in form of foils were used in this study (supplied by Goodfellow Ltd., UK): low density polyethylene (LDPE, density 0.92 g/cm^3 in the form of 30 μm thick foils), biaxially oriented polyethyleneterephthalate (PET, density 1.3 g/cm^3, 50 μm), polytetrafluoroethylene (PTFE, density 2.2 g/cm^3, of three thicknesses 100, 50 and 25 μm), biaxially oriented polystyrene (PS, density 1.05 g/cm^3, 30 μm), polypropylene (PP, density 0.9 g/cm^3, 50 μm), polyamides (PA6 and PA66, both of them 40 μm), poly L-lactic acid (PLLA, density 1.25 g/cm^3, 50 μm), polymethylpentene (PMP, density 0.84 g/cm^3, 50 μm), polyimides (Upilex R, Upilex S and Kapton, all of them 50 μm), polyaramide (PAr, 10 μm), polybutylenterephthalate (PBT, 25 μm), poly(ethylene-2,6-naphthalene) (PEN, 7 μm), polymethylmetacrylate (PMMA, 7 μm), polycarbonate (PC, 2 mm). Oriented high density polyethylene (HDPE, density 0.952 g/cm^3, 40 μm, supplied by Granitol Ltd., Czech Republic); polytetrafluoroethylene with perfluorovinyl pendant side chains ended by sulfonic acid groups (Nafion, total acid capacity: 0.95 to 1.01 meq/g, density 1.97 g/cm^3, 173 μm, manufacturer DuPont, USA).

2.3. Modifications

Polymer foils were studied in pristine form and also after modifications in the following variable ways: (i) plasma treatment, (ii) laser exposure, (iii) gold sputtering and consequent annealing, (iv) polyethylene glycol grafting, (v) dithiols grafting, (vi) methanol, dithiol and cysteamine grafting, (vii) steps (iv)-(vi) were supplemented by the subsequent Au nanoparticles/nanorods grafting.

2.3.1. Plasma treatment

Plasma treatment was performed on Balzers SCD 050 device under the following conditions: gas purity was 99.997%, flow rate 0.3 l/s, pressure 10 Pa, electrode distance 50 mm and its area 48 cm^2, chamber volume approx. 1000 cm^3, plasma volume 240 cm^3. Exposure times differ for individual tested polymer foils and individual consequent modifications from 15 to 500 s, discharge power was 3.8 or 8.3 W and the treatment was accomplished at room temperature [12].

2.3.2. Laser irradiation

For laser irradiation of samples we employed the F_2 laser (Lambda Physik LPF 202, wavelength of 157 nm, pulse duration of 15 ns). For the irradiation with the F_2 laser, the light was polarized linearly with a MgF_2 prism. For homogeneous illumination of the samples, we used only the central part of the beam profile by means of an aperture (10 mm - 3.5 mm). We performed the irradiation at fluences well below the ablation threshold of PET at 157 nm (29.6 mJ/cm^2). The samples were mounted onto a translation stage and scanned at a speed of 14 mm/s. At a repetition rate of the laser of 11 Hz. We carried out the experiments in a flow box purged with nitrogen at a pressure of 110 kPa [41].

2.3.3. Gold sputtering and subsequent annealing

We deposited the gold nanostructures or nanolayers from a gold target (99.999%) by diode sputtering (BAL-TEC SCD 050 equipment) onto pristine polymers. The deposition was performed at room deposition temperature with a different deposition time and a current of 20 mA.

Post-deposition annealing of Au-covered PTFE was carried out in air at 300°C (± 3°C) for 1 h using a thermostat Binder oven. The heating rate was 5°C/min and the annealed samples were left to cool in air to room temperature (RT) [42].

2.3.4. Polyethylene glycol grafting

Polymers (PE and PS) were used for modification in Ar^+ plasma discharge on a Balzers SCD 050 device. Immediately after the plasma treatment, plasma activated polymers were grafted by immersion into an aqueous solution (24 hours, room temperature, 2 wt.%) of polyethylene glycol (PEG, molecular weights (M_{PEG}) 300, 6000 and 20 000 g/mol). The non-bounded PEG was removed by rinsing the samples in distilled water for 24 hours [18].

2.3.5. Dithiols grafting and subsequent gold nanoparticles grafting

The plasma activated surface of PTFE was firstly grafted from methanol solution of biphenyl-4,4′-dithiol (BFD), 24 hours, room temperature, concentration $4 \cdot 10^{-3}$ mol/dm^3). Same samples was put in the colloidal solution of Au nanoparticles (24 hours, concentration $2.75 \cdot 10^{-9}$ mol/dm^3, size of nanoparticles was ca 15 nm [43]. This solution was prepared by citrate reduction of $K[AuCl_4]$ [44]. The non-bound chemicals were removed by immersion of the samples into distilled water for 24 hours [45].

2.3.6. Different thiols grafting and subsequent gold nanoobjects grafting

The samples were modified in direct (glow, diode) Ar^+ plasma on Balzers SCD 050 device under the conditions presented in 2.3.1. Immediately after the plasma treatment the samples were inserted into water solution (2 wt. %) of 2-mercaptoethanol (ME), methanol solution (5.10^{-3} mol/dm^3) of biphenyl-4,4'-dithiol (BPD) or into water solution (2 wt.%) of cysteamine (CYST) for 24 hours. To coat the polymers with the gold nanoobjects the plasma treated polymers with grafted thiols were immersed for 24 hours into freshly prepared colloidal citrate stabilized solution of Au nanoparticles (AuNPs), [44,46] or Au nanorods (AuNR), 0.1 mol/dm^3 water solution of cetyltrimethylammonium bromide).

2.4. Diagnostic methods

The ζ-potential measurement is advantageously used as one of methods for characterization of various materials [12-15,17,18]. The ζ-potential measurements often supplement other analyses characterized surface properties, as X-ray photoelectron spectroscopy (XPS) [12,17], goniometry (contact angle) [12-18,26,28], atomic force microscopy (AFM) [12,14,17,27,28] or other measurements [14,17,28].

2.4.1. *Electrokinetic analysis*

Zeta potential of all samples was determined by SurPASS Instrument (Anton Paar, Graz Austria). The SurPASS instrument is an electrokinetic analyzer for the investigation of the ζ-potential of macroscopic solids based on the streaming potential and/or the streaming current method. Samples were studied inside the adjustable gap cell in contact with the electrolyte 0.001 mol/dm^3 KCl. Only for investigation of electrolyte and its concentration we have also tested the 0.005 mol/dm^3 KCl or 0.001 mol/dm^3 KNO_3 [40]. All measurements were accomplished in temperature range of 22–24 °C. Due to strong dependence of ζ-potential on pH, we studied all polymers at pH=6.0-6.2 with the relative error of ± 10%. Only for study of isoelectric points of polymer foils we used pH dependence of zeta potential in range of 7.0 to 2.5. For each measurement a pair of polymer foils with the same top layer was fixed on two sample holders (with a cross section of 20x10 mm^2 and gap between 100 μm) [14,40].

The samples to be studied by streaming potential/current measurements have to be mechanically and chemically stable in the aqueous solutions used for the experiment. First, the geometry of the plug must be consolidated in the measuring cell. This can be checked by rinsing with the equilibrium liquid through repeatedly applying Δp in both directions until finding a constant signal [35]. Capacitance at the electrodes can lead to a long time constant for the system to reach equilibrium. Measurements must be conducted by ramping the pressure up and down to confirm that no hysteresis is observed [30]. Other serious problem of ζ-potential determination are bubbles, not only at suspensions but also at flat samples measured in a very small channels [35]. When bubbles are present, whole sample surface is not covered by electrolyte and results for both directions vary, and the values obtained are not stable. Due to these it is necessary to rinse the samples before experiment and repeat the measurement to exclude all adverse effects. Therefore all samples were rinsed several times before measurement and every experiment comprises 4 ramps. If ζ-potential values are constant during this multiple measurement, the surface is stable and values obtained are correct.

Because the ζ-potential of polymer foils is strongly affected by their surface roughness and other surface properties of measured samples [13,32,40], two approaches to determine ζ-potential were used, streaming current and streaming potential. For calculation of ζ-potential we used eqs. (1) and (2) presented above, to investigate and to disclose possible effects of the surface roughness or other surface properties.

2.4.2. *Contact angle*

Contact angles of distilled water were measured at room temperature at two samples and at seven positions using a Surface Energy Evolution System (SEES, Masaryk University, Czech Republic). The „static" contact angle was measured for all samples immediately after the plasma treatment (<10 min delay). Drops of 8.0±0.2 μl volume were deposited using automatic pipette (Transferpette Electronic Brand, Germany) and their images were taken with 5 s delay. Then the contact angles were evaluated using SEES code [12,13].

2.4.3. *Surface roughness and morphology*

The surface morphology and surface roughness R_a were examined by AFM microscopy. The AFM images were taken under ambient condition of a Digital Instruments VEECO CP II set-up. "Tapping mode" was chosen in preference to "Contact mode" to minimize damage to the samples surfaces. Si probe RTESPA-CP with the spring constant 20-80 N/m was used. By repeated measurements of the same region (2x2 μm^2) we proved that the surface morphology does not change after three consecutive scans. The mean roughness value (R_a) represents the arithmetic average of the deviations from the centre plane of the sample [12,13].

2.4.4. *X-ray photoelectron spectroscopy*

Concentrations of individual elements (O, C, F, N, Au and S) in the pristine and modified surface layer was measured by X-ray photoelectron spectroscopy (XPS). Omicron Nanotechnology ESCAProbeP spectrometer was used to measure photoelectron spectra (error of 10%). Exposed and analyzed area had dimension 2x3 mm^2. X-ray source was monochromated at 1486.7 eV with step size 0.05 event. The spectra evaluation was carried out by CasaXPS programme [46].

2.4.5. *Sheet resistance*

Electrical sheet resistance (R_s) is a suitable method for characterization of conductive surfaces. R_s of the gold nanostructures and nanolayers was determined by a standard two-point technique using KEITHLEY 487 pico-ampermeter. For this measurement additional Au contacts, about 50 nm thick, were created by sputtering. The electrical measurements were performed at a pressure of about 10 Pa to minimise the influence of atmospheric humidity. The typical error of the sheet resistance measurement did not exceed ±5% [47].

2.4.6. *Infrared spectroscopy*

In some cases the changes of chemical structure were examined by Fourier Transform Infrared Spectroscopy (FTIR) on Bruker ISF 66/V spectrometer equipped with a Hyperion microscope with ATR (Ge) objective.

3. Results

The ζ-potential measurement is advantageously used as one of fast and easy methods for characterization of various material surfaces. We used this for characterization of polymer surface and their variable modifications mentioned above. ζ-potential measurements is always in our papers supplemented with other analyses characterized surface properties, as X-ray photoelectron spectroscopy (XPS) [12,17], goniometry (contact angle) [12-18,26,28], with atomic force microscopy (AFM) determination [12,14,17,27,28] or with other measurements [14,17,28]. In this chapter we presented selected and interesting results for pristine polymers and their surface modified by variable ways.

3.1. Pristine polymers

ζ-potential of 21 polymer foils was determined for investigation of electrolyte type and concentration effect [40]. Fig. 1 presents ζ-potential of polymer foils in 0.001 mol/dm^3 KCl (left) and 0.005 mol/dm^3 KCl (right) electrolytes determined by streaming current and streaming potential approaches and calculated by both of equations, eq. 1 (HS, black columns) and eq. 2 (FM, red columns). Obtained results correspond with others presented previously [30,43].

Firstly, it is evident ζ-potential is a suitable characteristic for characterization and distinction of different polymer foils, generally variable solid substrates. Results for both electrolytes, even for both of concentrations, have the same trend, but the absolute values differ for particular electrolytes and individual concentrations [40]. It is in agreement with previously presented studies for silica, glass and some polymers [30,31], ζ-potential is an increasing function of electrolyte concentration, which corresponds to results presented previously for other systems [2,21,22,30,31,39,48]. It is necessary all zeta potential results supplement with information about type and concentration of applied electrolyte.

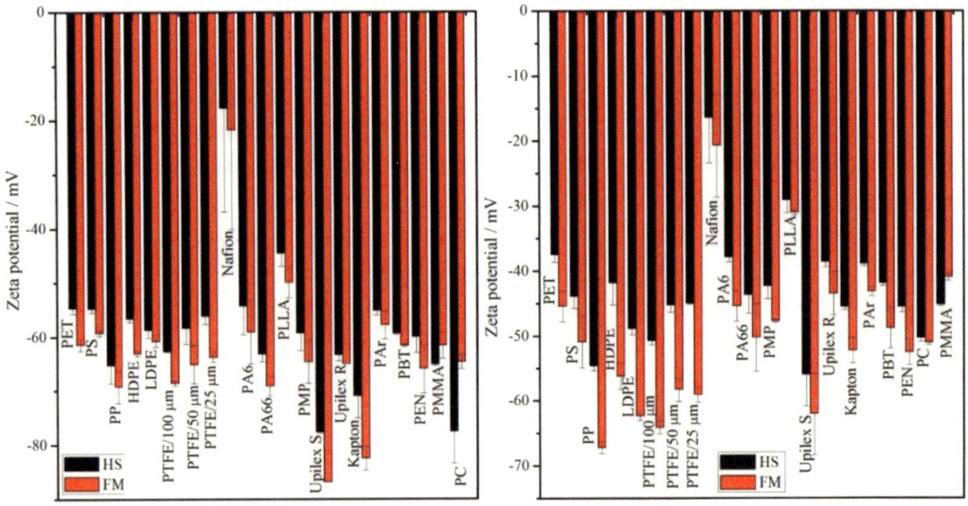

Figure 1. Zeta potential of 21 polymer foils under our study measured in 0.001 mol·dm^{-3} (left) and in 0.005 mol·dm^{-3} (right) KCl electrolyte and calculated by eq. 1 (HS, black columns) and by eq. 2 (FM, red columns) [40]

Fig. 1 shows that most of studied polymers are thought to be chemically inert, without reactive surface groups and displaying negative ζ-potential at neutral pH. The ζ-potential of these polymer foils, except Nafion, is generated only by the preferential adsorption OH^- or H_3O^+ ions on the surface. Zeta potential value and the differences between individual polymers are given only by the distinct adsorption range depending on polarity or non-polarity of individual polymers [49]. Interfacial charge at the non-polar polymer is caused by the preferential adsorption of OH^- ions and less strongly adsorbed H_3O^+ ions in KCl solution. K^+ and Cl^- ions behave indifferently at these non-polar polymer surfaces in comparison with polar ones. This effect is apparent from Fig. 1. The non-polar polymers have the lower ζ-potential values. In some cases ζ-potentials as polar and non-polar polymers (e.g. PTFE, PP, PE) seem to be the same, e.g. for polar PET and non-polar PE (HDPE or LDPE), but as it is clear from Fig. 1, at these polymers are different difference between zeta potential determined and calculated by both of methods and equations. Possible explanation could be found in our previous study [50] and it can be connected with oxygen containing molecular segments oriented toward the polymer bulk and also with surface roughness.

The higher ζ-potentials are found on polar polymers without any functional surface groups (e.g. PLLA). Polymers of similar surface chemistry exhibit similar ζ-potentials, e.g. three samples of PTFE foils differing only in the thickness, or HDPE and LDPE. Only one of used polymers, Nafion, contains sulfonic acid groups on the surface disposed to dissociation in contact with electrolyte. This dissociation is a key action to zeta potential generation at thie polymer. Due to this dissociation zeta potential of Nafion is the lowest. In some cases it can be even positive due to this [40].

Zeta potential of polymer foils is strongly affected by their surface roughness R_a and other surface properties as well [12,32]. Surface roughness R_a of selected polymers determined by AFM is summarized in Table 1. For confirmation of surface roughness effect we used two approaches to determine ζ-potential (streaming current and streaming potential) and two equations (eqs. 1 and 2) for calculation. For polymers, with higher surface roughness R_a, larger differences between both calculated values of ζ-potential (ζ and $\zeta_{apparent}$, see eqs. (1) and (2)) are observed (e.g. on PTFE foils known to have the highest surface roughness). Also HDPE with higher roughness (see Table 1) exhibits more significant differences in ζ-potentials in comparison with LDPE of lower surface roughness [40].

Polymer	R_a (nm)	Polymer	R_a (nm)	Polymer	R_a (nm)
PET	0.5	PTFE/25μm	17.2	Kapton	0.7
PS	1.3	Nafion	1.9	PAr	0.8
PP	2.1	PA6	7.4	PBT	1.6
HDPE	11.1	PA66	4.1	PEN	4.8
LDPE	2.5	PLLA	1.2	PMMA	0.4
PTFE/100μm	12.3	Upilex S	4.1	PC	1.7
PTFE/50μm	13.9	Upilex R	4.4	PMP	5.0

Table 1. Surface roughness (R_a) in nm of polymers under our study determined by atomic force microscopy [40]

We compare our results with those presented previously by other authors. As we mentioned above, many results cannot be compared in principle since the measurements were done using different techniques and quoted errors are often of opposite sign [30] or at different experimental conditions. Data on ζ-potential for polydimethylsiloxane (PDMS), PC, PET, PMMA, PE, PS, polyvinylchloride PVC, PTFE obtained by many authors and by different techniques were summarized in [31]. Values of ζ-potential for PC and PMMA, -70.0 mV and

-45.0 mV respectively, obtained with 0.001 mol/dm^3 KCl solution (for PMMA the higher concentration of electrolyte was used), pH=6 and at room temperature and reported in [31] are in reasonable agreement with ours, -77.6 and -65.0 mV respectively, with a knowledge about concentration effect for PMMA. Also ζ-potential for some type of PTFE -58 mV in 0.01 mol/dm^3 KCl reported in [49] is in good agreement with our presented values varying from -56.1 to -62.6 mV depending on the foil thickness.

Other comparison is difficult due to the authors examined the samples, combining polymer membrane foils, polymer fibres and colloidal dispersions for which the ζ-potential values vary strongly in dependence on their surface properties. Another chance is to compare results for PET [37], but the previous results were obtained for pH=5 and the ζ-potential = -33.6 mV is lower than presented here (see Fig. 1) due to the pH dependence mentioned above. Zeta potential for PET was determined [50] to be about -40 mV (only approximate estimation from figure [50]) using the same instrument as we used (the same cell, electrolyte, concentration, pH=6), this value agrees well with the present one of -54.7 mV, obtained for pH=6.2. Some polymers inclusive PC and PS have also been measured in 0.001 mol/dm^3 KCl, but in different analyser [51] and the data are not comparable with present ones. In addition PS was not in form of foil and other polymers were used without purification and the results can be affected by contaminants (stabilizers) in the polymers.

Zeta potential values are often correlated with contact angle measurements and/or surface roughness determination. As we discussed previously, that is not correct in all cases [40]. Most of polymers used in this study have no "reactive" surface groups in water solution and the surface charge only arises due to preferential adsorption of hydroxide ions in water solutions. The adsorption on the polymer surface is a drive force for ζ-potential initiation. Because both of contact angle and ζ-potential are connected to surface chemistry, some authors correlated only these two characteristics. At polymers with "active" functional groups the most important is "surface chemistry" and the correlation of the ζ-potential with contact angle measurement is better. For polymers, for which the adsorption of hydroxide ions is predominant, the surface roughness plays the more important role in ζ-potential due to a rough surface and not homogenous electrical double layer. Therefore, for correlation of zeta potential must be taken both contact angle and also surface roughness. We supplemented this study of polymers with contact angle values and also with surface roughness. Fig. 2 shows contact angle values for all polymers studied in this work measured in distilled water. Surface roughness R_a of samples examined is presented in Table 1. A general trend is that the "more" polar polymers have lower contact angle (e.g. PMMA, PLLA, PA), that means the higher hydrophilicity and wettability, the non-polar polymers the higher contact angle (PTFE, HDPE, LDPE, PMP) corresponding to their higher hydrophobicity. Polymers of the same surface chemistry embody the similar contact angle, as e.g. HDPE and LDPE or all PTFE samples (PTFE/25μm, PTFE/50μm, PTFE/100μm), slight differences between them is caused by distinct surface roughness. From the comparison of the results of surface roughness presented in Table 1 and of contact angle presented in Fig. 2 follows that the polymers with lower R_a (e.g. PET, PS and PP) exhibit higher wettability and vice versa. The same conclusion has been done for zeta potential determination.

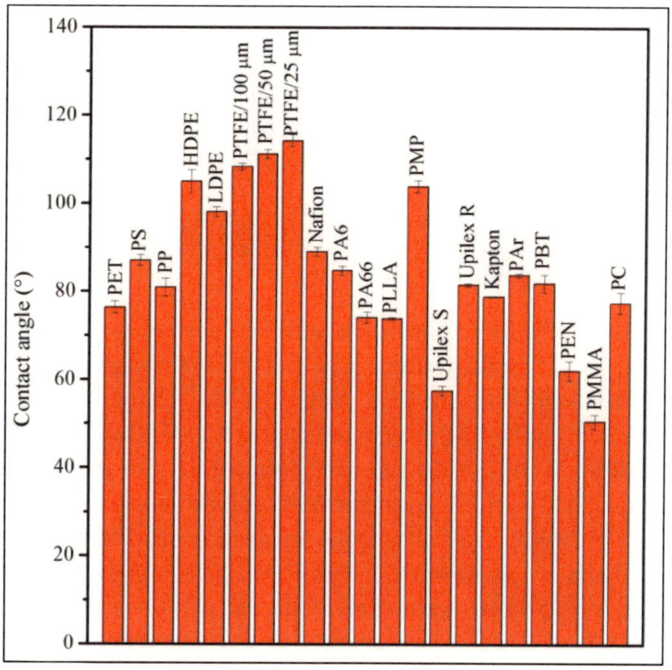

Figure 2. Contact angle for 21 polymer foils under study determined by goniometry [40]

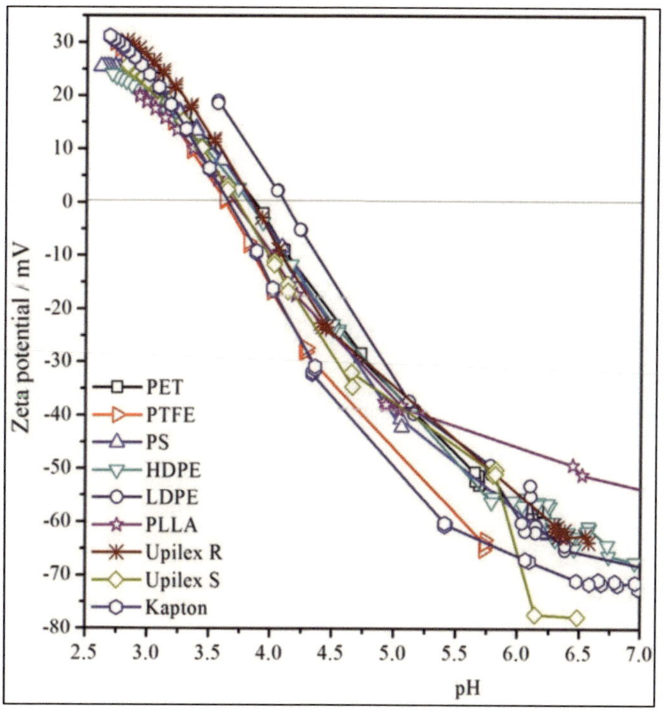

Figure 3. pH dependence of zeta potential of selected polymer foils. Isoelectric point is the point of zero value of zeta potential [52]

Zeta potential also serves for determination of isoelectric point (IEP), which is defined as the point at which the electrokinetic potential equals zero. It is also important characteristic for wrappers, for material for study of living cell adhesion, etc. Fig. 3 present zeta potential of several polymer foils as pH dependent. As it is clear, IEP of all polymer samples are obtained at pH ca 4. For this determination we have titrated samples by 0.1 mol/dm^3 HCl in pH range from 7.0 to 2.5 [52].

3.2. Plasma and laser treatment of polymers

As it is known, plasma treatment affects the polymer surface to the depths from several hundred to several thousand angstroms but the bulk properties of polymers remain unchanged [12]. Therefore only surface properties are changed dramatically by this treatment. Generally, during the plasma treatment the macromolecular chains are degraded, cleavaged and oxidized. Due to this, plasma treatment results in a dramatic increase of polymer wettability. The wettability increase is caused by creation of polar groups, exhibiting enhanced hydrophilicity, at the polymer surface [12]. The similar effect on surface properties has also the laser irradiation [53]. Fig. 4 shows the effect of plasma treatment (left) and laser irradiation (right) on zeta potential of selected polymer foils. For more transparent presentation only results obtained by streaming current method (eq. 1) are shown.

As it is clear from Fig. 4, plasma treatment results in a dramatic increase of the zeta-potential due to that the plasma modification of polymer surfaces leads to dramatic change of surface chemistry (e.g. chains and bonds scission, ablation, oxidation and cross-linking) and creation of polar groups [12]. The effect of laser irradiation is not too strong due to that this treatment way influences surface chemistry and surface properties more weak [12,53]. Even in some causes zeta potential decreases after laser irradiation. It can be explained by the fact the cleavage of original bonds is most significant than surface oxidation after laser exposure in comparison with the plasma one. These effects strongly depend on plasma power or number of laser pulses.

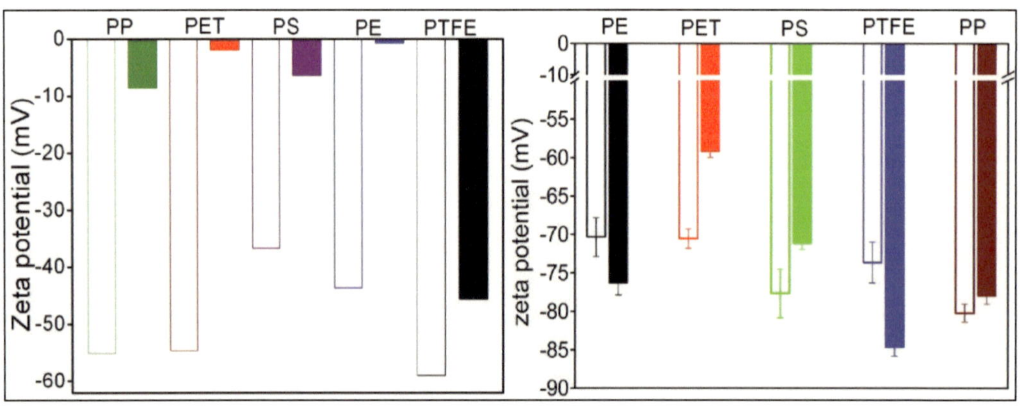

Figure 4. Zeta potential of selected polymer foils and effect of plasma (left, [12]) or laser (right, [53]) treatment determined by electrokinetic analysis. Empty columns presents pristine polymers, full columns treated ones

For explanation of zeta potential behaviour on pristine and treated polymer surfaces we analysed surface chemistry by other techniques. Oxygen concentration in the surface layer of the laser modified polymers was determined from XPS spectra, the surface wettability and polarity were characterized by the measurement of contact angle of distilled water. These results are presented in Figs. 5 and 6. As it is clear form Fig. 5, after plasma or laser treatment increase the presence of oxygen groups on polymer surface. This increase in the oxygen content is much lower (Fig. 5, left) in comparison with that observed on the same polymers treated in plasma discharge (Fig. 5, right) [12,53].

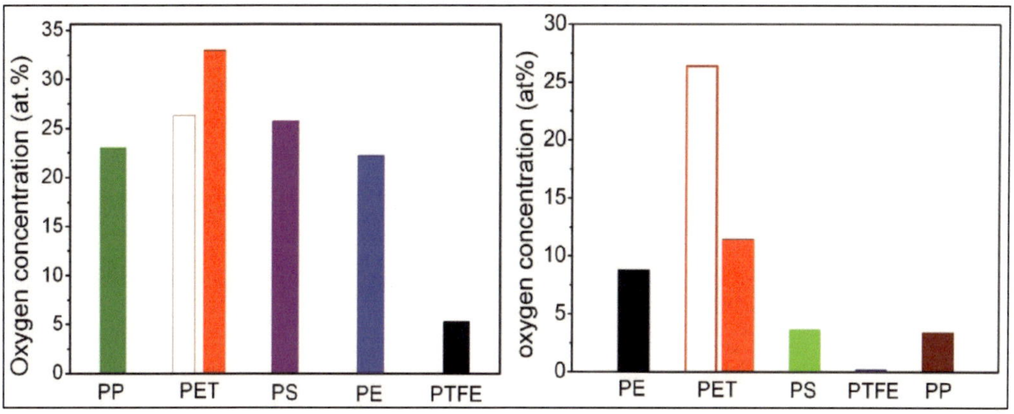

Figure 5. Concentration of oxygen groups on surface of selected polymer foils after plasma (left, [12]) or laser (right, [53]) treatment determined by XPS. Empty columns are pristine polymers (only at PET), full columns treated ones

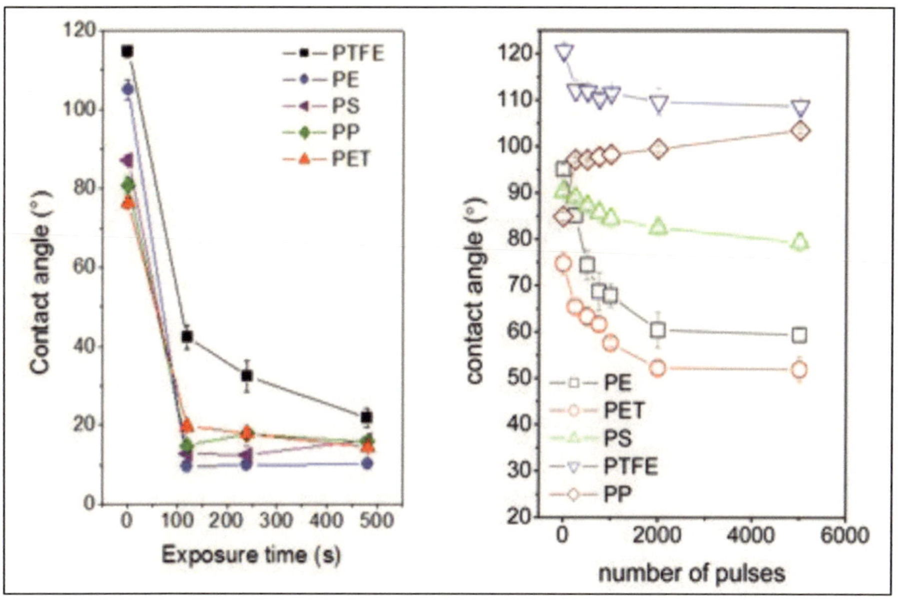

Figure 6. Contact angle on surface of selected polymer foils after plasma (left, [12]) or laser (right, [53]) treatment determined by goniometry

Plasma treatment or laser irradiation result in a dramatic decline of the contact angle and corresponding increase of polymer wettability as it shown in Fig. 6. As is presented above, also contact angle analyses compare the result that laser irradiation effects surface chemistry and surface properties much less (Fig. 6, left) in comparison with plasma treatment (Fig. 6, right).

3.3. Gold sputtering and consequent annealing

In this part of work we have studied the changes of surface morphology and some other physico-chemical properties of sputtered gold nanostructures and nanolayers on polymer surface (PTFE) induced by post-deposition annealing [42]. Fig. 7 presents zeta potential (left) and sheet resistance (right) of pristine PTFE (Fig. 7, thickness = 0) and Au coated PTFE in Au thicknesses dependences for room temperature (RT) and the same samples after annealing by different temperatures (100°C red line, 200°C green line, 300°C, blue line). Zeta potential was determined only for samples deposited by gold at RT and at 300°C.

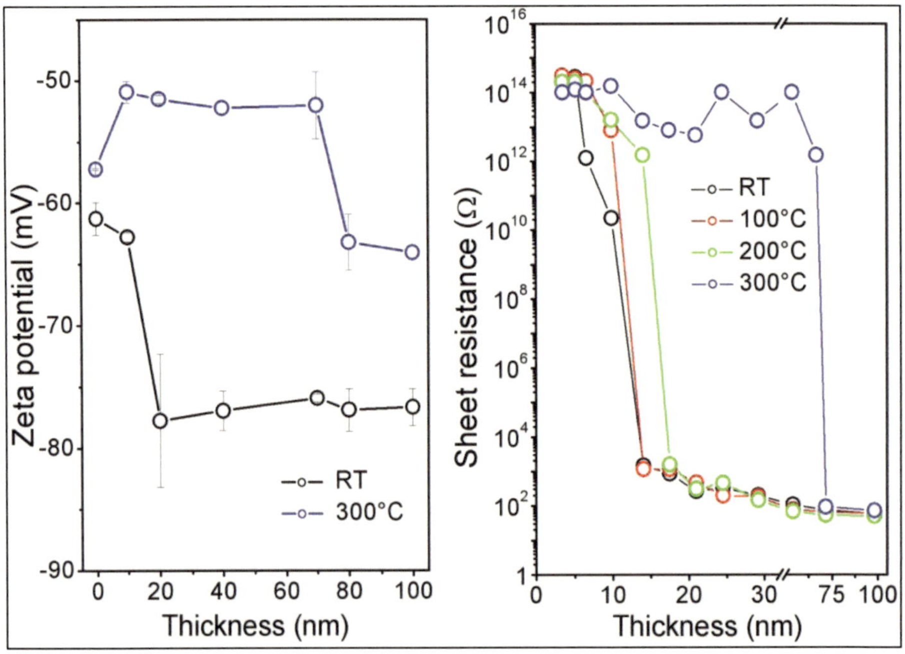

Figure 7. Zeta potential (left) and sheet resistance (right) of gold nanostructures and annealed ones [42]

We can see, that after deposition of metal layer zeta potential dramatically decrease due to increasing presence of surface coverage by metal and accompanying surface conductivity. For as-sputtered samples and very thin gold nanolayers, the zeta potential is close to that of pristine PTFE due to the discontinuous gold coverage since the PTFE surface plays dominant role in zeta potential value. Then, for thicker layers, where the gold coverage prevails over the original substrate surface, the zeta potential decreases rapidly and for the thicknesses above 20 nm remains nearly unchanged, indicating total coverage of original substrate by gold. For annealed samples, the dependence on the layer thickness is quite different. It is seen that the annealing leads to a significant increase of the zeta potential for thin nanolayers. This increase may be due to thermal degradation of the PTFE accompanied by production of excessive polar groups on the polymer surface, which plays the important role when the gold coverage is discontinuous [42].

These results on zeta potential are in good agreement with sheet resistance measurement and other surface analyses [42]. As its obvious from Fig. 7 (right), for the as-sputtered samples the sheet resistance decreases rapidly in the narrow thickness from 10 to 15 nm when an electrical continuous gold coverage is formed. After annealing at 300°C a dramatic change in the resistance curve is observed. The annealed layers are electrically discontinuous up to the Au effective thickness of 70 nm above which the continuous coverage is created and a percolation limit is overcome. These results confirmed zeta potential analysis well as it is presented in Fig. 7, blue lines of both diagrams.

3.4. Polyethylene glycol grafting

As a representative of polymers grafted by polyethylene glycol we present PS grafted by polyethylene glycol (PEG) of different molecular weight 300 (PEG 300), 6000 (PEG 6000) and 20000 (PEG 20000) g/mol in Fig. 8. Modification of polymer surface by PEG plays important role in research of living cell adhesion and proliferation. Due to improving grafting process of PEG on PS it is better to treated polymer surface by plasma. We exposed PS surface by plasma for 50, 100 and 300 s and subsequently we grafted these surfaces by PEGs of mentioned molecular weights. These results have usage in tissue engineering due to positive effect to living cell adhesion and proliferation discussed below in Fig. 9.

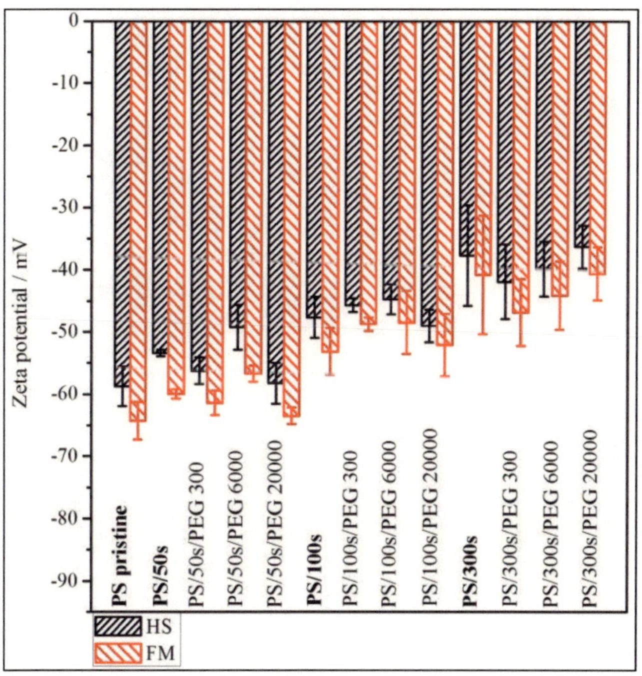

Figure 8. Zeta potential of PS treated by plasma for 50, 100 or 300 s and subsequently grafted with PEG of different molecular weight 300 (PEG 300), 6000 (PEG 6000) and 20000 (PEG 20000) g/mol)

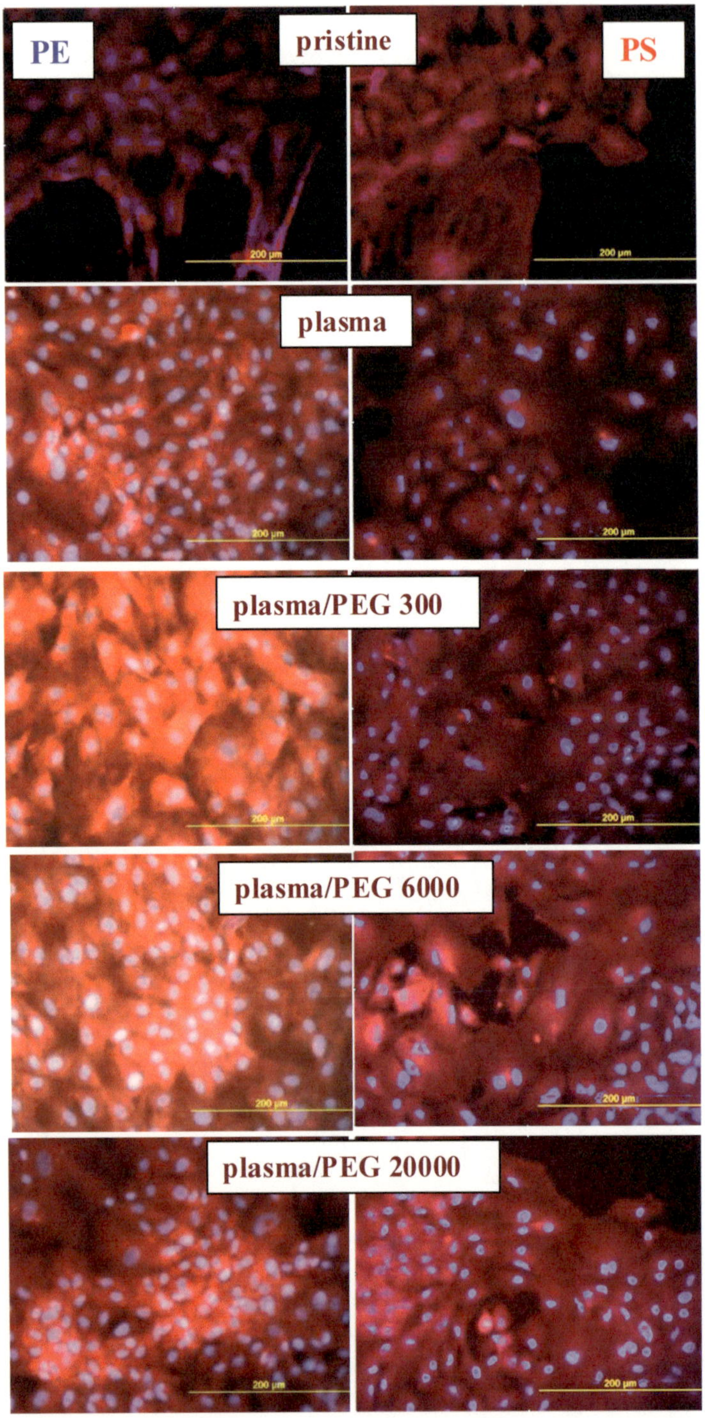

Figure 9. Photographs of cells (VSMC) proliferated (7th day) on pristine PE and PS, polymers treated with plasma (plasma) and plasma modified and subsequently grafted with PEG (plasma/PEG ...). The molecular weights of PEG were M_{PEG} =300, 6000 and 20000 [18]

As it is clear from Fig. 8, plasma treatment leads to increase of zeta potential due to creation of more polar groups on surface and this effect depends on exposure time positively. Subsequent grafting of PEG influences surface chemistry dependently on its molecular weight. While PEG 300 and PEG 20000 in most cases causes a slight decrease of zeta potential, PEG 6000 for shorter exposure time causes increase of zeta potential. It indicates quite different surface chemistry and nevertheless these changes are quiet slight, they play important role in process of adhesion and proliferation of living cells (see Fig. 9).

Adhesion and proliferation of rat vascular smooth muscle cells (VSMC) were studied on the pristine, plasma modified and subsequently PEG grafted PE and PS samples. It is supposed that the PEG bonds to several positions on the same macromolecular chain, limits chain mobility and in this way facilitates adsorption and colonization of the cells. Moreover, the oxygen-containing groups present in the PEG molecules might also enhance the colonization of the material by VSMCs as indicated by the highest final cell numbers on the PEG grafted PE observed on the 7th day after cell seeding [18].

Attractiveness of a substrate for cells is manifested not only by the number of growing VSMC cells but also by the homogeneity of their spreading over substrate surface. In Fig. 9 images of the cell cultures cultivated on PE and PS substrates subjected to different modification steps are shown. On pristine polymers the coverage with proliferating cells is not homogenous. It could be explained by the tendency of cells to compensate the insufficient cell-material adhesion by cell-to-cell adhesion, which leads to formation of cell clusters (aggregates) on the material surface. Plasma modification and especially the PEG grafting improves the coverage homogeneity substantially on both polymers.

3.5. Biphenyldithiol grafting and subsequent Au nanoparticles grafting

We tested several polymers (PE, PS, PET) for influence of surface properties affected by grafting of dithiols (methanol solution of biphenyl-4, 4′-dithiol (BFD) or 1, 2-ethanedithiol (ED)) and subsequent grafting by gold nanoparticles [42,54]. Before this grafting has also been polymer surface exposed to plasma discharge. As it is clear from Fig. 10 (left) for PET and BFD grafting, the zeta potential dramatically increases after plasma treatment due to creation of polar oxygen groups, which has been discussed above several times. The grafting of BFD leads to other zeta potential changes caused by thiol groups SH^- bonded at polymer surface, which results in dissociation of this group in liquid surrounding during zeta potential measurement. This process leads to creation of negative charge on polymer surface and more presence of positive charge in electrolyte following by decrease of zeta potential. Subsequent Au nanoparticles grafting results to other decrease of zeta potential due to presence of metal gold nanostructures on polymer surface. From data obtained by both of methods and eqs. (1) and (2) (Fig. 10, left, HS, FM data resp.) it is clear the higher difference between HS and FM data for surface with Au nanoparticles which is caused by presence of some gold nanostrucrures and due to this increasing conductivity and also surface roughness.

In this case, the results on zeta potential were compared with infrared spectroscopy (Fig. 10, right), where the blue curve indicates the sulphur presence and the red curve indicates the gold presence in sample, confirming both of successful chemical graftings. FTIR spectroscopy was used for the characterization of chemical composition of modified PET samples. In the Fig. 10 (right) the differential FTIR spectra of the PET samples plasma treated and grafted in biphenyldithiol (BFD) and with Au nanoparticles are shown. The band at 790 cm^{-1} corresponds to absorption of the S-C group and the band at 761 cm^{-1} is assigned to the S-Au group. That is confirmation that both of graftings were successful as has been established by zeta potential analysis.

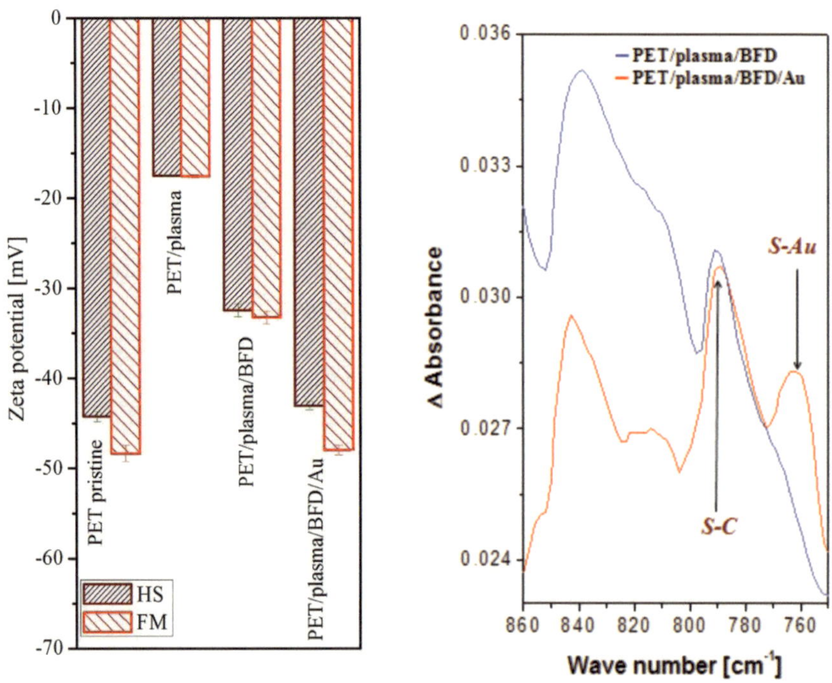

Figure 10. Zeta potential (left) and infrared spectroscopy (right) results for pristine PET, PET treated by plasma and consequently grafted by BFD and subsequent grafting by Au nanoparticles [54]

On the other hand, grafting of ED on PET was not successful. After the grafting of plasma treated PET with ED and Au nanoparticles the peak at 761 cm^{-1} (S-Au) in FTIR spectra was not detected. This finding supports the conclusion that no Au nanoparticles are bonded to the PET treated in ED [54].

3.6. Different thiols grafting and subsequent Au nanoobjects grafting

Other attempt to modify polymer surfaces was treatment of PET and PTFE by plasma discharge and subsequent grafting with different thiols (mercaptoethanol, ME, biphenyldithiol, BFD and cysteamine, CYST). Thiols are expected to be fixed via one of their functional groups –OH, –SH or –NH_2 to reactive places created by the preceding plasma treatment. The remaining „free" –SH group is then allowed to interact with gold nanoparticles or gold nanorods. The main goal of this study is to examine the effect of the plasma treatment and the thiol grafting on the binding of gold nanoobjects on the polymer surface. Chemical structure of the modified polymer films is expected to influence substantially theirs elektrokinetic potential in comparison with pristine polymers. Zeta potentials results for PET (left) and PTFE (right) modified by mentioned ways are presented in Fig. 11. Zeta potential of pristine PET, resp. PTFE, plasma treated PET, resp. PTFE, subsequently grafted with thiols and then with gold nanoobjects (nanoparticles, AuNP or nanorods, AuNR) are presented in Fig. 11. Zeta potential is affected by several factors, such as surface morphology, chemical composition (e.g. polarity, wetability) and electrical conductivity of surface. As it is clear from Fig. 11, zeta potential increases after plasma treatment. Due to higher ablation PTFE this increase is more significant at PET. After grafting of thiols in most cases zeta potential decreases due to presence of dissociative groups on surface. The Au nanoobjects grafting results in decrease of zeta potential values due to presence of gold nanostructures on surfaces.

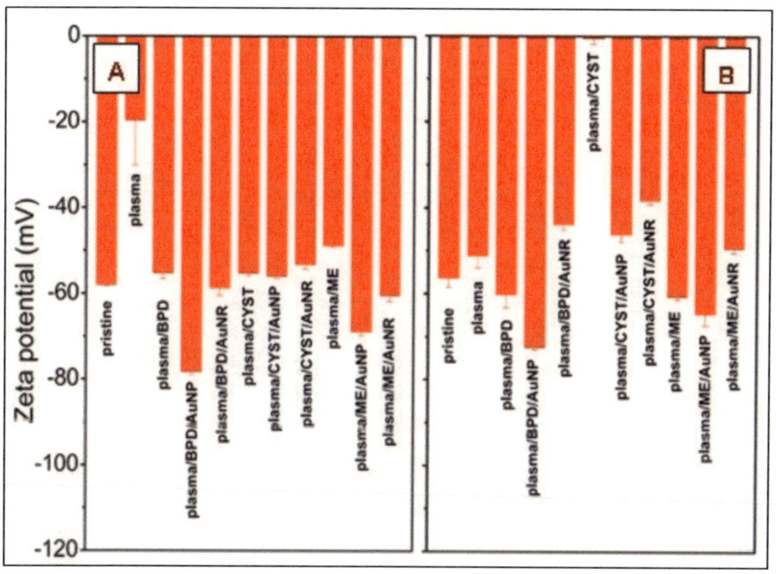

Figure 11. Zeta potential determined by SurPASS of pristine (pristine) and modified (A) PET and (B) PTFE. The polymer foils were plasma treated (plasma), plasma treated and grafted with (i) thiols (biphenyldithiol-BPD, cysteamine-CYST and mercaptoethanol-ME) and then (ii) grafted with Au nanoobjects (nanoparticles-AuNP and nanorods-AuNR).

Very interesting results we obtained after CYST exposition on PTFE, which results in dramatic increase of zeta potential. It can be explained by presence of "free" NH_2- groups on surface caused probably by not preferential binding of cysteamine to plasma treated PTFE. This compound can be bonded probably not only via NH- group on activated surface but also via SH- group. This result was obtained only at PTFE, not at PET. At both of polymers the highest decrease of zeta potential after gold nanoobjects was found out after BPD grafting and subsequent grafting of Au nanoparticles.

Also in previous cases, zeta potential sensitively indicates all changes in surface chemistry and surface properties after any modification. These chemical changes were studied by XPS analysis. Elements (C(1s), O(1s), F (1s), S(2p), N (1s) and Au (4f)) concentrations (in at.%) determined by XPS measurements for PET and PTFE pristine and plasma treated (for 120 s) and grafted with (i) thiols (biphenyldithiol-BPD, cysteamine-CYST and mercaptoethanol-ME) and then (ii) grafted with Au nanoobjects (nanoparticles-AuNP and nanorods-AuNR) are presented in Table 2. The interesting results is for cysteamine on plasma treated PTFE, where nitrogen contents is not significantly different from the same sample of PET, but contents of bonded gold nanoparticles embody significantly lower value. It can confirm the zeta potential results discussed above. At both polymers obtained results clearly indicate the gold nanoparticles grafted at these substrates more easily than gold nanorods.

PET	**C(1s)**	**O(1s)**	**F(1s)**	**S(2p)**	**N(1s)**	**Au(4f)**
Pristine	73.7	26.3	-	-	-	-
plasma treated	67.0	33.0	-	-	-	-
plasma/BPD	73.8	18.5	-	7.7	-	-
plasma/BPD/AuNP	81.1	12.7	-	6.1	-	0.1
plasma/BPD/AuNR	82.2	12.1	-	5.0	-	0.7
plasma/CYST	71.6	20.9	-	3.1	4.4	-
plasma/CYST/AuNP	70.1	21.4	-	2.3	2.9	3.3
plasma/CYST/AuNR	75.5	19.9	-	1.8	2.7	0.1
plasma/ME	72.0	26.8	-	1.2	-	-
plasma/ME/AuNP	77.4	21.7	-	0.6	-	0.3
plasma/ME/AuNR	80.5	16.4	-	1.0	-	2.1
PTFE	**C(1s)**	**O(1s)**	**F(1s)**	**S(2p)**	**N(1s)**	**Au(4f)**
pristine	33.4	-	66.6	-	-	-
plasma treated	39.8	5.2	55.0	-	-	-
plasma/BPD	78.0	6.2	4.7	11.1	-	-
plasma/BPD/AuNP	60.5	4.9	29.7	4.4	-	0.5
plasma/BPD/AuNR	58.8	4.5	33.0	3.0	-	0.6
plasma/CYST	48.9	5.4	39.2	3.4	3.1	-
plasma/CYST/AuNP	44.3	5.1	45.4	2.7	2.3	0.2
plasma/CYST/AuNR	57.4	5.9	31.2	2.4	3.0	0.1
plasma/ME	41.2	8.0	49.0	1.8	-	-
plasma/ME/AuNP	52.8	6.7	35.1	0.3	-	5.1
plasma/ME/AuNR	50.1	5.6	40.6	1.0	-	2.6

Table 2. Elements (C(1s), O(1s), F (1s), S(2p), N (1s) and Au (4f)) concentrations (in at.%) determined by XPS measurements in pristine PET and PTFE and plasma treated (for 120 s) and grafted with (i) thiols (biphenyldithiol-BPD, cysteamine-CYST and mercaptoethanol-ME) and then (ii) grafted with Au nanoobjects (nanoparticles-AuNP and nanorods-AuNR).

4. Conclusion

The ζ-potential of many polymer foils was determined and the effect of different modification approaches was studied to confirm that this analysis is sensitive to indicate all changes in surface chemistry. Zeta potential can distinguish individual pristine polymer foils due to their different polarity, which causes that ζ-potential is generated by the preferential adsorption OH^- or H_3O^+ions on the surface. Zeta potential of planar samples of polymer foils depends on many surface properties of samples (surface chemistry, wettability, polarity, surface roughness and morphology), as well as on electrolyte solution properties (counter-ion type and concentration and pH). As was confirmed, the best electrolyte for study of planar polymer foils was 0.001 mol/dm^3 KCl. Zeta potential is also strongly dependent on measurement conditions, temperature, pH, etc. All zeta potential must be supplemented by this experimental information.

In our research all zeta potential values are always complemented with contact angle measurement by goniometry and determination of surface roughness and surface morphology by AFM, by spectroscopy analyses or other techniques.

Pristine polymer foils can be distinguished by electrokinetic analysis, as same as all modifications of polymer surface were indicated by changes in zeta potential. Due to this electrokinetic potential can serve as a fast and easy characteristic for study of changes in surface chemistry. Elektrokinetic analysis is the fast, the easy and very sensitive method for characterization solid substrates.

Author details

Zdeňka Kolská,
Faculty of Science, J. E. Purkyně University, Ústí nad Labem, Czech Republic,

Zuzana Makajová, Kateřina Kolářová, Nikola Kasálková Slepičková,
Simona Trostová, Alena Řezníčková, Jakub Siegel and Václav Švorčík
Department of Solid State Engineering, Institute of Chemical Technology, Prague, Czech Republic

Acknowledgement

This work was supported by the grants of the GACR projects Nos. P108/10/1106, P108/12/1168 and P503/12/1424.

5. References

[1] Hiemenz P, Rajagopalan R (1997) The electrical double layer and double-layer interactions. In: Principles of colloid and surface chemistry, Marcel Dekker, Inc., New York, pp. 499-533.

[2] Beattie JK (2006) The intrinsic charge on hydrophobic microfluidic substrates. Lab. Chip 6: 1409-1411.

[3] Kosmulski M (2001) Chemical properties of material surfaces. Vol. 102, Marcel Dekker, Inc., New York, 2001.
[4] Marinova KG, Alargova RG, Denkov ND, Velev OD, Petsev DN, Ivanov IB, Borwankar RP (1996) Charging of oil-water interfaces due to spontaneous adsorption of hydroxyl ions. Langmuir 12: 2045-2051.
[5] Stana-Kleinschek K, Kreze T, Ribitsch V, Strnad S (2001) Reactivity and electrokinetical properties of different type sof regenerated celulose fibres. Colloids Surf. A 195: 275-284.
[6] Stana-Kleinschek K, Ribitsch V (1998) Electrokinetic properties of processed celulose fibers. Colloids Surf. A 140: 127-138.
[7] Grancaric AM, Tarbuk A, Pusic T (2005) Electrokinetic properties of textile fabrics. Color. Technol. 121: 221-227.
[8] Stana-Kleinschek K, Strnad S, Ribitsch V (1999) The influence of structural and morphological changes on the electrokinetic properties of PA 6 fibres. Colloids Surf. A 159: 321-330.
[9] Moritz T, Benfer S, Arki P, Tomandl G (2001) Influence of the surface charge on permeate flux in the dead-end filtration with ceramic membrane. Sep. Purif. Technol. 25: 501-508.
[10] Elimelech M, Chen WH, Waypa JJ (1994) Measuring the zeta (electrokinetic) potential of reverse-osmosis membranes by a streaming potential analyser. Desalination 95: 269-286.
[11] Childress AE, Elimelech M (1996) Effect of solution chemistry on the surface charge of polymeric reverse osmosis and nanofiltration membranes. J. Membrane Sci. 119: 253-268.
[12] Řezníčková A, Kolská Z, Hnatowicz V, Stopka P, Švorčík V (2011) Comparison of glow argon plasma-induced surface changes of thermoplastic polymers. Nucl. Instrum. Meth. B 269: 83-88.
[13] Švorčík V, Řezníčková A, Kolská Z, Slepička P, Hnatowicz V (2010) Variable surface properties of PTFE foils. e-Polymers 133: 1-6.
[14] Švorčík V, Kolská Z, Luxbacher T, Mistrík J (2010) Properties of Au nano-layer sputtered on PET. Mater. Lett. 64: 611-613.
[15] Slepička P, Vasina A, Kolská Z, Luxbacher T, Malinský P, Macková A, Švorčík V (2010) Argon plasma irradiation of polypropylene. Nucl. Instrum. Meth. B 268: 2111-2114.
[16] Bratskaya S, Marinin D, Nitschke M, Pleuel D, Schwarz S, Simon F (2004) Polypropylene surface functionalization with chitosan. J. Adhes. Sci. Technol. 18: 1173-1186.
[17] Slepičková Kasálková N, Slepička P, Kolská Z, Sajdl P, Bačáková L, Rimpelová S, Švorčík V (2012) Cell adhesion and proliferation on polyethylene grafted with Au nanoparticles. Nucl. Instrum. Meth. B, 272: 391 - 395.
[18] Švorčík V, Makajová Z, Kasálková N, Kolská Z, Bačáková L (accepted) Plasma-modified and polyethylene glycol-grafted polymers for potential tissue engineering applications. J. Nanosci. Nanotechnol.
[19] Jachowicz J, Berthiaume M, Garcia M (1985) The effect of the amphiprotic nature of human-hair keratin on the adsorption of high charge-density cationic polyelectrolytes. Colloid. Polym. Sci. 263: 847-858.
[20] Werner C, Konig U, Augsburg A (1999) Electrokinetic surface characterization of biomedical polymers - a survey. Colloids Surf. A 159: 519-529.

[21] Volkova AV, Ermakova LE, Sidorova MP, Antropova TV, Drozdova IA (2005) The Effect of thermal treatment on the structural and electrokinetic properties of porous glass membranes. Colloid J. 67: 263-270.

[22] Ermakova LE, Medvedeva SV, Volkova AV, Sidorova MP, Antropova TV (2005) Structure and electrosurface properties of porous glasses with different compositions in KCl and NaCl Solutions. Colloid J. 67: 304-312.

[23] Young A, Smistad G, Karlsen J, Rölla G, Rykke M (1997) Zeta Potentials of human enamel and hydroxyapatite as measured by the Coulter® DELSA 440. Adv. Dental Res. 11: 560-565.

[24] Hiemenz P, Rajagopalan R (1997) Electrophoresis and other electrokinetic phenomena. In: Principles of colloid and surface chemistry. Vol. 3, Marcel Dekker, Inc., New York, pp. 534-574.

[25] Goddard JM, Hotchkiss JH (2007) Polymer surface modification for the attachment of bioactive compounds. Prog. Polym. Sci. 32: 698-725.

[26] Jelínek M, Smetana K, Kocourek T, Dvořánková B, Zemek J, Remsa J, Luxbacher T (2010) Biocompatibility and sp3/sp2 ratio of laser created DLC films. Mater. Sci. Eng. B-Solid 169: 89-93.

[27] Jelínek M, Kocourek T, Remsa J, Mikšovský J, Zemek J, Smetana K Jr, Dvořánková B, Luxbacher Z (2010) Diamond/graphite content and biocompatibility of DLC films fabricated by PLD. Appl. Phys. A 101: 579-583.

[28] Jelínek M, Kocourek T, Jurek K, Remsa J, Mikšovský J, Weiserová M, Strnad J, Luxbacher T (2010) Antibacterial properties of Ag-doped hydroxyapatite layers prepared by PLD method. Appl. Phys. A 101: 615-620.

[29] Rance GA, Khlobystov AN (2010) Nanoparticle-nanotube electrostatic interactions in solution: the effect of pH and ionic strength. Phys. Chem. Chem. Phys. 12: 10775-10780.

[30] Kirby BJ, Hasselbrink EF Jr (2004), Zeta potential of microfludic substrates: 1. Theory, experimental techniques, and effects on separation. Electrophoresis 25: 187-202.

[31] Kirby BJ, Hasselbrink EF Jr (2004) Zeta potential of microfludic substrates: 2. Data for polymers. Electrophoresis 25: 203-213.

[32] Zimmermann R, Bickert O, Gauglitz G, Werner C (2003) Electrosurface phenomena at polymer films for biosensor application. Chem. Phys. Chem. 4: 509-514.

[33] Kohler HH (1993) Surface charge and surface potential. In: Coagulation and flocculation: Theory and application. Marcel Dekker, Inc., New York, pp. 43-69.

[34] Luxbacher T, Bukšek H, Petrinić I, Pušić T (2009) Zeta potential determination of flat solid surfaces using a SurPASS electrokinetic analyser. Tekstil 58: 393-400.

[35] Delgado AV, González-Caballero F, Hunter RJ, Koopal LK, Lyklema J (2007) Measurement and interpretation of electrokinetic phenomena. J. Colloid Interf. Sci. 309: 194-224.

[36] Yaroshchuk A, Luxbacher T (2010) Interpretation of electrokinetic measurements with porous films: role of electric conductance and streaming current within porous structure. Langmuir 26: 10882-10889.

[37] Bukšek H, Luxbacher T, Petrinić I (2010) Zeta potential determination of Polymeric materials using two differently designed measuring cells of an electrokinetic analyser. Acta Chim. Slov. 57: 700-706.

[38] Slepička P, Kolská Z, Náhlík J, Hnatowicz V, Švorčík V (2009) Properties of Au nanolayers on polyethyleneterephtalate and polytetrafluorethylene. Surf. Interface Anal. 41: 741-745.

[39] Afonso MD, Hagmeyer G, Gimbel R (2011) Streaming potential measurements to assess the variation of nanofiltration membranes surface charge with the concentration of salt solution. Sep. Purif. Technol. 22-23: 529-541.

[40] Kolská Z, Řezníčková A, Švorčík V (accepted) Surface characterization of polymer foils e-Polymer.

[41] Siegel J, Slepička P, Heitz J, Kolská Z, Sajdl P, Švorčík V (2010) Gold nano-wires and nano-layers at laser-induced nano-ripples on PET. Appl. Surf. Sci. 256: 2205–2209.

[42] Siegel J, Krajcar R, Kolská Z, Sajdl P, Švorčík V (2011) Annealing of gold nano-structures sputtered on PTFE. Nanoscale Res. Lett. 6:588

[43] Švorčík V, Řezníčková A, Sajdl P, Kolská Z, Makajová Z, Slepička P (2011) Au nanoparticles grafted on plasma treated polymers. J Mater Sci 46: 7917-7922.

[44] Řezanka P, Záruba K, Král V (2008) A change in nucleotide selectivity pattern of porphyrin derivatives after immobilization on gold nanoparticles. Tertahedron Lett. 49:6448-6453.

[45] Ročková-Hlaváčková K, Švorčík V, Bačáková L, Dvořánková B, Heitz J, Hnatowicz V (2004) Bio-compatibility of ion beam-modified and RGD-grafted polyethylene. Nucl Instrum Meth B 225:275-282.

[46] Švorčík V, Chaloupka A, Záruba K, Král V, Bláhová O, Macková A, Hnatowicz V (2009) Nucl. Instrum. Meth. B 267: 2484-2488.

[47] Švorčík V, Kvítek O, Lyutakov O, Siegel J, Kolská Z (2011) Annealing of sputtered gold nano-structures. Appl. Phys. A 102: 747–751.

[48] Min Y, Pesika N, Zasadzinski J, Israelachvili J (2010) Studies of bilayers and vesicle adsorption to solid substrates: Development of miniature streaming potential apparatus (SPA). Langmuir 26: 8684-8689.

[49] Zimmermann R, Dukhin S, Werner C (2001) Electrokinetic measurement reveal interfacial charge at polymer films caused by simple electrolyte ions. J. Phys. Chem. B 105: 8544-8549.

[50] Škvarla J, Luxbacher T, Nagy M, Sisol M (2010) Relationship of surface hydrophilicity, charge, and roughness of PET foils stimulated by incipient alkaline hydrolysis. ACS Appl. Mater. Inter. 2: 2116-2127.

[51] Bismarck A, Kumru ME, Springer J (1999) Characterization of several polymer surfaces by streaming potential and wetting measurements: some reflections on acid-base interactions. J. Colloid Interf. Sci. 217: 377-387.

[52] Kolská Z, Makajová Z, Trostová S, Slepičková Kasálková N, Siegel J, Švorčík V, Electrokinetic Potential for Characterization of Nanosctructured Solid Flat Surfaces. Submitted to J. Nano Res.

[53] Řezníčková A, Chaloupka A, Heitz J, Kolská Z, Švorčík V (2012) Surface properties of polymers treated with F2 laser. Surf. Interface Anal. 44: 296–300.

[54] Švorčík V, Kolská Z, Kvítek O, Siegel J, Řezníčková A, Řezanka P, Záruba K (2011) "Soft and rigid" dithiols and Au nanoparticles grafting on plasma-treated polyethyleneterephthalate. Nanoscale Res. Lett. 6: 607.

Chapter 9

Polymerization of Peptide Polymers for Biomaterial Applications

Peter James Baker and Keiji Numata

Additional information is available at the end of the chapter

http://dx.doi.org/10.5772/46141

1. Introduction

The term biomaterial, as defined by a consensus conference on Definitions in Biomaterial Science, is a material intended to interface with biological systems to evaluate, treat, augment or replace any tissue, organ of function of the body. [1] Biomaterials range from the simple embedded material to complex functional devices. There are three primary types of biomaterials: **1)** *metallic*, based on metallic bonds, **2)** *ceramic*, based on ionic bonds, and **3)** *polymeric*, based on covalent bonds. [2] Polymers that are mechanical resistance, degradable, permeable, soluble and transparent, have been used in both simple and complex biomaterial applications. [3-5] The mechanical properties of poly(esters), poly(amides), poly(amidoamines), poly(methyl methacrylate), and poly(ethyleneimine), are particularly well suited **(Table 1)**. These polymeric materials have given rise to first and second-generation biomaterials. [6] 'Next'-generation biomaterials will have less toxic degradation products, undergo hierarchal assemble to form supramolecular structures, and maintain a sustainable design. The degradation products of synthetic polymers are of acidic and cannot be metabolized biological systems. This can result in a bioaccumulation of these products offsetting the homeostatic balance of the system. The production of synthetic polymers requires bulk separation and crystallization, which often inhibits the formation of higher ordered structures. Finally, the source materials for these polymers are petrochemicals. The dependency on petrochemicals presents both environmental and sustainability concerns. The development of materials that overcome these limitations would be a significant advancement in the field of biomaterials.

Biological materials have capabilities that far exceed those which are synthesized chemically. [21, 22] Biopolymers including polypeptides, can serve as replacement materials for synthetic polymers. Polypeptides such as collagen, elastin, and silk, are currently being sought as next-generation biomaterials **(Table 2)**. [23-28] Collagen, a major constituent of bone, cartilage, tendon, skin and muscle, are the most abundant proteins in the human body. [29] Several different types of collagen have been identified; these proteins are distinguished by their triple-helical structure. Type I collagen forms supramolecular assemblies. This assembly is controlled by environmental parameters such as concentration, pH, and ionic strength, making it of particular interest as a biomaterial. [29] Type I collagen is approximately 1000 amino acids long and contains a tripeptide (-Pro-Hyp-Gly-)$_n$ tandem repeat, where Hyp is a postranslationally modified hydroxyproline. Elastin is an extracellular matrix protein responsible for the extensibility and elastic recoil of blood vessels, ligaments and skin. Elastin is approximately 70 kDa protein composed of crosslinking domains and elastin domains. Elastin domains are composed poly(Gly-Val-Gly-Val-Pro)$_n$ and poly(Gly-Val-Gly-Val-Ala-Pro)$_n$ of repeating sequences. These domains undergo an inverse temperature transitions where the protein forms a crystalline state on raising the temperature and redissolve on lowering the temperature. [30, 31] Silks are fibrous proteins spun by silkworms and spiders. They have a range of functions, including cocoons to protecting eggs or larvae, draglines to support spiders, and the formation of webs that can withstand high impacts of insect prey. [32] The mechanical strength of dragline spider silks has led to its employment in several biomedical applications. Dragline spider silk is around 300 kDa protein composed of two repeating domains: poly(Ala)$_n$ and poly(Gly-Gly-Gly-Xaa-Gln-Tyr)$_n$, where Xaa can be any amino acid. The strength of this polymer is a result of the highly crystalline structure of the poly(Ala) domains and the amorphous poly(Gly-Gly-Gly-Xaa-Gln-Tyr) repeat is amorphous in structure which allows for flexibility. [33, 34] All of three of these high molecular weight polypeptides are composed of highly repetitive amino acid sequences. The present review will be concerned with the synthesis of polypeptides and how they polypeptides will be used in the production of new biomaterials.

Polymer	Biomaterial Application	Ref
Synthetic Polymer		
Polyester	Drug Delivery, Sutures, Stents, Nanoparticles	[7-10]
Polyamides	Sutures, Wound Dressings	[11]
Polyamidoamine	Biomedical Imaging	[12]
Poly(methyl methacrylate)	Contact Lens	[4]
Polyethyleneimine	Gene Delivery, Tissue Engineering	[13-15]
Natural Polymer		
Polypeptides	Gene Delivery, Biomedical Imaging, Drug Delivery,	[16-18]
Polysaccharides	Tissue Engineering, Nanoparticles	[19]
Polynucleotide	Biomedical Imaging	[20]

Table 1. Survey of synthetic and natural polymers used as biomaterials

Polypeptide	Repeating Sequences	MW (kDa)	Ref
Collagen	(-Pro-Hyp-Gly-)$_n$	~ 400	[29]
Elastin	(-Gly-Val-Gly-Val-Pro-)$_n$, or (-Gly-Val-Gly-Ala-Pro-)$_n$	~ 70	[30,31]
Spider Silk	(-Ala-)$_n$ or (-Gly-Gly-Gly-Xaa-Gln-Tyr-)$_n$	~ 300	[33,34]

Table 2. Amino acid sequence of polypeptides being used as biomaterials

The synthesis of polypeptides means formation of amide bonds between amino acid monomers. Amide bonds are formed by a condensation reaction between a carboxylic acid and an amine. The conventional chemical method for amide bond synthesis requires the activation of the carboxyl group followed by a nucleophilic attack by a free amine. This requires the presence of coupling reagents, base and solvents **(Figure 1A)**. [35] When this condensation reaction occurs between two amino acids the resulting amide bond is a peptide bond. Solid phase peptide synthesis (SPPS), developed by Merrifield, provides a fast an efficient manner to synthesis polypeptides. In SPPS, the N-terminal amino acid is attached to a solid matrix with the carboxyl group and the amine group is protected **(Figure 1B)**. Amine group undergoes a deprotection step revealing an N-terminal amine. This is followed by a coupling reaction between the activated carboxyl group of the next amino acid and the amine group of the immobilized residue. Side chains of several amino acids contain functional groups, which may interfere with the formation of amide bonds and must be protected. This process may continue through iterative cycles until the polypeptide has reached its desired chain length. Polypeptide is then cleaved off the resin and purified. [35-37] SPPS has enabled the synthesis of ogliopeptides ~40 to 50 amino acids residues. However, limitations in chemical coupling efficiency have made it impractical to synthesize longer polypeptides with reasonable yields.

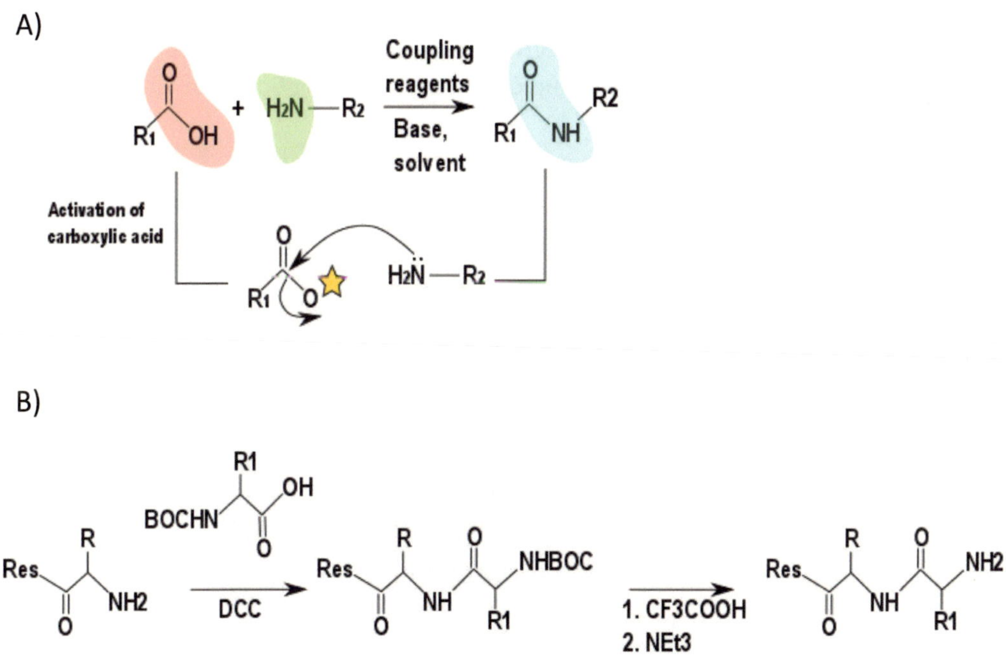

Figure 1. Chemical synthesis of polypeptides. A) The chemical synthesis of an amide bond requires the activation of the carboxylic acid B) Solid phase peptide synthesis.

Peptide bonds, the key chemical linkage in proteins, are synthesized not only chemically but also biologically. In this process the primary structure of a protein is encoded in a gene, a short sequence of deoxyribonucleic acid (DNA). A two-step process of transcription and translation are required to biosynthetic extract this information from a gene. Transcription is the step where the genetic information encoded from the DNA is transcribed into messenger ribonucleic acid (mRNA) in the form of an overlapping degenerate triplet code. This process occurs in three stages: initiation, chain elongation, and termination. Initiation occurs when the RNA polymerase binds to the promoter gene sequence on the DNA strand. Elongation begins as the RNA polymerase is guided along the template DNA unwinding the double-stranded DNA molecule as well as synthesizing a complementary single-stranded RNA molecule. Finally, transcription is terminated with the release of RNA polymerase from template DNA. Translation involves the decoding of the mRNA to specifically and sequentially link together amino acids in a growing polypeptide chain. Decoding of mRNA occurs in the ribosome, a macromolecular complex composed of nucleic acids and proteins. mRNA is read in three-nucleotide increments called codons; each codon specifies for a particular amino acid in the growing polypeptide chain **(Figure 2)**. Ribosomes bind to mRNA at the start codon (AUG) that is recognized by initiator tRNA. Ribosomes proceed to elongation phase of protein synthesis. During this stage, complexes composed of an amino acid linked to tRNA sequentially bind to the appropriate codon in mRNA by forming complementary base pairs with the tRNA anticodon. Ribosome moves from codon to codon along the mRNA. Amino acids are added one by one, translated into polypeptide sequences dictated by DNA and represented by mRNA. At the end, a release factor binds to the stop codon, terminating translation and releasing the complete polypeptide from the ribosome.

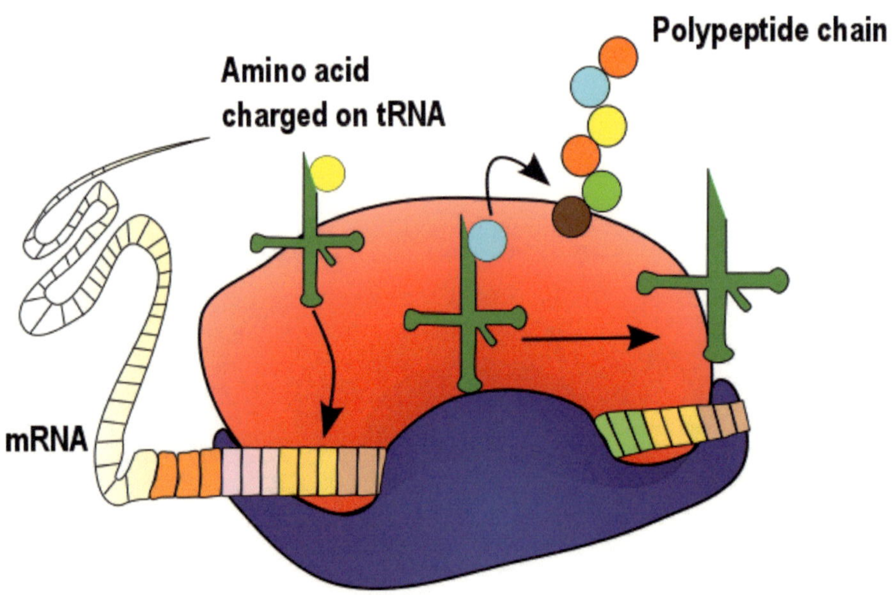

Figure 2. Cartoon representation of polypeptide synthesis on the ribosome.

The advent of recombinant DNA technology has allowed the facile, large-scale production of many polypeptide sequences. However, polypeptides containing highly repetitive sequences are difficult to express recombinantly due to undesirable elements in the secondary structure of the mRNA. [38-41] The development of techniques to synthesize high molecular weight polypeptide sequences in a high yield, low impact manner will significantly advance the field of biomaterials **(Figure 3)**.

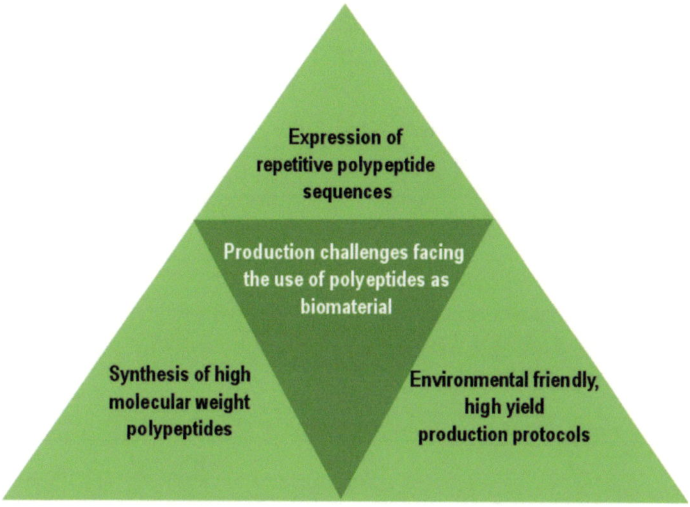

Figure 3. Synthesis challenges facing the use of polypeptides as biomaterials.

These current methods in polypeptides synthesis, namely, SPPS and the recombinant DNA technique, have significantly improved the field of biomaterials. The 'next' generation of biomaterials will require high yields of high molecular weight, repetitive sequences bearing modified amino acids. Below we will review the recent advances of 1) chemoenzymatic synthesis by protease, 2) peptide synthesis by NRPS 3) oligomerization of peptides by amino acid ligase, and 4) native chemical ligation, in the context of biomaterial production.

2. Chemoenzymatic synthesis of polypeptides

In 1898 vant' Hoff postulated that based on the principle of reversibility of chemical reactions, proteases would catalyze peptide synthesis. Bergman *et al.* actualized this in 1938 by successfully demonstrating the protease-mediated synthesis of Leu-Leu and Leu-Gly dipeptides. [42-44] Under standard conditions proteases hydrolyze peptide bonds (**Figure 4**). The active sites of proteases are composed of subsites located on either side of the catalytic site. The geometry and electrostatic potential of these subsites influence the substrate specificity of these enzymes (**Table 3**). Protease mediated polypeptide synthesis proceeds through either a thermodynamically controlled synthesis (TCS) or a kinetically controlled synthesis (KCS). [45] TCS is a reversal of hydrolysis and requires conditions, which shift the equilibrium towards synthesis. Any protease is suitable for TCS, the protease increases the rate at which equilibrium is established but does not alter the final equilibrium.

[46] TCS undergoes a two-step process **(Figure 5)**. The first step is an exergonic process where a proton is transferred to –COOH and –NH_2. The second step is an endergonic condensation. Reaction conditions should be selected that ensure for optimal catalytic activity. Manipulation of reaction conditions is required to increase the product yield. Strategies such as, product precipitation, introduction of organic solvents or water immiscible solvents, have been employed to favor synthesis.

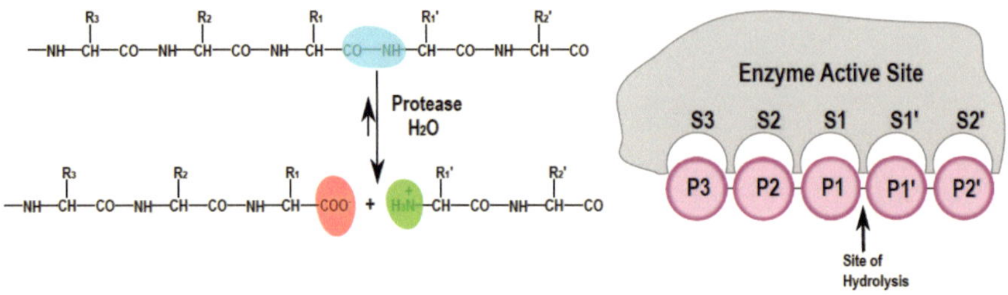

Figure 4. Protease mediated hydrolysis of peptide bonds. A) Hydrolytic reaction scheme B) Proteases active site is composed of subsites (S). Each S has an affinity for residues (P). This "lock and Key" mechanism dictates protease specificity.

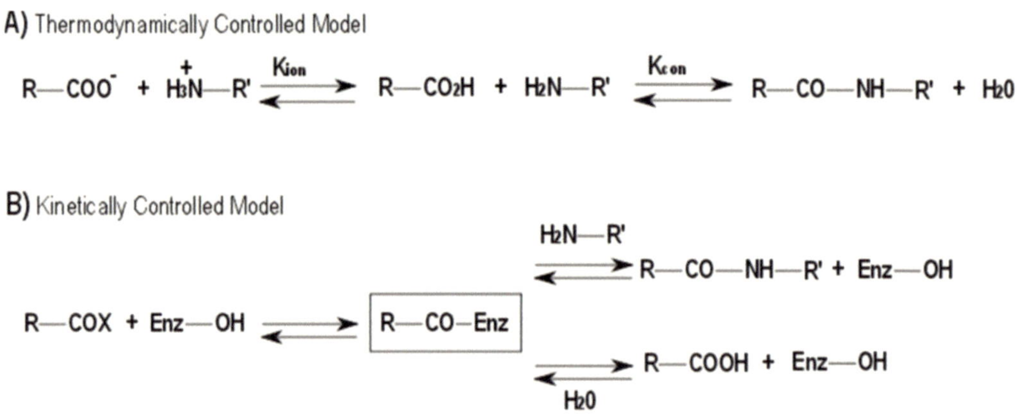

Figure 5. Protease mediated polypeptide synthesis occurs either through a thermodynamically controlled mechanism or a kinetically controlled mechanism.

Protease	Class	Cleavage Sequence
α-Chymotrypsin	Serine	-(Trp, Tyr, Phe, Leu, Met)🡻Xaa-
Trypsin	Serine	-(Arg, Lys)🡻Xaa-
Papain	Cysteine	-Phe-(Leu, Val)🡻Xaa-Xaa-
Bromelain	Cysteine	-Phe-(Leu, Val)🡻Xaa-Xaa-
Pepsin	Aspartly	-Phe-(Tyr, Leu)🡻(Leu, Phe)-
Thermolysin	Metalloproteases	-Phe-(Gly, Leu)🡻(Leu, Phe)-

Table 3. Protease specificity. Arrows indicate cleavage site and Xaa represents an amino acid.

In KCS the protease acts as a transferase; it mediates the transfer of an acyl group to an amino acid of peptide-derived nucleophile **(Figure 5)**. The reaction requires an activated C-terminal ester of the substrate and a protease containing a Cys or Ser residue in the catalytic site.

In aqueous conditions, homopolymers of L-amino acids and modified polypeptides homopolymers and heteropolymers, have been synthesized using cysteine or serine proteases. [47-54] This reaction is initiated by the formation of an acyl-enzyme intermediate between the Cys or Ser residue in the active site and the ester group on modified carboxylic acid. In presence of high concentration of the substrate, the free amine group of the L-amino acids acts as the nucleophile resulting in propagation. Water can also serve as the nucleophile resulting in the termination reaction **(Figure 6).** Under aqueous environments, competition between synthesis and hydrolysis is significant and reaction parameters such as protease activity, pH, buffer capacity, substrate concentration, and reaction time, in influence product formation. Several proteases hydrolyze the peptide bond between –Lys–Xaa, where Xaa represents any amino acid. These proteases were surveyed for optimal conditions for oglio(L-Lys) synthesis. At pH 7.0 bromelain demonstrated a 76% monomer conversion rate and an average chain length (DP_{avg}) of 3.5. Under basic conditions both the monomer conversion and the DP_{avg} were reduced, 10% and 3.0, respectively. pH 10.0 was the optimal condition for trypsin-mediated synthesis of oglio(L-Lys); the monomer conversion rate was 65% and the DP_{avg} of 2.25. At neutral pH the monomer conversion rate was 10% and the DP_{avg} was below 2.0.

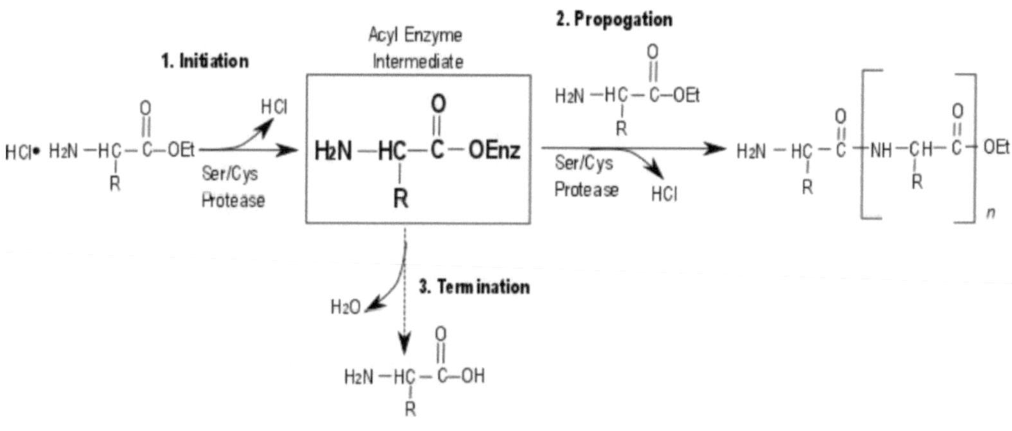

Figure 6. Ser/Cys protease mediated polypeptide synthesis. A carboxy terminal modified amino acid ethyl ester serves as the monomer. Initiation occurs upon the formation of the acyl enzyme intermediate. In the presence of high concentrations of the substrate, the amino terminal of the monomer will act as the nucleophile resulting in chain length propagation. A water molecule can also act as the nucleophile resulting in termination of the reaction.

Upon synthesis insoluble homopolymers of poly(L-Tyr) and poly(L-Ala) underwent self-assembly to form macromolecular structures. [51, 54] Polypeptide crystals have been observed in high molecular weight α-polypeptides synthesized by ring-opening polymerization. Papain-mediated polymerization of L-Tyr ethyl ester at pH 7.0 resulted in a homopolymer with molecular weight greater than 1,000. After 100 minutes, 100% of the monomer was converted to either the polypeptide or an oligomeric state. The resulting polymer precipitated in globular highly crystalline state. Wide-angle X-ray diffraction (WAXD) and scanning electron microscopy (SEM) of the crystal demonstrated rod-like crystal structures originated in the center and radially grew to a diameter larger than 50 μM. [51] Under alkaline pH L-Ala ethyl ester was polymerized into poly(L-Ala) resulting in higher molecular weight product as compared to those synthesized at neutral pH. These higher molecular polymers showed distinct β-sheet formation and were capable of fibril assembly. This was in stark difference to the lower molecular weight polymers that were composed of mostly random coils and formed submicron aggregates.

Chemoenzymatic synthesis of low molecular weight polypeptides have been successful in the production of cholecystokinin and aspartame in non-aqueous media [55, 56] However the bulk production of high molecular weight homo- or heteropolypeptides has yet to be realized. Advances in enzyme and media engineering will significantly advance this field.

3. Nonribosomal peptide synthetase

Nonribosomal peptide synthetases (NRPSs) represent another enzymatic approach to synthesize polypeptides. NRPSs are multimodular complexes of enzymes found in lower organisms that assemble secondary metabolites such as polypeptides, polyketides, and fatty acids **(Figure 7)**. [57] Each NRPSs module is subdivided into four catalytic domains 1) Adenylation domain (A-domain) 2) Peptidyl carrier protein (PCP-domain) 3) Condensation domain (C-domain) and 4) Thioester domain (TE-domain) **(Figure 7)**. [57] Initiation occurs in the A- and PCP-domains. A-domains serve as 'gatekeepers', recognizing specific amino acids (or hydroxy acids) and activating their carboxyl group in an ATP-dependent manner. The A domain has broad substrate recognition and allows for the facile incorporation of modified amino acids. [58, 59] The activated amino acids are transferred to the PCP-domain where they are covalently tethered to the 4′-phosphopantetheinyl cofactor. [60] Propagation takes place at the C-domain, where the amino acids are linked via a condensation reaction. Termination occurs at the TE domain through either a hydrolysis or a cyclization reaction, resulting in a linear or cyclic polypeptide, respectively. [61]

NRPSs have made significant advances in the synthesis of cyclic polypeptides and the ability to introduce modified amino acids will assist in the development of scaffolds for biomaterials. However, NRPSs produce low molecular weight polypeptides and require large enzymatic complexes, which are difficult to use in large-scale production of polypeptides.

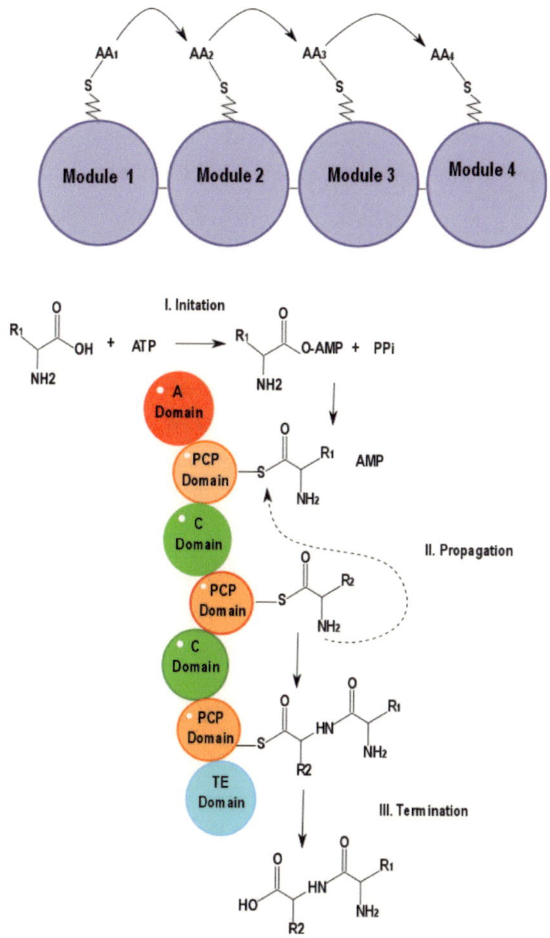

Figure 7. General scheme of nonribosomal peptide synthesis (NRPS). Each NRPS module incorporates one amino acid into the growing peptide chain. The modules are composed of several domains: Adenylation domain (red) is responsible for substrate selectivity, peptidyl carrier protein domain (orange) and condensation domain (green) work synergistically to form the peptide bond, and thioester domain (blue) which terminates the reaction, resulting in either a linear or cyclic polypeptide.

4. Amino acid ligase

Biological synthesis of polypeptides also occurs independent of genetic information. Several enzymes including, folylpolyglutamate synthetase, poly-γ-glutamate synthetase, and D-alanine: D-alanine ligase (DDL), mediate polypeptide synthesis. [62-64] These synthetases either produce non-α-linked amino acids or use non-natural amino acids as their substrates. [65-67] The reaction mechanism of these synthetases requires an ATP-dependent ligation of a carboxyl of one amino acid with an amino- or imino group of a second. These enzymes all belong to the carboxylate-amine/thiol ligase superfamily. [68] This superfamily is identified by structural motifs corresponding to the phosphate-binding loop and an Mg^{2+} binding site located within the ATP-binding domain. [68]

Cell-free biochemical studies suggest the existence of dipeptide synthetases, which form α-peptide linkages between L-amino acids. [73, 74] Tabata *et al.* using *in silico* screening identified the L-amino acid ligase gene (*YwfE*) in *Bacillus subtilis*. [69] Recombinant YwfE (also reported as BacD, in the literature) protein demonstrated α-dipeptides synthesis activity, using from unprotected L-amino acids as the substrate. YwfE contains an ATP binding domain however this domain showed no sequence homology with aminoacyl-tRNA synthetase or A-domains of NRPSs. The crystal structure of YwfE was refined to 2.5 Å and is superimposable on DDL from *E. coli* (PDB ID: 2DLN). [75] YwfE is divided into three domains: an N-terminal domain, a central domain and a C-terminal domain. The N-terminal and central domains are structurally similar to DDL while the C-terminal domain is 100 residues longer and contains additional antiparallel β-sheets. [75] Critical differences were observed in the active site cavity of YwfE when compared to DDL: binding mode of dipeptide moiety, and the size and electrostatic potential of the active site. These differences contribute to the substrate preference for L-amino acids and play a critical role in the stabilization of the tetrahedral intermediate state as proposed for DDL mechanism. [75, 76]. The specificity of YwfE is currently limited to non-bulky, neutral residues at the N-terminus and bulky, neutral residues at the C-terminus.

Using YwfE as a template sequence new L-amino acids ligases (L-AAL) have been identified *in silico* **(Table 4)**. [70-72] RizB from *B. subtilis* NBRC3134 mediated the synthesis of branch chained L-amino acids and L-Met homo-polypeptides. [77-81] RizB also synthesized heteropeptides with high specificity at the N-terminal and relaxed specificity towards the C-terminal. [70] RizB polypeptide synthesis is similar to other L-AAL however RizB uses amino acid monomers as well as ogliopeptides as their substrates while other L-AAL use only amino acids monomers.

Ligase	Species	Preference	Length	Composition	Ref
YwfE	*B. subtilis*	Neutral	Dimer	Hetero-	[69]
RizB	*B. subtilis* NBRC3134	Branched	Oligomer	Hetero-	[70]
spr0969	*S. pneumoniae*	Branched	Oligomer	Hetero-	[71]
BAD_1200	*B. adolescentis*	Aromatic	Oligomer	Hetero-	[71]
RimK	*E. coli*	Glutamic Acid	Oligomer	Homo-	[72]

Table 4. Amino acid ligases homologs, substrate specificity and products

In a subsequent study, RizB was used as a template sequence to identify additional L-AAL that synthesis high molecular polypeptides. [71] spr0969 and BAD_1200 from *Streptococcus pneumoniae* and *Bifidobacterium adolescentis*, respectively, were identified as RizB homologs. spr0969 showed a modest improvement in polypeptide chain length. spr0969 polymerization of Val resulted in six repeat units while RizB polymerization showed only four repeats. BAD_1200 was more promiscuous than the other L-AAL investigated. The activity towards branched amino acids was lower than RizB but BAD_1200 also polymerized aromatic amino acids. [71]

Ribosomal protein S6 from *Escherichia coli* undergoes a unique post-translation modification where up to six glutamic acid residues are ligated to the C-terminus. [82-84] RimK, also a member of carboxylate-amine/thiol ligase superfamily, mediated this post-translational modification. [84] In vitro analysis of RimK synthesis resulted in 46-mer (maximum length) of α-poly (L-Glu) at pH 9.0, 30 °C. The maximum chain length was pH dependent. Furthermore, RimK demonstrated strict substrate specificity for Glu. [72]

The data reported in these manuscripts demonstrates the enzymatic synthesis of homo- and heteropolypeptides. Data mining has expanded both the diversity and chain length of the resultant polypeptides. Recently, the crystal structure YwfE has been reported which will assist in the development of protein design studies. Engineering the reaction media for enhanced solubility of the polypeptides may influence the molecular weight.

5. Native chemical ligation

Native chemical ligation (NCL) allows the combination of two unprotected peptide segments by the reaction of a C-terminal thioester with an N-terminal cysteine peptide in a two-step process (**Figure 8**). The first step is a transthioesterification where the thiolate group of peptide 2 attacks the C-terminal thioester of peptide 1 under mild reaction conditions (aqueous buffer, pH 7.0, 20 °C), resulting in a thioester intermediate. This intermediate rearranges by an intramolecular S→N acyl shift that results in a peptide bond at the ligation site. [85] NCL has been employed for the synthesis of biomaterials such as type I collagen and hydrogels.

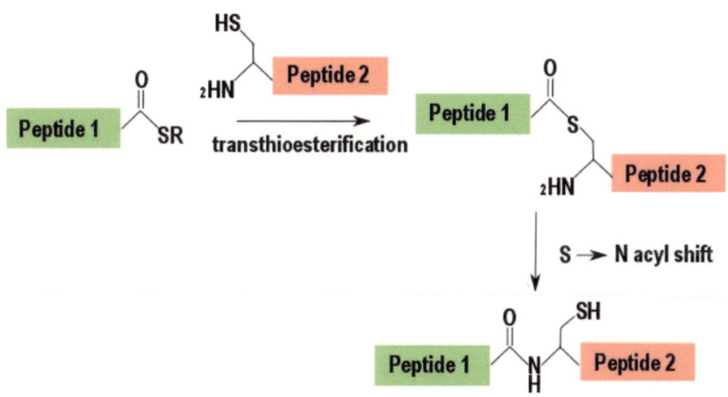

Figure 8. Native chemical ligation, the sulfur atom of the N-terminal Cys residue of peptide 2 attacks the C-terminal thiol group of peptide 1 producing a thioester intermediate that rearranges to yield a peptide bond.

The length of type I collagen far exceeds the length which can be synthesized by SPPS. [86] Recombinant expression of this protein is limited due to the high number of hydroxyproline residues. [87] Paramonov *et. al.* was able to overcome these limitations by employing an NCL strategy. [88] Low molecular weight (Pro-Hyp-Gly)$_n$ repeats bearing an N-terminal Cys residue and a C-terminal thioester were prepared by SPPS. Following NCL polymeric material with M_w = 28,000, M_n = 12,000, and PDI = 2.3, were observed. Circular dichroism studies demonstrated that the secondary structure of the ligated collagen was in agreement with the native collagen. TEM studies of the ligated collagen revealed a dense network of fibers with diameter in the nanometer range and microns in length. These high molecular weight products indicated that the collagen was self-assembling in a fashion similar to natural collagen.

Hydrogels are hydrophilic cross-linked polymer networks that can retain water while maintaining a distinct three-dimensional shape. [89] 'Smart' hydrogels have been designed which can shrink, swell or degrade based on environmental factors. These hydrogels are used for tissue engineering, surgical adhesives and drug delivery. [89-91] Polypeptide hydrogels utilize protein scaffolds such as coiled coil domains and four-helix bundles. [92] These scaffolds do not meet the required mechanical robustness for many biomaterial applications. To enhance these mechanical properties, peptide forming hydrogels were modulated with β-sheet forming peptides by NCL. [93] The storage modulus of the hydrogel with the ligated β-sheet was 48.5 kPa, five times higher than the unmodified gel. The hydrogels with the increased stiffness were able to support higher proliferation rates of primary human umbilical vein endothelial (HUVE) cells as compared to the more elastic hydrogels.

NCL has been useful tool in the development of these biomaterials. However, to further expand this technique for the development of higher molecular weight polypeptides significant hurdles must be address. High molecular weight polypeptides would require iterative ligations with peptides bearing both an N-terminal cysteine and a C-terminal thioester. Under typical NCL conditions the expected product would be the cyclic peptide preventing the formation of larger products. [94]

6. Conclusion and future perspectives

As earlier noted, amide bonds are the key chemical linkage in polypeptides. In 2007 the American Chemical Society of Green Chemistry Institute named amide bond formation as a top challenge for organic chemistry. [95] Since then, several new amide bond synthesis reactions have been developed. These methods are less expensive and friendly to the environment. The further development of these reactions and the transitioning of them from small molecules to macromolecules will be a future prospect of developing polypeptides as biomaterials.

Polypeptides represent a class of molecules, which are uniquely qualified to serve as biomaterials. They undergo self-assembly to form macroscopic structures and are synthesized from renewable resources. Chemoenzymatic synthesis, identification of new enzyme sequences and native chemical ligation has advanced the more traditional routes of polypeptide production. Despite the successes outlined above, these techniques have been modest in their production of new biomaterials. Progress in the development of 'next'-generation biomaterials will require media and protein engineering as well as combining these methods reviewed above. One of the major limitations in the chemoenzymatic synthesis of high molecular polypeptides is the solubility of the product. Short chain polypeptides composed of hydrophobic amino acids precipitate out of aqueous solution, yielding a phase separation between enzyme and substrate. Engineering the media to enhance the polypeptide solubility will help achieve higher molecular weight products. Screening organic solvents as well as different immobilization techniques needs to be investigated.

Author details

Peter James Baker and Keiji Numata*
Enzyme Research Team, RIKEN Biomass Engineering Program,
RIKEN, 2-1 Hirosawa, Wako-shi, Saitama, Japan

7. References

[1] Williams DF (1999) The Williams Dictionary of Biomaterials. Liverpool: Liverpool University Press. 343 p.
[2] Williams DF (2009) On the Nature of Biomaterials. Biomaterials. 30: p. 5897-909.
[3] Angelova N, Hunkeler D (1999) Rationalizing the Design of Polymeric Biomaterials. Trends biotechnol. 17 p. 409-21.
[4] Refojo MF (1982) Current Status of Biomaterials in Ophthalmology. Survey of ophthalmology. 26: p. 257-65.
[5] Dumitriu S (2002) Polymeric Biomaterials. In: New Concepts in Polymer Science. Boca Raton, FL: CRC/Taylor & Francis. 1168 p.
[6] Hench L, Polak JM (2002) Third-Generation Biomedical Materials. Science. 295: p. 1014-7.
[7] Seyednejad H, et al. (2011) Functional Aliphatic Polyesters for Biomedical and Pharmaceutical Applications. J. controlled release. 152: p. 168-76.
[8] Postlethwait RW, Dillon ML, Reeves JW (1961) Experimental Study of Polyester Fiber Suture. American journal of surgery. 102: p. 706-9.
[9] Waksman R, Pakala R (2010) Biodegradable and Bioabsorbable Stents. Curr. pharm. des. 16: p. 4041-51.
[10] Lenz RW, Marchessault RH (2005) Bacterial Polyesters: Biosynthesis, Biodegradable Plastics and Biotechnology. Biomacromolecules. 6: p. 1-8.
[11] Shtil'man MI (2003) Polymeric Biomaterials. In: New Concepts in Polymer Science. Boca Raton, FL: CRC/Taylor & Francis. 52 p.
[12] Esfand R, Tomalia DA (2001) Poly(amidoamine) (PAMAM) Dendrimers: from Biomimicry to Drug Delivery and Biomedical Applications. Drug discovery today. 6: p. 427-436.
[13] Chung YC, et al. (2010) A Variable Gene Delivery Carrier-Biotinylated Chitosan/Polyethyleneimine. Biomedical materials. 5: p 650.

* Corresponding Author

[14] Men K, et al. (2010) *A* Novel Drug and Gene Co-Delivery System Based on Poly(epsilon-caprolactone)-Poly(ethylene glycol)-Poly(epsilon-caprolactone) Grafted Polyethyleneimine Micelle. J. nanosci. nanotechnol. 10: p. 7958-7964.
[15] Kuo YC, Ku IN (2009) Application of Polyethyleneimine-Modified Scaffolds to the Regeneration of Cartilaginous Tissue. Biotechnol. progr. 25: p. 1459-67.
[16] Martin ME, Rice KG (2007) Peptide-Guided Gene Delivery. The AAPS journal. 9: p. E18-29.
[17] Nie S, et al. (2007) Nanotechnology Applications in Cancer. Annu. rev. biomed. eng. 9: p. 257-88.
[18] Choe UJ, et al. (2012) Self-assembled Polypeptide and Polypeptide Hybrid Vesicles: from Synthesis to Application. Top. curr. chem. 310: p. 117-34.
[19] Calvo P, et al. (1997) Chitosan and Chitosan/Ethylene Oxide-Propylene Oxide Block Copolymer Nanoparticles as Novel Carriers for Proteins and Vaccines. Pharm. res. 14: p. 1431-6.
[20] Lee KY, et al. (2010) Bioimaging of Nucleolin Aptamer-Containing 5-(N-benzylcarboxyamide)-2'-Deoxyuridine more capable of Specific Binding to Targets in Cancer Cells. J. biomed. biotechnol. p. 168306.
[21] Bar-Cohen Y (2006) Biomimetics: Biologically Inspired Technologies. Boca Raton, FL: CRC/Taylor & Francis. 32 p.
[22] Dumitriu S (2002) Polymeric Biomaterials. New York: Marcel Dekker, Inc. 1168 p.
[23] Dai M, et al. (2011) Artificial Protein Block Polymer Libraries Bearing Two SADs: Effects of Elastin Domain Repeats. Biomacromolecules. 12: p. 4240-6.
[24] Haghpanah JS, et al. (2009) Artificial Protein Block Copolymers Blocks Comprising Two Distinct Self-Assembling Domains. Chembiochem. 10: p. 2733-5.
[25] Kim W, Chaikof EL (2010) Recombinant Elastin-Mimetic Biomaterials: Emerging Applications in Medicine. Adv. drug delivery rev. 62: p. 1468-78.
[26] Kyle S, et al. (2009) Production of Self-Assembling Biomaterials for Tissue Engineering. Trends biotechnol. 27: p. 423-33.
[27] Numata K, Yamazaki S, Naga N (2012) Biocompatible and Biodegradable Dual-Drug Release System Based on Silk Hydrogel Containing Silk Nanoparticles. Biomacromolecules. *in press*
[28] Pritchard EM, Kaplan DL (2011) Silk Fibroin Biomaterials for Controlled Release Drug Delivery. Expert opinion on drug delivery. 8: p. 797-811.
[29] Cen L, et al. (2008) Collagen Tissue Engineering: Development of Novel Biomaterials and Applications. Pediatric research. 63: p. 492-6.
[30] Urry DW (1990) Aqueous Interfacial Driving Forces in the Folding and Assembly of Protein (Elastin)-Based Polymers. J. am. chem. soc. 199: p. 36.
[31] Urry DW (2002) Elastin: a Representative Ideal Protein Elastomer. Philos. trans. r. soc. london, ser. B. 357: p. 169-184.
[32] Altman GH, et al. (2003) Silk-Based Biomaterials. Biomaterials. 24: p. 401-416.
[33] Vollrath F, Knight DP (2001) Liquid Crystalline Spinning of Spider Silk. Nature. 410: p. 541-548.

[34] Rising A, et al. (2005) Spider Silk Proteins--Mechanical Property and Gene Sequence. Zoological science. 22: p. 273-81.
[35] Merrifield RB (1964) Solid-Phase Peptide Synthesis. Biochemistry. 3: p. 1385-90.
[36] Arlinghaus R, Shaefer J, Schweet R (1964) Mechanism of Peptide Bond Formation in Polypeptide Synthesis. PNAS. 51: p. 1291-9.
[37] Chang CD, Meienhofer J (1978) Solid-Phase Peptide Synthesis using Mild Base Cleavage of N Alpha-Fluorenylmethyloxycarbonylamino Acids. Int. j. pept. protein res. 11: p. 246-9.
[38] Arcidiacono S, et al. (1998) Purification and Characterization of Recombinant Spider Silk Expressed in *Escherichia coli*. Appl. microbiol. biotechnol. 49: p. 31-8.
[39] Prince JT, et al. (1995) Construction, Cloning, and Expression of Synthetic Genes Encoding Spider Dragline Silk. Biochemistry. 34: p. 10879-85.
[40] Winkler S, et al. (1999) Designing Recombinant Spider Silk Proteins to Control Assembly. Int. j. biol. macromol. 24: p. 265-70.
[41] Rising A, et al. (2011) Spider Silk Proteins: Recent Advances in Recombinant Production, Structure-Function Relationships and Biomedical Applications. Cell. mol. life sci. 68: p. 169-84.
[42] Bergmann M, Fruton JS (1941) The Specificity of Proteinases. Adv. enzymol. relat. areas mol. biol. 1: p. 63-98.
[43] Bergmann M, Fruton JS (1937) The Nature of Papain Activation. Science, 1937. 86(2239): p. 496-7.
[44] Bergmann M, Fruton JS (1938) Some Synthetic and Hydrolytic Experiments with Chymotrypsin. J. Biol. Chem. 124: p. 321-329.
[45] Lombard C, Saulnier J, Wallach JM (2005) Recent Trends in Protease-Catalyzed Peptide Synthesis. Protein pept lett. 12: p. 621-9.
[46] Jakubke HD (1994) Protease-Catalyzed Peptide-Synthesis - Basic Principles, New Synthesis Strategies and Medium Engineering. J. Chinese Chem. Soc. 41: p. 355-370.
[47] Soeda Y, Toshima K, Matsumura S (2003) Sustainable Enzymatic Preparation of Polyaspartate Using a Bacterial Protease. Biomacromolecules. 4: p. 196-203.
[48] Li G, et al. (2006) Rapid Regioselective Oligomerization of L-Glutamic Acid Diethyl Ester Catalyzed by Papain. Macromolecules. 39: p. 7915-7921.
[49] Viswanathan K, et al. (2010) Protease-Catalyzed Oligomerization of Hydrophobic Amino Acid Ethyl Esters in Homogeneous Reaction Media Using L-Phenylalanine as a Model System. Biomacromolecules. 11: p. 2152-60.
[50] Qin X, et al. (2011) Protease-Catalyzed Oligomerization of L-Lysine Ethyl Ester in Aqueous Solution. ACS Catalysis. 1: p. 1022-1034.
[51] Fukuoka T, et al. (2002) Enzymatic Polymerization of Tyrosine Derivatives. Biomacromolecules. 3: p. 768-74.
[52] Li G, et al. (2008) Protease-Catalyzed Co-Oligomerizations of L-Leucine Ethyl Ester with L-Glutamic Acid Diethyl Ester. Macromolecules. 41: p. 7003-7012.
[53] Aso K, Kodaka H (1992) Trypsin-Catalyzed Oligomerization of L-Lysine Esters. Biosci. Biotech. Biochem. 56: p. 755-758.

[54] Baker PJ, Numata K (2012) Chemoenzymatic Synthesis of Poly(l-alanine) in Aqueous Environment. Biomacromolecules, 13: p. 947-951.
[55] Ran N, et al. (2009) Chemoenzymatic Synthesis of Small Molecule Human Therapeutics. Curr. pharm. design. 15: p. 134-52.
[56] Yagasaki M, Hashimoto S (2008) Synthesis and Application of Dipeptides. Appl. micro. and biotech. 81: p. 13-22.
[57] Kopp F, Marahiel MA (2007) Macrocyclization Strategies in Polyketide and Nonribosomal Peptide Biosynthesis. Nat. prod. rep. 24: p. 735-49.
[58] Conti E, et al. (1997) Structural Basis for the Activation of Phenylalanine in the Non-ribosomal Biosynthesis of Gramicidin S. EMBO. 16: p. 4174-83.
[59] Lautru S, Challis GL (2004) Substrate Recognition by Nonribosomal Peptide Synthetase Multi-enzymes. Microbiology. 150: p. 1629-36.
[60] Koglin A, Walsh CT (2009) Structural Insights into Nonribosomal Peptide Enzymatic Assembly Lines. Nat. prod. rep. 26: p. 987-1000.
[61] Samel SA, et al. (2006) The Thioesterase Domain of the Fengycin Biosynthesis Cluster. J. mol. biol. 359: p. 876-89.
[62] Ritari SJ, et al. (1975) The Determination of Folylpolyglutamate Synthetase. Analytical biochem. 63: p. 118-29.
[63] Scrimgeour KG (1986) Biosynthesis of Polyglutamates of Folates. Biochem. cell biol. 64: p. 667-74.
[64] Carpenter CV, Neuhaus FC (1972) Enzymatic Synthesis of D-alanyl-D-alanine. Two Binding Modes for Product on D-alanine: D-alanine Ligase (ADP). Biochemistry. 11: p. 2594-8.
[65] Evers S, et al. (1996) Evolution of Structure and Substrate Specificity in D-alanine:D-alanine Ligases and Related Enzymes. J. mol. evol. 42: p. 706-12.
[66] Urushibata Y, Tokuyama S, Tahara Y (2002) Characterization of the *Bacillus subtilis* ywsC Gene, Involved in Gamma-Polyglutamic Acid Production. J. bacteriol. 184: p. 337-43.
[67] Sun X, et al. (1998) Structural Homologies with ATP- and Folate-Binding Enzymes in the Crystal Structure of Folylpolyglutamate Synthetase. PNAS. 95: p. 6647-52.
[68] Galperin MY, Koonin EV (1997) A Diverse Superfamily of Enzymes with ATP-Dependent Carboxylate-Amine/Thiol Ligase Activity. Protein sci. 6: p. 2639-43.
[69] Tabata K, Ikeda H, Hashimoto S (2005) ywfE in *Bacillus subtilis* Codes for a Novel Enzyme, L-Amino Acid Ligase. J. bacteriol. 187: p. 5195-202.
[70] Kino K, Arai T, Tateiwa D (2010) A Novel L-Amino Acid Ligase from Bacillus subtilis NBRC3134 Catalysed Oligopeptide Synthesis. Biosci. biotech. biochem. 74: p. 129-34.
[71] Arai T, Kino K (2010) New L-Amino Acid Ligases Catalyzing Oligopeptide Synthesis from Various Microorganisms. Biosci. biotech. biochem. 74: p. 1572-7.
[72] Kino K, Arai T, Arimura Y (2011) Poly-Alpha-Glutamic Acid Synthesis Using a Novel Catalytic Activity of RimK from *Escherichia coli* K-12. Appl. environ. microbiol. 77: p. 2019-25.
[73] Sakajoh M, Solomon NA, Demain AL (1987) Cell-Free Synthesis of the Dipeptide Antibiotic Bacilysin. J. Indust. Microbiol. 2: p. 201-208.

[74] Ueda H, et al. (1987) Kyotorphin (Tyrosine-Arginine) Synthetase in Rat-Brain Synaptosomes. J. Biol. Chem. 262: p. 8165-8173.

[75] Shomura Y, et al. (2012) Structural and Enzymatic Characterization of BacD, an L-Amino Acid Dipeptide Ligase from *Bacillus subtilis*. Protein science 21: p. 707-716.

[76] Fan C, et al. (1994) Vancomycin Resistance: Structure of D-alanine:D-alanine Ligase at 2.3 A resolution. Science. 266: p. 439-43.

[77] Senoo A, et al. (2010) Identification of Novel L-amino Acid Alpha-Ligases through Hidden Markov Model-based Profile Analysis. Biosci. biotech. biochem. 74: p. 415-8.

[78] Kino K, Nakazawa Y, Yagasaki M (2008) Dipeptide Synthesis by L-amino Acid Ligase from Ralstonia Solanacearum. Biochem. biophys res. comm. 371: p. 536-40.

[79] Kino K, et al. (2010) Identification and Characterization a Novel L-Amino Acid Ligase from Photorhabdus luminescens subsp. laumondii TT01. J. biosci. bioengin. 110: p. 39-41.

[80] Arai T, Kino K (2008) A Novel L-Amino Acid Ligase is Encoded by a Gene in the Phaseolotoxin Biosynthetic Gene Cluster from Pseudomonas syringae pv. phaseolicola 1448A. Biosci. biotech. biochem. 72: p. 3048-50.

[81] Kino K, et al. (2008) A Novel l-Amino Acid Ligase from *Bacillus licheniformis*. J. biosci. bioeng. 106: p. 313-5.

[82] Kade B, Dabbs ER, Wittmannliebold B (1980) Protein Chemical Studies on *Escherichia coli* Mutants with Altered Ribosomal Protein-S6 and Protein-S7. FEBS. 121: p. 313-316.

[83] Reeh S, Pedersen S (1979) Post-Translational Modification of *Escherichia Coli* Ribosomal-Protein S6. MGG. 173: p. 183-187.

[84] Kang WK, et al. (1989) Characterization of the Gene rimK Responsible for the Addition of Glutamic Acid Residues to the C-terminus of Ribosomal Protein S6 in Escherichia coli Ribosomal-Protein S6. 217: p. 281-8.

[85] Dawson PE, et al. (1994) Synthesis of Proteins by Native Chemical Ligation. Science. 266: p. 776-9.

[86] Tickler AK, Wade JD (2007) Overview of Solid Phase Synthesis of "Difficult Peptide" Sequences. In: Coligan, JE, editor. Current protocols in protein science. Chapter 18.

[87] Myllyharju J (2009) Recombinant Collagen Trimers from Insect Cells and Yeast. Methods mol. biol. 522: p. 51-62.

[88] Paramonov SE, Gauba V, Hartgerink JD (2005) Synthesis of Collagen-like Peptide Polymers by Native Chemical Ligation. Macromolecules. 38: p. 7555-7561.

[89] Kopecek J, Yang J (2009) Peptide-Directed Self-Assembly of Hydrogels. ACTA biomaterialia. 5: p. 805-16.

[90] Cushing MC, Anseth KS (2007) Hydrogel Cell Cultures. Science. 316: p. 1133-1134.

[91] Jackson MR (1996) Tissue sealants: Current Status, Future Potential. Nature medicine. 2: p. 637-8.

[92] Banta S, Wheeldon IR, Blenner M (2010) Protein Engineering in the Development of Functional Hydrogels. Ann. rev. of biomed. engin. 12: p. 167-86.

[93] Jung JP, et al. (2008) Modulating the Mechanical Properties of Self-Assembled Peptide Hydrogels via Native Chemical Ligation. Biomaterials. 29: p. 2143-51.

[94] Tam JP, Xu J, Eom KD (2001) Methods and Strategies of Peptide Ligation. Biopolymers. 60: p. 194-205.

[95] Constable DJC, et al. (2007) Key Green Chemistry Research Areas – A Perspective from Pharmaceutical Manufactures. Green chem. 9: p. 411-420.